Climate Change, Sustainable Development, and Human Security

Climate Change, Sustainable Development, and Human Security

A Comparative Analysis

Edited by Dhirendra K. Vajpeyi

LEXINGTON BOOKS
Lanham • Boulder • New York • Toronto • Plymouth, UK

Published by Lexington Books
A wholly owned subsidiary of Rowman & Littlefield
4501 Forbes Boulevard, Suite 200, Lanham, Maryland 20706
www.rowman.com

10 Thornbury Road, Plymouth PL6 7PP, United Kingdom

British Library Cataloguing in Publication Information Available

Library of Congress Cataloging-in-Publication Data

Climate change, sustainable development, and human security: a comparative analysis / edited by Dhirendra K. Vajpeyi.
pages cm.
Includes bibliographical references and index.
ISBN 978-0-7391-8146-1 (cloth : alkaline paper) — ISBN 978-0-7391-8147-8 (electronic) 1. Climatic changes. 2. Climatic changes—Social aspects. 3. Climatic changes—Government policy. 4. Sustainable development. 5. Human security. I. Vajpeyi, Dhirendra K.
QC903.C553 2013
363.738'74—dc23

2013023061

∞™ The paper used in this publication meets the minimum requirements of American National Standard for Information Sciences—Permanence of Paper for Printed Library Materials, ANSI/NISO Z39.48-1992.

Printed in the United States of America

Dedicated to

Sarla and Uma Shankar Awasthi,
Cathy and Edward J. Gallagher, Jr.,
Linda and Steve Segebarth

Contents

List of Figures, Tables, and Maps ix

Preface xv

1 Introduction 1
Dhirendra K. Vajpeyi

2 Global Warming and Climate Extremes: The Impact of Carbon Dioxide (CO_2) on the Ozone 27
Clifton C. White and Rebecca McGuire

3 Health, Human Security and Climate Change 51
Laura L. Janik

4 Accommodating "Climate Refugees:" Models of Sovereignty and Security in the International Climate Regime 83
Craig A. Johnson

5 China 111

Part I. China and Climate Change: Environmental Impacts, Human Security, and Mitigation Policies and Actions 111
Jian Li

Part II. Climate Change and Sustainable: Development in Western China's Minqin Oasis—Joining Forces with Civil Society 139
Maria Bondes and Ding Li

6 Climate Change: Challenges for Brazil 169
Eduardo Assad, António Márcio Buainain, Hilton Pinto, Miguel Rocha de Sousa, and Vanessa Duarte

7 Climate Change Vulnerabilities and Human Insecurities—Understanding the Impact on India 201
Roopinder Oberoi and M. P. Singh

8 Climate Change and its Impacts: Domestic and International Responses of Japan 227
Cheng Fang-Ting

9 Domestic Debates on Climate in Russia 249
Leonid Grigoryev, Igor Makarov, and Alla Salmina

10 Climate Change, Its Effects, and the Political Economy of Adaptation and Mitigation: Turkey and the Eastern Mediterranean Region 281
Paul A. Williams

11 The Ecological Paradox in Russia: Political, Social, and Economic Issues and Challenges 307
Natalia Eremina and Igor N. Barygin

12 Climate Change in Bangladesh and Nepal: Issues Challenges and Strategies 325
R. K. Mishra and P. S. Janaki Krishna

13 Conclusion 337
Dhirendra K. Vajpeyi

Selected Bibliography 341

Index 355

About the Contributors 357

List of Figures, Tables, and Maps

FIGURES

1.1 Global Land-Ocean Temperature Index 4

1.2 Effects of Changes in Temperature Distribution 5

2.1 U.S. Greenhouse Gas Emissions 2009 40

2.2 Per Capita GHG Emissions 2009 41

2.3 Atmospheric CO_2 Levels 42

2.4 Global Surface Temperature Trends 44

2.5 Atmospheric Carbon Dioxide and Global Surface Temperature Trends 45

5.1 Water Supply from the Upper Reaches of the Shiyang River and Area of Cultivated Land in Minquin 1956–2010 149

6.1 Agroecological Zoning Flowchart 173

6.2 Flowchart of Components and Biophysical, Climatic and Plant Growth Processes Used 176

6.3 Results—Federal Capital, Brasilia—to the temperature calculated to all models 179

6.4 Monthly maximum and minimum temperatures estimated from seven models for the weather station "DFUNBFAL" located in Brasilia, DF, Brazil 180

6.5 Absolute values of reducing deforestation in the Amazon in km_2 since 2002 188

7.1 Climate Change, Vulnerability, and Impact on Development 205

7.2 Assessment of Vulnerability of Agriculture to Climate Change 212

8.1 Observed results of the concentration of greenhouse gases in Japan, 1987–2009 (CO_2) 228

8.2 Observed results of the concentration of greenhouse gases in Japan, 1994–2009 (CH_4) 228

8.3 Observed results of the concentration of greenhouse gases in Japan, 1994–2009 (N_2O) 229

8.4 Annual Temperature in Japan 1890–2010 229

8.5 Temperature of the largest cities in Japan, the average temperature in Japan, and changes in Sea surface temperature of the peripheral sea area 1900–2010 230

8.6 Observation and Projections of Changes in Temperature in Japan 1900–2010 231

8.7 Observations and Projections of hot days, extremely hot days, sultry nights, icy days during summer and winter in Japan 1900–2100 232

8.8 The number of occurrences of typhoons, the number of typhoons that reached near Japan, and the number of typhoons that landed in Japan, 1950–2010 232

8.9 Number of occurrences of "strong" typhoons and their percentage to the total number of typhoons 1975–2010 233

8.10 Observations and Projections of changes in days of heavy rains (100mm) during summer in Japan 1900–2100 235

8.11 Observations and Projections of Annual average precipitation in Japan 1900–2100 235

8.12 Number of casualties and deaths caused by Heat strokes during the summer (July–September 2008–2012) 238

8.13 Number of prefectures with at least one problem caused by global warming 240

9.1 The number of hydro-meteorological disasters causing significant damage in Russia between 1996–2011 252

9.2 Russian GDP (right scale) and volume of carbon emissions (without—LULUCF) (left scale) 266

9.3 Atmosphere (left scale) and water (right scale) Pollution in Russia 266

9.4 Russian Greenhouse gas emissions in comparison to other countries 267

9.5 Russian Greenhouse gas emissions by sectors (without LULUCF); Mt CO_2e 268

9.6 Structure of Energy Savings in areas of the state Program in Energy Savings and Efficiency of the period of up to 2020 272

9.7 CO_2 emissions by countries Gt CO_2 276

TABLES

1.1 Information on Climate Extreme across Spatial Scales 13

2.1 United States production of First Generation ODS phase out schedule 38

4.1 Support for Migration in the National Adaptation Programs of Action 91, 92

5.1 Top 8 CO_2 Emitting Countries in 2011 112

6.1 Impacts of Climate Change on Brazilian Crops—Optimistic and Pessimistic Scenarios for 2020 and 2030 181

6.2 Values of reducing emissions of greenhouse gases in Brazil by 2020 by industry 187

6.3 Reduction of deforestation by land use 189

7.1 Potential indicators for assessing human vulnerability to Environmental Change 205

7.2 Number of reported natural disaster events and fatalities in selected countries 1990–1999 207

8.1 Deaths and missing persons due to disaster related to weather events, 1945–2011 234

8.2 Measures for the occurrence of chalky grains 243

8.3 Measures for the Satsuma mandarin adversely affected by global warming 243

9.1 Climate characteristics of Russian regions 250

9.2 Estimates of the linear trends in average annual and seasonal temperature rise 1976–2011 251

9.3 Additional mortality due to climate change and carbon emissions—Annual average 256

9.4 Deaths in Moscow in Summer 2010 258

9.5 "Which local environmental problems concern you most?" 262

9.6 Russian Public Opinion Research Center (VCIOM) Poll of April 2007—"Which of the following Opinions about global warming do you most agree to?" 263

9.7 Russian Public Opinion Research Center Poll of April 2007—"What consequences of global warming for Russia will be in your opinion?" 264

9.8 Russian Public Opinion Research Center 2007 Opinion Poll—"What effects will global warming cause?" 264

12.1 Summary of Climate Risks in South Asia 326

12.2 Scope of study in the context of Programs and Activities included in the Bangladesh Climate Change Strategy and Action Plan 331

12.3 Outputs of GCM exercise using GFD 01 Transient 332

MAPS

1.1 Climate Change, Precipitation and Dry Days 7

5.1 Map of China: Provinces, Municipalities, Autonomous Regions, and Major Rivers 115

5.2 Geographic Distributions of Minquin County and Minquin Oasis 142

5.3 Extension of Minquin Oasis 1960–1987 147

5.4 Extension of Minquin Oasis 1987–1998 147

5.5 Extension of Minquin Oasis 1998–2001 147

8.1 Spatial distribution of favorable regions for satsuma mandarin production, assuming that the annual mean temperature appropriate for satsuma manjorin ranges 15–18°C 241

8.2 Projections of Changes in the yield of Paddy rice in Japan, 2046–2065; 2081–2011 242

Preface

Despite strong differences of opinions between the supporters and the deniers of climate change and global warming, a large majority of people around the world perceive it to affect the environmental well-being and global human security. The prominent place President Barak H. Obama gave in his Second Inaugural Address on January 21, 2013 was a clarion call to one of the most pressing problems of the twenty-first century.

The present collection of scholarly essays analyzes the climate change/ global warming issues in major countries of the world, and its impact on human health and habitats. The book does not purport any side of the climate change discourse. It simply aims to provide an objective, data based discussion on this crucial economic, political and national security issue. It is to be noted that none of these chapters were published before. They were specifically prepared for this collection. Several of them were presented at the International Political Science Association (IPSA) World Congress, Madrid, Spain (July 8–12, 2013), and at the three day (January 8–12, 2012) regional conference on "Climate Change and Sustainable Development in a Global Perspective" jointly organized by IPSA Research Committee 35 "Technology and Development," and the Advanced Centre for American Studies, Osmania University, Hyderabad, India. Besides the input provided at these conferences the papers were peer reviewed and reformulated in chapter lengths for publication.

During the preparation of the manuscript I have benefitted from several of my colleagues and friends' suggestions and advice. At the University of Northern Iowa my thanks to Dr. Brenda L. Bass, Interim Dean, College of Social and Behavioral Sciences for her support and encouragement, and Professor Donna R. Hoffman, Head of the Department of Political Science for providing a conducive academic environment to work in. Special thanks to Clifton C. White for showing great care, patience, and humor in going through a diversity of styles and expressions of written English language from different continents and countries. I am also thankful to Dr. Sanghamitra Patnaik presently Associate Professor at KIIT School of Law, KIIT University, Bhubaneswar, Odisha, India for her excellent leadership in organizing the conference in India. Last but not least my thanks to Mr. Justin Race, Associate Editor for Political Science and International Relations, and his Assistant Editor Alissa Parra at Lexington Books for their much needed help and advice in preparing the manuscript. Of course I am most indebted to my colleagues—authors for contributing their papers. The book is dedicated to my sister Sarla and brother-in-law Uma Shankar Awasthi for their constant support, unselfish love, and emotional support throughout my adult life, my friends Cathy and Edward J. Gallagher, Jr., and Linda and Steve Segebarth for their *forty-plus* years of friendship, moral support, and for keeping faith in me.

1

Introduction

Dhirendra K. Vajpeyi

Since the adoption of the United Nations Framework Convention on climate change at the Earth Summit in Rio (1992) the discourse on climate change/global warming has become one of the most controversial environmental issues. Passionate cheerleaders and the true believers have churned immense amounts of "scientific" data to support these claims while equally strident skeptic deniers have called it a "hoax," a "fake science," and a "crock" ever perpetuated on the American public by "political hacks and leftists" (Nordhaus 2012: 32). This uncertainty is mainly due to a variety of reasons including contradictory information and views on the nature, causes, and consequences of climate change on the earth (Singer 2012: 85; Nordhaus 2012: 85).

Following analysis presents the ongoing debate on the issue by addressing three questions:

1. Is the climate changing? And the factors responsible for the change,
2. Impact of climate changes on environment in general and human life in particular,
3. What is to be done?

IS THE CLIMATE CHANGING?

Despite the absence of perfectly reliable data about the extent and geographical distribution (Raleigh and Urdal 2007: 647) of climate change on Earth's ecosystem and the quality of human life there is emerging an "overwhelming consensus of more than 1,250 authors and 2,000 scientific experts reviewers from the Nobel Prize-winning Intergovernmental Panel on Climate Change (IPCC), as well as eighteen American Scientific Associations, is that global warming is observably happening and growing threat to our world" (*UCS Fact Check*, November 2011). According to the IPCC "our collective security in a fragile and increasingly interdependent world" would be affected by human induced climate changes. Rainfall patterns may be affected contributing to water scarcity, floods, droughts, and overall food and health sectors (*The Economist*, 2011, May 7: 86–87).

THE CLIMATE MODELS:

Over several years scientists have used global climate change models (GCM) to measure and predict global warming/climate change. Most of these "climate models are based on well-established physical principles and have demonstrated to reproduce observed features of recent climate and past climate changes" (Randall, et al. 2007: 59), and there is considerable confidence that climate models provide credible quantitative estimates for future climate change" (Randall, et al. 2007: 600). This confidence in models emanates from the following factors:

- Physical bases of the models,
- Their skill in representing observed climate, and past climate changes as far back as 6,000 years warm mid-Holocene or the last glacial maximum of 21,000 years.

However, this "considerable confidence" placed by supporters of GCM has been challenged. Serious doubts and misgivings have been expressed about their ability to predict climate responses. Models are beautiful to look at but can we live with them. Reconciling ever increasing divergent models has been a daunting task for climate change scientists and policy makers (Reifen and Toumi 2009). These models are not perfect and often suffer from usual technological (computing power), and human (scientific knowledge) limitations even if they are becoming "increasingly more comprehensive and useful," and "over several decades of model development, they have consistently provided a robust and unambiguous picture of significant climate warming in response to increasing greenhouse gasses" (Randall et al. 2007: 601). On the other hand the critics of GCMs posit that "the real world data provide little or no support . . . The global warming of the past few decades was part of the much longer warming trend that began in many places throughout the world a little over three centuries ago (about 1680) with the dramatic" "beginning of the end," of the Little Ice Age well before there was any significant increase in the air CO_2 content" (Soloman et al. 2009). Soloman et al. (2009) also point out several "intrinsic" problems and limitations with GCMs by setting forth *three* major criteria for modeled climatic parameters:

1. Observed changes are already occurring and there is evidence for anthropogenic contributions to these changes.
2. The Earth's current temperature is no higher now (and maybe slightly less) than it was during the peak warmth of the Medieval Warm Period (MWP), when there was more than 100 ppm less CO_2 in the air than there is today."
3. "The phenomon (a) (are) based upon physical principles thought to be well understood. There are non-modeled chemical and biological principles that may be equally as important as the physical principles employed in the models."
4. "Projections are available and are broadly robust across models. But these models often diverge so greatly in their assumptions and in their specific spatial and temporal findings that they cannot be said to validate each other, nor can such discordant projections be combined to produce meaningful averages. . . ."According to studies "real world data contradict the models. To say such models are 'robust' is a wishful thinking" (Woolings 2010: 37333756).
5. On the basis of seventeen climate models' results the Fourth Assessment Report of the IPCC found "no evidence of future prediction skill delivered by past performance-based model selection . . . favoring no particular model or groups of models consistently" (Reifen and Toumi 2009: 10).
6. Whether these models are perfectly capable of predicting climate change, and its short and long range impact on Earth's ecosystem it seems that there is an emerging broad consensus that climate "warming is projected to continue through the twenty-first century under all models and all emissions scenarios, and by the end of the twenty-first century, the global average temperature will be several degrees celsius warmer than the present" (Dessler and Parson 2006: 81–82).

It should be noted here that "climate is not just temperature, but also includes such factors as humidity, precipitation, cloudiness, and winds etc." (Dessler and Parson 2006: 47). These factors do affect changes in raising sea levels, melting/retreat of glaciers, flow of rivers, and water level of lakes, etc.

Several models show that over a period of time the Earth's temperature has risen. During the twentieth century the global average surface temperature of the Earth has increased by 0.4–0.8 degrees celsius mainly between 1910 and 1945, and then from 1976 to date (Dessler and Parson 2006: 47), and 1990s have been the warmest years since climate warming data was kept in the mid-nineteenth century (Figure 1.1).

IPCC 2012 Summary for Policy makers points out that "it is virtually certain that increases in the frequency and magnitude of warm daily temperature extremes and decreases in cold extremes will occur in the twenty-first century at the global scale." It further states that "a changing climate leads to changes in the frequency, intensity, spatial extent, duration, and timing of extreme weather and climate events, and can result in unprecedented extreme weather and climate events" (Figure 1.2).

However, the climate change "deniers" believe that some of these predictions are purely scare tactics. They hold that "a modest warming is likely to be beneficial. "We do not see any significant warming of the atmosphere . . . the oceans, covering 71 percent of the Earth's surface, show no appreciable warming either. Also, non-thermometer data from so called "proxies" (tree rings, stalagmites, etc.) show mostly no warming

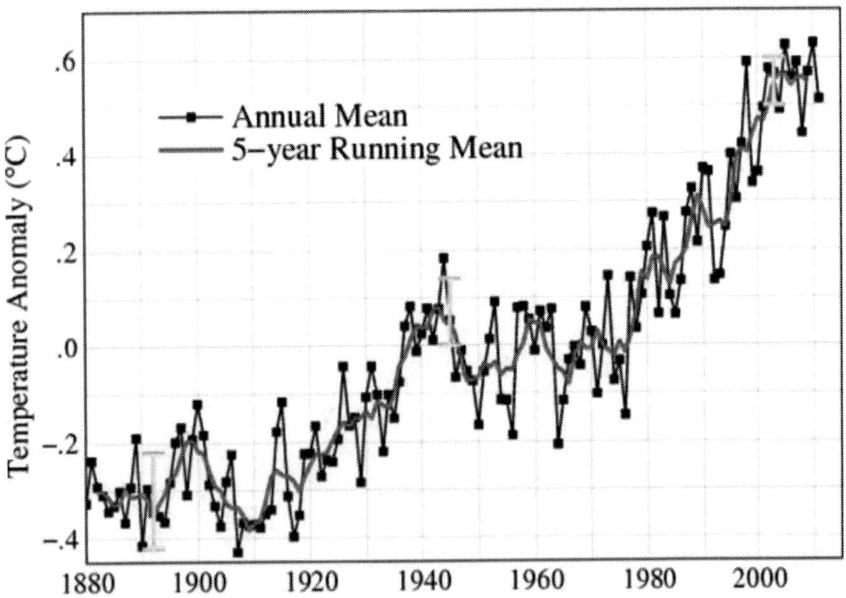

Figure 1.1. Global Land-Ocean Temperature Index
Source: NASA 2011:1.

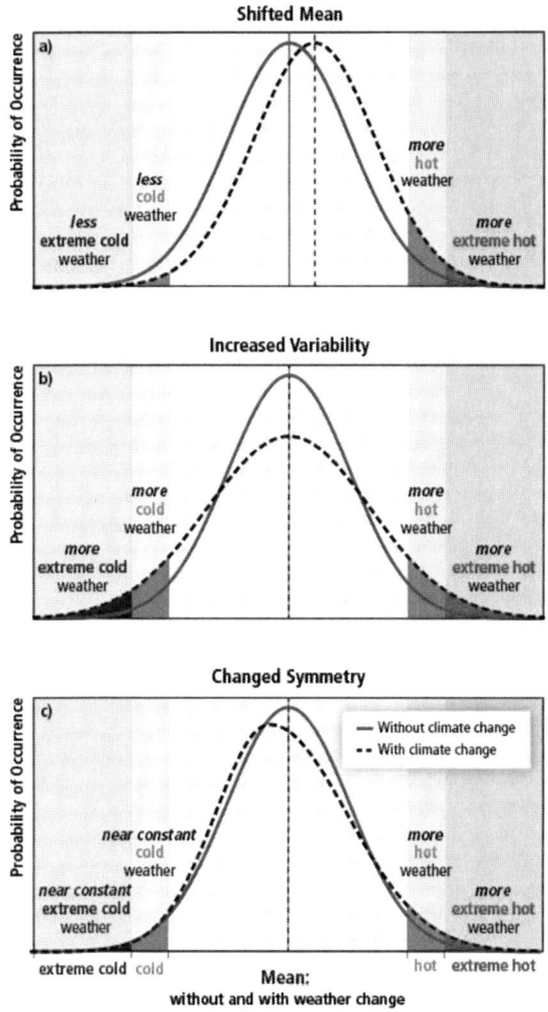

Figure 1.2. Effects of Changes in Temperature Distribution[1]

Source: Special Report, IPCC 2012: 5

during this period (1978–2000). So in spite of rising CO levels, undoubtedly anthropogenic, most observations show no warming trends and that additional warming will be beneficial (Singer 2012: 85).

Glaciers are one of the major sources of water. They cover about 10 percent of the Earth's total land area. They are mostly in Antarctica, Greenland, and the Himalayas. The rising temperature due to global warming will melt (retreat) these glaciers depriving water but will also occasionally raise the sea levels, and contribute to floods. It is reported that between 1860 and 1900 thirty-six glaciers were monitored. Thirty-five of them were affected by rising temperatures. Between 1900 and 1980 one hundred and forty-four glaciers were observed. A total of one hundred and forty two retreated due to warmer temperatures. "The Arctic is warming roughly twice as fast as the rest of the planet. Since the 1950s the lower atmosphere has warmed by a global average of 0.7 degrees Celsius; Greenland's air has warmed by 1.5 degrees. The main reason appears to be catalytic warming effect triggered by global warming" (Astill 2012: 13) The melting glaciers in general and Arctic in particular will have serious economic and geostrategic consequences such as disruptive effect of new trade routes (Astill 2012: 5) impacting cultural and social lives of humans who have lived there for a long period of time. It has been pointed out that melting of Antarctic snow will provide immense natural resources, however, the environmental cost to mine these resources will be too high. According to U.S. Geological Survey 30 percent of the world's undiscovered reserve of natural gas, and 13 percent of oil is under the Arctic Snow. "It also contains coal, iron, uranium, gold, copper, rare earths, gemstones, and much more" (Banyan 2012: 46). According to the World Bank the cost of "adapting" to climate change between 2010 and 2050 will be around $75 billion to $100 billion per year.

Oceans are very good indicators of Earth's health. It has been documented that in recent decades the sea levels have risen mainly due to global warming. During the twentieth century global average sea level rose by 1.5 mm per year or 15 cm over the last century (Dessler and Parson 2006: 53 and 65). Since the 1970s the sea has retreated by around 12 percent each decade. A rise in the sea level could have serious impact on low lying areas such as Bangladesh, Maldive, and even some states in the United States (*The New York Times*, 2012: March 15; 2012 September 11) and could affect an estimated 145 million people around the world (Hertsgaard 2011: 33).

A vast majority of people in developing countries in Asia, Africa, and Latin/South America depend on regular rainfall for their agriculture, and everyday water related needs. Any change, even if it is not severe, in temperature will adversely affect their lives. Droughts and floods will evaporate moistures from the soil, and will produce more dry days (Map 1.1) (Rosenberg 2010: 113–117).

"It is likely that the frequency of heavy precipitation or the proportion of total rainfalls will increase in the twenty-first century over many areas of the globe. This is particularly the case in the high latitudes and tropical regions, and in winter in the northern mid-latitudes. Heavy rainfalls associated with tropical cyclones are likely to increase with continued warming." the droughts will intensify in the twenty-first century in some seasons and areas due to reduced precipitation and/or increased evapotranspiration," including Southern Europe, the Mediterraneans, Central

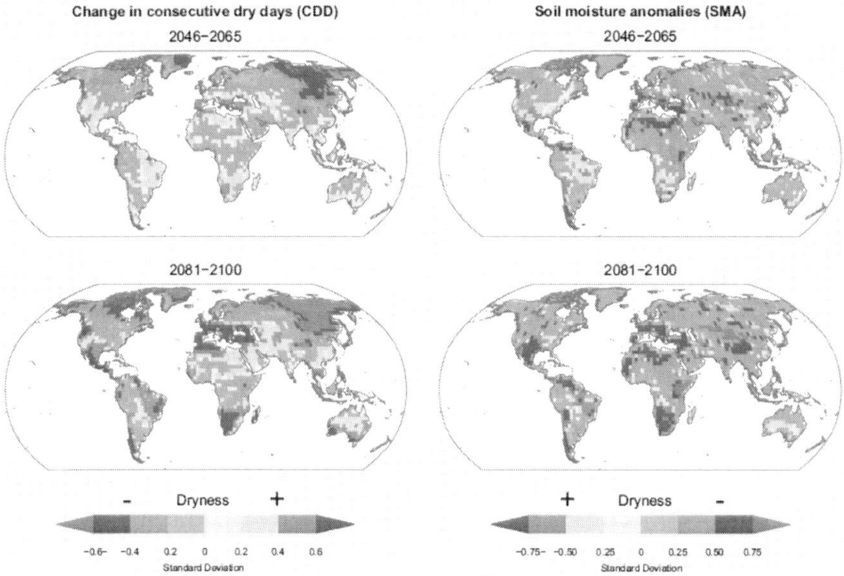

Map 1.1. Climate Change, Precipitation and Dry Days[2]
Source: Special Report, IPCC 2012: 13

Europe, Central North America, Central America and Mexico, northeast Brazil, and Southern Africa (IPCC 2012: 11).

One of the most contentious and complex point in the climate change discourse has been about who is to blame, and what factors have been responsible for global warming. Many scholars and scientists hold that nature shifts and adjusts with time. It has been doing it for millions of years. Others maintain that besides normal shifting and adjusting there is another very important player—the humans—who has contributed immensely to the unprecedented levels of global warming in past two centuries. Observed data show that "human activities have been increasing the concentration of CO_2 and other greenhouse gases in the atmosphere for at least the past century or two" (Dessler and Parson 2006: 73). A survey reported in the "Proceedings of the National Academy of Sciences" observed that 97 percent of the scientists out of a sample of 1,372 believed that climate change is "very likely" caused by human activity. The report was released as the Earth continues to sizzle in 2010 (Rice 2010 *USA Today*, June 22). Overuse of fossil fuels, deforestation, and consumption habits of the rich countries are mainly responsible for increased presence of CO_2 and greenhouse gasses in the ecosystem. Eighty percent of global greenhouse emission is from the richest 2 percent of the world population. While atmospheric CO_2 is considered to be the principal control knob governing Earth's temperature" (Lacis et al. 2010: 356), there are other greenhouse gasses emitted by humans. They include methane (CH_4) which is released by agricultural production, mainly rice, livestocks, biomass burning, chlorofluorocarbons (CFCs) used as refrigerants in various industrial products, and the most powerful greenhouse gas in the atmosphere is water vapor which is very difficult to regulate or control by humans (Dessler and Parson 2006: 9).

HUMAN SECURITY: A PARADIGM SHIFT

The quest to establish a stable society—utopian dystopian—which could provide basic necessities of life is as old as humankind itself. To achieve this goal (El Dorado) humans have taken a variety of economic, political and spiritual paths. In recent times this journey has become more arduous and complicated due to new challenges posed by ever changing global problems—ethnic, ecological, religious, and political. Twentieth century has not only witnessed the explosive inventions and adoption of new technologies and their impact on all aspects of human as well as nonhuman lives but it has also suffered from two brutal World Wars, famines, genocide, tremendous population explosion in poorer countries, environmental degradation, weaponization of space, and nuclear proliferation. The end of World II was followed by the emergence of an era characterized by a Cold War between the U.S. led Western Europe and Soviet Union led Eastern Europe. In ideological terms a competition and confrontation ensued between capitalism, free market, democracy on one hand, and planned (command) economy (Marxism-Communism) nondemocratic system on the other hand. During this period the two super powers tried to manage international affairs by entering and creating military and economic alliances—NATO, and the Warsaw Pact. While both super powers avoided a direct military confrontation, they tested their military weapons and political wills. The world witnessed an unprecedented arms race both nuclear and conventional—which deprived hundreds of millions of people both in advanced countries (North) and in poorer countries (South) of resources necessary for human survival—food, shelter, education, health, and a decent environment. Both super powers demonstrated an amoral, cynical attitude towards democratization, human rights and the economic welfare of the people. Neither individual security nor international security received broad based attention (Danapolous and Vajpeyi 2004: 1–2). The years of Cold War were dominated by the dictates of realist statecraft that the state is the prime actor in international political life and the force is widely available and frequently used to adjust relations on the basis of power (military). While "Realism" never became the sole paradigm but it did tremendously influence the foreign policies of both superpowers. They squandered trillions of dollars and rubles to destroy millions of lives in Korea, Vietnam, Cambodia, and Afghanistan to achieve their narrow national security interests. By late 1980s this race to outbid each other started to exhaust both the United States and the Soviet Union. Domestic needs required greater attention and larger allocation of resources at home. International pressure also demanded "new thinking" to solve soaring problems related to quality of human life. The fall of the Berlin Wall in 1989 symbolically heralded the end of communism, and drastically changed the political and security map of Europe and proved to be of tremendous importance to the entire world overturned an international system that had dominated the world since the end of World War II. An alternative paradigm to the Realist Theory, a more comprehensive holistic definition of security to include broader well-being of humankind started to take shape in a new global environment with a realization that "Many of the dangers the world now faces stem not from conflicts between countries but from conflicts within them. Many States today cause trouble and spread chaos not because of their strength but because of their internal weaknesses—the ethnic, religious, or political conflicts that tear apart their flimsy structure or the private public gangsterism that erodes their civil society" (Hoffmann 2000: 6). A major shift in security paradigm was on its way, the idea of "Comprehensive Security" was first propounded by Olaf Palme. It did not get sufficient attention during the Cold War preoccupation with realist concept of security. Palme's framework of Comprehensive Security included freedom from hunger, disease, repression, environmental security, and personal security. Despite slow progress serious debate among policymakers and academic researchers kept the need for redefinition alive and moving. In 1987 the publication of the World Commission on Environment and Development (WCED) Report *Our Common Future* highlighted issues related to sustainable development and the well being of the environment. In 1989, the United Nations General Assembly by a special resolution called on the world leaders to meet in three years to explore new ways to sustainable development. On June 13, 1992 one of the "largest and most complex conference . . . bigger than the momentous meetings at Versailles, Yalta, and Potsdam" (Elmer-Dewitt: 42) was held in Rio, Brazil. One of the most important contributions of the Rio Conference was that about 178 countries agreed on a goal of "Stabilization of greenhouse gas concentrations in the atmosphere at a level that would prevent dangerous anthropogenic interference with the climate system" (Nordhaus, 2012: 85). Two major treaties were approved and signed: (1) The UN Framework Convention on Climate Change which addressed global warming issues, and (2) The Convention on Biological Diversity to protect endangered species and biodiversity, however the most important product—the crown jewel—was a 400 page document popularly known as Agenda 21—"a document of hope and promise, and a blue print to prevent earth's environmental degradation" (Sitarz 1993: 6). The Agenda 21 included several sub-agendas such as Social Development. The Social Development sub-agenda focused on crucial issues related to basic human security needs such as food, shelter, health care, education, culture, and human rights. Special attention was given to the welfare of children, women, and indigenous people, the role of NGOs, democracy, and the capacity building of the institutions. It also discussed (chapter 9) the protection of global atmosphere and the need to rely more on efficient use of energy resources by increasing dependence on renewable nonpolluting energy for industries and transportation, paying more attention to urban air pollution from cars and trucks. Ozone depletion and global warming are potential threats to human well-being. The signatory states also agreed to study their greenhouse gas emissions and ways to reduce them. One of the major criticisms of Agenda 21 in general, and the Social Development sub-agenda in particular has been that while they provided us a road map to sustainable development and human security, they fell short of tackling many related problems, and a viable operational definition of human security for several reasons:

1. Too ambitious and often times unrealistic
2. Too abstract and broad

Several researchers (Paris 2001: 88; Deudney 1991: 23) have expressed their reservations about human security concept as "practical guide" for academic research. The concept lacks precise definition. Most of the definitions are considered "extraordinarily expansive and vague, encompassing everything from physical security to psychological well-being" (Paris 2001: 88; Dwivedi 2008: 60–75). Its cultivated ambiguity renders human security an effective campaign slogan, but it also diminishes as a guide for academic research or policy making. Despite these critical reservations there have emerged several definitions of human security. Following is a brief summary of major human security components accepted and propounded by scholars, policy makers and international agencies. According to King and Murray (2001: 598–590) there are four main characteristics of human security: (1) it is universal, (2) its components are interdependent, (3) it is best ensured through prevention, and (4) is people centered. Many other definitions which are not as broad as King and Murray's provide us a better understanding of the concept. Myers (1993:31) observes that "in essence . . . human security applies most at the level of the citizen. It amounts to human well-being not only protection from harm and injury but access to water, food, shelter, health, employment, and other basic requisites that are due to every person on Earth. It is the collectivity of these citizen needs—overall safety and quality of life—that should figure prominently in the nations view of security" Dwivedi (2008: 65) contends that human well-being also includes "social regrettables" (pollution, crime, and divorce) social cohesion (Social Capital), better health and the quality of the environment. Matthews (1989: 162–177) definition of human security involves global issues such as stratospheric ozone depletion, greenhouse warming, and deforestation. Brown's (2006: xxiii–xxvii) list of human security components are health, climate, governance, energy, human rights, and ozone layer. Nef (2008: 141–169) postulates an alternative fivefold classification on the basis of the elements of world system: (1) ecology/environment, (2) economy, (3) society, (4) polity, (5) culture. Several reports issued by UNDP on Human Development have emphasized the overall improvement of human life by providing security "from the threat of disease, hunger, unemployment, crime, social conflict, political repression, and environmental hazards" (1994: 22). Many other conceptualizations of human security were presented by diverse national and international organizations such as Human Security Network bilaterally agreed between Canada and Norway in 1998, and later joined by other countries (Austria, Chile, Greece, Ireland, Jordan, Mali, the Netherlands, Switzerland, Slovenia, Thailand) and Barcelona Report (2004) published by the European Union. Above definitions and conceptualizations provide a workable framework of human security. They reflect categories presented by UNDP's Human Development Report (1994) summarized in seven specific categories (Paris 2001: 90; Henk 2005: 91):

1. Economic Security (freedom from poverty)
2. Food Security (access to basic food)
3. Health Security (minimum protection from diseases)
4. Environmental Security (short and long-range ravages of nature, resource depletion, etc.)
5. Personal Security (protection from violence, torture, domestic violence, drug use, suicide)
6. Community security (protection of traditional cultures, values, and ethnic violence)
7. Political Security (civil and political rights, freedom from repression, human rights)

It has also been recognized that human security concerns are global and therefore they require global commitment and solution. They transcend ethnic, religious, economic, and political boundaries.

Linkages between human security, national security, and environmental security: As discussed above that the shift in human security paradigm encompasses a broad arena of physical and emotional well-being and is closely linked with crucial issues related to national, and environmental security including climate change. "The linkages between environmental degradation and security are fairly explicit whether you think of security narrowly in terms of protecting our physical safety or more broadly, in terms of defending our nation's economic strength, values, and way of life" (Claussen 1995: 44), however, there are skeptics who seem to deny these linkages (Deltraz 2009: 320; Deudney and Matthews 1995; Bruynincx 1993; Conca 1994) by proclaiming that "environmental stress is not likely to cause interstate war" (Homer-Dixon 1994: 5–40) and "keep a military based definition of security" (Deudney 1991: 461) and 'environmental security is a bureaucratic tactic' (Finger 1991: 220–225; Le Prestre 1993). Recent climate change data reflects these contentions. First and foremost the new paradigm of human security does not deny the military aspect of security. It advocates shifting of resources—excessive military buildup at the cost of ecological and other human problems. "Issues related to human security today are inextricably linked to the ecological and social problems . . . weapons do not necessarily provide security . . ." (Flavin 2005: XIX; Vajpeyi 2012).

IMPACT OF CLIMATE CHANGE

There is a broad consensus among climate scientists and social policy makers that "simultaneous impact of climate change and globalization" will expose "certain regions, sectors, ecosystems, and social groups" (O'Brian, and Robin M. Leichenko 2000: 221). However, the physical and social impacts of these changes will be uneven. While many areas could experience temperature increases by 1.5–4.5 degrees celsius the other areas may even get cold. Some social groups and ecosystems are more vulnerable than others due to their location, system of government, spatial differences in the magnitude, and the direction of climate change (Bohle 1994: 37–48; Table 1.1).

There has been a lively and robust debate among climate scholars about the impact of these changes on rich and poor countries—the North and the South (Mendelshon et al. 2006: 159–178; O'Brien et al. 2000: 221–232, Grubb 1995: 463–496; Parry and Carter 1998). Mendelson et al. (2006) used three models—the Partial Climate Model (PCM), The Center for Climate Research Studies (CCSR), and the Canadian General Circulation Model (CGCM1) to analyze "the relative impact of climate across countries, impact per capita, and impact per GDP" on the Distributional

Table 1.1. Information on Climate Extreme Across Spatial Scales

Example	Exposure and vulnerability at scale of risk management in the example	Information on Climate Extreme Across Spatial Scales			Options for risk management and adaptation in the example
		GLOBAL Observed (since 1950) and projected (to 2100) global changes	**REGIONAL** Observed (since 1950) and projected (to 2100) changes in the example	**SCALE OF RISK MANAGEMENT** Available information for the example	
Inundation related to **extreme sea levels** in tropical small island developing states	Small island states in the Pacific, Indian, and Atlantic Oceans, often with low elevation, are particularly vulnerable to rising sea levels and impacts such as erosion, inundation, shoreline change, and saltwater intrusion into coastal aquifers. These impacts can result in ecosystem disruption, decreased agricultural productivity, changes in disease patterns, economic losses such as in tourism industries, and population displacement – all of which reinforce vulnerability to extreme weather events. [3.5.5, Box 3-4, 4.3.5, 4.4.10, 9.2.9]	<u>**Observed:**</u> *Likely* increase in extreme coastal high water worldwide related to increases in mean sea level. <u>**Projected:**</u> *Very likely* that mean sea level rise will contribute to upward trends in extreme coastal high water levels. *High confidence* that locations currently experiencing coastal erosion and inundation will continue to do so due to increasing sea level, in the absence of changes in other contributing factors. *Likely* that the global frequency of tropical cyclones will either decrease or remain essentially unchanged. *Likely* increase in average tropical cyclone maximum wind speed, although increases may not occur in all ocean basins. [Table 3-1, 3.4.4, 3.5.3, 3.5.5]	<u>**Observed:**</u> Tides and El Niño–Southern Oscillation have contributed to the more frequent occurrence of extreme coastal high water levels and associated flooding experienced on some Pacific Islands in recent years. <u>**Projected:**</u> The *very likely* contribution of mean sea level rise to increased extreme coastal high water levels, coupled with the *likely* increase in tropical cyclone maximum wind speed, is a specific issue for tropical small island states. See global changes column for information on global projections for tropical cyclones. [Box 3-4, 3.4.4, 3.5.3]	Sparse regional and temporal coverage of terrestrial-based observation networks and limited in situ ocean observing network, but with improved satellite-based observations in recent decades. While changes in storminess may contribute to changes in extreme coastal high water levels, the limited geographical coverage of studies to date and the uncertainties associated with storminess changes overall mean that a general assessment of the effects of storminess changes on storm surge is not possible at this time. [Box 3-4, 3.5.3]	Low-regrets options that reduce exposure and vulnerability across a range of hazard trends: • Maintenance of drainage systems • Well technologies to limit saltwater contamination of groundwater • Improved early warning systems • Regional risk pooling • Mangrove conservation, restoration, and replanting Specific adaptation options include, for instance, rendering national economies more climate-independent and adaptive management involving iterative learning. In some cases there may be a need to consider relocation, for example, for atolls where storm surges may completely inundate them. [4.3.5, 4.4.10, 5.2.2, 6.3.2, 6.5.2, 6.6.2, 7.4.4, 9.2.9, 9.2.11, 9.2.13]
Flash floods in informal settlements in Nairobi, Kenya	Rapid expansion of poor people living in informal settlements around Nairobi has led to houses of weak building materials being constructed immediately adjacent to rivers and to blockage of natural drainage areas, increasing exposure and vulnerability. [6.4.2, Box 6-2]	<u>**Observed:**</u> *Low confidence* at global scale regarding (climate-driven) observed changes in the magnitude and frequency of floods. <u>**Projected:**</u> *Low confidence* in projections of changes in floods because of limited evidence and because the causes of regional changes are complex. However, *medium confidence* (based on physical reasoning) that projected increases in heavy precipitation will contribute to rain-generated local flooding in some catchments or regions. [Table 3-1, 3.5.2]	<u>**Observed.**</u> *Low confidence* regarding trends in heavy precipitation in East Africa, because of insufficient evidence. <u>**Projected.**</u> *Likely* increase in heavy precipitation indicators in East Africa. [Table 3-2, Table 3-3, 3.3.2]	Limited ability to provide local flash flood projections. [3.5.2]	Low-regrets options that reduce exposure and vulnerability across a range of hazard trends: • Strengthening building design and regulation • Poverty reduction schemes • City-wide drainage and sewerage improvements The Nairobi Rivers Rehabilitation and Restoration Programme includes installation of riparian buffers, canals, and drainage channels and clearance of existing channels; attention to climate variability and change in the location and design of wastewater infrastructure; and environmental monitoring for flood early warning. [6.3, 6.4.2, Box 6-2, Box 6-6]
Impacts of **heat waves** in urban areas in Europe	Factors affecting exposure and vulnerability include age, pre-existing health status, level of outdoor activity, socioeconomic factors including poverty and social isolation, access to and use of cooling, physiological and behavioral adaptation of the population, and urban infrastructure. [2.5.2, 4.3.5, 4.3.6, 4.4.5, 9.2.1]	<u>**Observed:**</u> *Medium confidence* that the length or number of warm spells or heat waves has increased since the middle of the 20th century, in many (but not all) regions over the globe. *Very likely* increase in number of warm days and nights at the global scale. <u>**Projected.**</u> *Very likely* increase in length, frequency, and/or intensity of warm spells or heat waves over most land areas. *Virtually certain* increase in frequency and magnitude of warm days and nights at the global scale. [Table 3-1, 3.3.1]	<u>**Observed:**</u> *Medium confidence* in increase in heat waves or warm spells in Europe. *Likely* overall increase in warm days and nights over most of the continent. <u>**Projected:**</u> *Likely* more frequent, longer and/or more intense heat waves or warm spells in Europe. *Very likely* increase in warm days and nights. [Table 3-2, Table 3-3, 3.3.1]	Observations and projections can provide information for specific urban areas in the region, with increased heat waves expected due to regional trends and urban heat island effects. [3.3.1, 4.4.5]	Low-regrets options that reduce exposure and vulnerability across a range of hazard trends: • Early warning systems that reach particularly vulnerable groups (e.g., the elderly) • Vulnerability mapping and corresponding measures • Public information on what to do during heat waves, including behavioral advice • Use of social care networks to reach vulnerable groups Specific adjustments in strategies, policies, and measures informed by trends in heat waves include awareness raising of heat waves as a public health concern; changes in urban infrastructure and land use planning, for example, increasing urban green space; changes in approaches to cooling for public facilities; and adjustments in energy generation and transmission infrastructure. [Table 6-1, 9.2.1]
Increasing losses from **hurricanes** in the USA and the Caribbean	Exposure and vulnerability are increasing due to growth in population and increase in property values, particularly along the Gulf and Atlantic coasts of the United States. Some of this increase has been offset by improved building codes. [4.4.6]	<u>**Observed.**</u> *Low confidence* in any observed long-term (i.e., 40 years or more) increases in tropical cyclone activity, after accounting for past changes in observing capabilities. <u>**Projected.**</u> *Likely* that the global frequency of tropical cyclones will either decrease or remain essentially unchanged. *Likely* increase in average tropical cyclone maximum wind speed, although increases may not occur in all ocean basins. Heavy rainfalls associated with tropical cyclones are *likely* to increase. Projected sea level rise is expected to further compound tropical cyclone surge impacts. [Table 3-1, 3.4.4]	See global changes column for global projections.	Limited model capability to project changes relevant to specific settlements or other locations, due to the inability of global models to accurately simulate factors relevant to tropical cyclone genesis, track, and intensity evolution. [3.4.4]	Low-regrets options that reduce exposure and vulnerability across a range of hazard trends: • Adoption and enforcement of improved building codes • Improved forecasting capacity and implementation of improved early warning systems (including evacuation plans and infrastructure) • Regional risk pooling In the context of high underlying variability and uncertainty regarding trends, options can include emphasizing adaptive management involving learning and flexibility (e.g., Cayman Islands National Hurricane Committee). [5.5.3, 6.5.2, 6.6.2, Box 6-7, Table 6-1, 7.4.4, 9.2.5, 9.2.11, 9.2.13]
Droughts in the context of food security in West Africa	Less advanced agricultural practices render region vulnerable to increasing variability in seasonal rainfall, drought, and weather extremes. Vulnerability is exacerbated by population growth, degradation of ecosystems, and overuse of natural resources, as well as poor standards for health, education, and governance. [2.2.2, 2.3, 2.5, 4.4.2, 9.2.3]	<u>**Observed:**</u> *Medium confidence* that some regions of the world have experienced more intense and longer droughts, but in some regions droughts have become less frequent, less intense, or shorter. <u>**Projected:**</u> *Medium confidence* in projected intensification of drought in some seasons and areas. Elsewhere there is overall *low confidence* because of inconsistent projections. [Table 3-1, 3.5.1]	<u>**Observed:**</u> *Medium confidence* in an increase in dryness. Recent years characterized by greater interannual variability than previous 40 years, with the western Sahel remaining dry and the eastern Sahel returning to wetter conditions. <u>**Projected:**</u> *Low confidence* due to inconsistent signal in model projections. [Table 3-2, Table 3-3, 3.5.1]	Sub-seasonal, seasonal, and interannual forecasts with increasing uncertainty over longer time scales. Improved monitoring, instrumentation, and data associated with early warning systems, but with limited participation and dissemination to at-risk populations. [5.3.1, 5.5.3, 7.3.1, 9.2.3, 9.2.11]	Low-regrets options that reduce exposure and vulnerability across a range of hazard trends: • Traditional rain and groundwater harvesting and storage systems • Water demand management and improved irrigation efficiency measures • Conservation agriculture, crop rotation, and livelihood diversification • Increasing use of drought-resistant crop varieties • Early warning systems integrating seasonal forecasts with drought projections, with improved communication involving extension services • Risk pooling at the regional or national level [2.5.4, 5.3.1, 5.3.3, 6.5, Table 6-3, 9.2.3, 9.2.11]

Source: Special Report, IPCC 2012: 17.

impact of climate change on rich and poor countries. They concluded that "poor countries bear a larger burden of climate damages because they are already hot" and the "damages rise with income" however, "what happens to some countries in aggregate does not necessarily indicate what will happen to the poor residents of a country (Mendelsohn et al. 2006: 163, 167–170). Grubb (1995: 467) concurs with above observations for following reasons:

- Developing countries will suffer more than developed ones because most the developing countries are in tropical regions,
- Their economies are more dependent upon agriculture and other natural resources,
- Their institutional and social structures tend to be weaker hence they are less prepared to cope with changes,
- They have fewer financial resources to invest in more robust infrastructure.
- Several other studies dispute above conclusions and maintain that damages due to climate change will not be confined only to poor tropical, countries.

The above discussion on the impact of climate change highlights the following:

- Higher temperatures and more hot days through the next century (record-breaking heat that would have been a once in twenty year high are likely to become a one-in-two year event).
- More frequent and heavier rains, especially in winter.
- Stronger hurricanes that will do more damage.
- Increased droughts, especially in the center of the country.
- Higher sea levels, resulting in more coastal erosion and other damages.
- All these changes will affect agriculture, water supplies, health, and even tourism, and global economies (Gaby 2011: 1).

In contrast to above several studies dispute these conclusions. They maintain that damages due to climate change will not be confined only to poor tropical regions alone. They counter argue that "Extreme climate events can impact the wealthy and poor alike particularly in high-risk environments" (O'Brien et al. 2000: 225). For example, the tsunami in Asia (December 26, 2004) was an equal opportunity leveler from Bangladesh to Indonesia and Thailand. The nuclear meltdown due to tsunami in Japan (March 2011) did not spare rich and powerful. It is, therefore, maintained that both "biophysical and social vulnerability should be taken into account (O'Brien 2000: 224). " . . . the most vulnerable people may not be in the most vulnerable places—poor people can live in productive biophysical environments and be vulnerable, and wealthy people can live in fragile physical environments and live relatively well" (Liverman 1994: 332). Coastal areas for example are the most vulnerable to climate change. These areas also have expensive properties especially in rich countries such as the United States of America and are very vulnerable to floods due to climate change induced sea level rise (*New York Times* 2012 March 14: A1 and 3). According to the IPCC many other countries such as China's coastal areas with approximately 70 million, both rich and poor, will be affected by climate change.

The counter arguments mainly take aim and scoff at several IPCC conclusions. The Fourth Assessment Report of the IPCC expressed "high confidence that climate change contributes to the global burden of disease and premature deaths" (2007—II: 393). It also predicts that changes in climate will increase malnutrition and consequent disorders . . . increase the number of people suffering from death, disease, and injury from heatwaves, floods, storms, fires, and droughts . . . continue to change the range of some infectious disease vectors . . . increase the burden of diarrheal diseases . . . increase cardiorespiratory morbidity and mortality associated with ground-level zones . . . and increase the number of people at risk of dengue (IPCC 2007—II: 393). According to the 2009 report of the Nongovernmental International Panel on Climate Change (NIPCC) above claims of the IPCC do not support the real world data. On the basis of studies (Idso and Singer 2009; Bayentin et al. 2010; Cao et al. 2009: 328–332; Confalonieri et al. 2007; Deschenes et al. 2009: 659–681; Johansson et al. 2009; Kyle and Harris 2008: 71–82) it has been posited by the counter argumentators that "clearly long term trends for hunger, disease, and deaths from droughts, floods, and other extreme weather events are not consistent with the IPCC's narrative regarding the impacts of global warming. Perhaps global warming is not happening after all, or if it is, its effects are relatively small and/or overwhelmed by improvements in human adaptive capacity or other factors. Or Perhaps the global warming narrative is simply based on false expectations, that warming's real impacts are more positive than negative. Whatever the explanation (or combination of explanations) is correct, the salient fact is that real-world data do not support claims that global warming is reducing human well-being" (*Climate Change Reconsidered*—2011 Interim Report: 388).

INTERNATIONAL RESPONSE:

Issues related to environment have been with us since the creation. Eruption of volcanoes, severe rainfalls, droughts, and erratic weather changes have generated philosophical debates, involvement of priests and shamans to explain mother nature's behavior, but it is only recently, due to scientific advancement and our concern to rationally and scientifically analyze these changes, that human efforts have accelerated. Climate change and its impact is a late comer in the environmental discourse. At Stockholm Conference in 1972 it was not even mentioned. Other factors related to poverty, health, and natural disasters were on the agenda. Climate change was not on the radar. Following is a short discussion of various international efforts to address the long- and short-range impact and nature of climate change.

It was at the Earth Summit (1992) in Rio that two major treaties were approved: (1) The U.N. Framework Convention on Climate Change which addressed global warming issues, and (2) The Convention on Biological Diversity to protect endangered species and biodiversity. Chapter 9 of Agenda 21 resolved to protect global climate ozone depletion and global warming which were perceived as potential threats to world environment. At Rio the 190 signatory states agreed to evaluate their greenhouse gas emissions and ways to reduce them. The Framework convention on Climate Change (FCCC) also urged the development of "international joint implementation" projects between developing (South) and the developed (North) states to reduce greenhouse gases. The Climate Treaty was diluted due to objections and reservations of several developed countries led by the United States.

Between Rio (1992) and Kyoto (2005) several international protocols were signed (Montreal Protocol, Berlin Mandate, Johannesburg Summit), but it was at Kyoto, Japan that a major conference with its legally binding emissions limits for industrial countries elevated the climate change discourse, exposing serious differences between the South and the North. Under heavy pressure from the petroleum industry the United States rejected the Kyoto Protocol outright. Europe, Canada, Japan, and Russia ultimately ratified the Kyoto Treaty in following years, but the unilateral actions of George W. Bush's administration almost crippled the Treaty. Despite its ratification, the fate of Kyoto is uncertain. Since major, and to some extent unequal, responsibilities to reduce gas emissions were assigned to developed countries there has been a weak enforcement of the agreement. In 2011, Canada withdrew from the Treaty further weakening the arrangement. The next major conference was held in Copenhagen in December 2009. The aim of the Conference/Summit was to cut emissions in half from 1990 levels by 2050. Serious differences between the high emitters from industrialized countries, and developing countries particularly from India and China surfaced. Despite the change in the American position on the greenhouse gas emission and initial strong support by U.S. president Obama differences could not be bridged. "International climate politics have faltered like a mortally wounded animal" observed Germany's Spiegel newspaper (http://www.spiegel.de/international/world/0,1518,druck-692861,oohtml). Copenhagen demonstrated a lack of political will and narrow national interests driven power politics. The next conference was scheduled in Cancun, Mexico in November 2010. By many indications "the Cancun Climate Change Summit played out surprisingly well, achieving a fair bit of what its overhyped and acrimonious predecessor (at Copenhagen) had fluffed" (*The Economist*, 2010, December 10: 16). In Cancun, the ill will faded, and several practical steps such as $100 billion a year for developing countries by 2020 as climate assistance, a climate fund, partly under the World Bank, and a much needed deal on the conditions under which countries may be paid to decrease the damage being done due to deforestation were taken. The details were not very clear and were left for future deliberations. "The UN climate process did quite well out of Cancun. The climate, not so well" (*The Economist* 2010, December 18: 122).

Differences between the climate change deniers, and ardent supporters will not fade away soon, however, much progress has taken place in the last several decades. New scientific data, and events in nature, hopefully, will help us make better policy decisions.

WHAT IS TO BE DONE?

Differences of opinions among scientists, policy makers and politicians will not go away. Climate change is too complicated an issue and requires political, economic, ethical, and managerial solutions. The topic of climate change needs to be disconnected from ideological and purely economic considerations. The debate has to go beyond international conferences, weak pious resolutions, and absence of transparent political will. Global solutions must be found "One of the greatest international hurdles to be surmounted in the process is the discontinuity between the crude North-South division and the longer term problem of defining emission entitlements" (Grubb 1995: 495), and failure to accept moral responsibility—individual and collective to achieve sustainable development. Societies and civilizations which fail to adapt and adopt new ideas and technologies collapse in due time (Diamond 2005). While it is widely realized that "eliminating all the risks of climate change is impossible because carbon dioxide emissions, the chief human contribution to global warming are unlike conventional air pollutants which stay in the atmosphere for only hours or days—remain over a hundred years" (Victor et al. 2009: 65). "Carbon dioxide concentration is largely irreversible for 1,000 years after emissions stop" (Solomon et al. 2009: 1750). Dessler and Parson (2006: 91–127) and several others (O'Brien et al. 2000: 221–232; Victor et al. 2009:

64–76) suggest following responses to deal with the ever increasing threat of climate change to Earth's ecosystem.

a. Adaptation:

Earth's climate has been changing for centuries, and despite advances made in technological and scientific knowledge humans have little control. Mother Nature never ceases to surprise us in many ways. Humans, therefore, have to accommodate and adapt to these changes. Excessive consumerism, wasteful exploitation of natural resources, unsustainable life styles, our politics, economic, political, and cultural institutions will have to make necessary changes. Beliefs that God made humans lord of this earth to enjoy its bounties are too promiscuous, because he did not give one generation more privileges to destroy Earth's gifts over the future ones. It is difficult but humans during their entire existence on this Earth have been exceptionally resilient, and they are capable of adaptation and change. To quote an African proverb that if "you can't turn the wind, so turn the sail."

b. Mitigation:

". . . adapting to the impacts of climate change is not by itself an adequate response, it is also necessary to target the causes of climate change through mitigation measures, thereby slowing the impacts that we have to adapt to" (Dessler and Parson 2006: 96). It will mean the reduced use of the largest source of human emissions CO_2, fossil fuels, coal, natural gas, and other gases generated by agriculture, and animals (methane CH_4). The economic and social costs to mitigation efforts will be high. Narrow national interest in the name of economic development will be insurmountable hurdles to overcome. "There is a tension between adaptation and mitigation to some, the former implies a disregard of the latter, as if society were giving up on trying to reduce green house-gas (GHG) emissions" (Henson 2011:322).

c. Integrated Assessment (IA):

Integrated assessment models suggest a combined approach—adaptation plus mitigation—to face the climate change threat. "Integrated assessment models" according to Dessler and Parson (2006:121) "represent the climate system, the socioeconomic factors that drive emissions, the impacts of climate change, and potential mitigation and adaptation responses in a consistent quantitative framework. While highly simplified, these models can be used to simulate the effects of different mitigation and adaptation strategies, and to calculate the costs and benefits of alternative scenarios and policies."

d. Geoengineering:

Scientists and policymakers have suggested another "most cost-effective" option to cope with climate changes by "manipulating" or fiddling with the climate "to offset the effects of increased greenhouse gases in the atmosphere. This approach, usually called geoengineering or "climate engineering using technology" includes a diverse collection of proposals and ideas to cool the earth. Some involve (i) solar-radiation management. It focuses on reducing the impact of the sun by seeding clouds, spreading giant mirrors in the desert, or injecting sulfates into the stratosphere . . . ; (ii) removing carbon directly from the atmosphere and burying it in vast ocean storage beds or deep inside the earth (Specter 2012: 99) The proponents of geoengineering do realize that the approach is not "since Qua non" to all climate change problems but it is an option, and it would be dangerous to ignore it. Hence "it is time to take geoengineering out of the closet" (Victor et al. 2009; 74–76). According to Specter, however, geoengineering is 'hubris' and many other suggested variations are not only costly but from "plausible to absurd.

CONCLUSION/SUMMARY:

The above discussion points out

1. That there is a broad consensus among scientists and policy makers about the threats climate change poses to humanity and the sustained survival of Earth's ecosystem.
2. There is—despite a robust debate about specific models and approaches—still uncertainty about the nature and extent of climate change.
3. More research and scientific data is required.
4. Climate change discourse and research should not fall victim to ideological and political partisanship.
5. Climate change is a global issue and it requires global solutions by private and public institutions. Let us remember that "What good is a house if you do not have a decent planet to put it on" (Henry D. Thoreau).

This anthology of essays analyzes the increasing challenge posed by climate change/global warming threatening the global human security and sustainable development. In chapter 1 Dhirendra Vajpeyi presents a broad framework related to climate change, and its potential to impact earth's ecosystem, and the ongoing debate about climate change itself among scientists and policy makers around the globe.

Clifton White and Rebecca McGuire in chapter 2 analyze the recent scientific data on the frequency of extreme climate events and natural disasters due to rise in the earth's temperature and the impact of CO_2 on the Ozone Layer.

In chapter 3 Laura L. Janik analyzes both direct and indirect consequences of climate change for human security and citizen well-being. She investigates the impact of climate change and environmental degradation on human health and strategies to assuage these problems by positively contributing to human conditions in the global north and global south in general and the human health consequences of ozone depletion, flooding, droughts, and rising temperatures in particular.

Craig A. Johnson in chapter 4 explores the possibilities and limitations of using adaptation to support the entitlements of populations displaced by the impacts of climate change. He examines the priorities outlined in UNFCCs National Adaptation Programs of Action (NAPAs) which assert that migration has been characterized almost universally by threats and vulnerabilities it creates for societies, ecological systems, and migrants. Johnson further draws upon the "social protection" literature to identify the ways in which cash transfers, asset transfers, and other forms of social protection may be used to break the cycle of vulnerability distribution, and distress migrants suffer during the time of severe environmental stress.

Chapter 5 is divided into two parts. In part I Li Jian analyzes. China's environmental problems related to climate change in a broader framework of human security and sustainable development. He also discusses China's policies and measures mitigation, adaptation, role of technology, public awareness, and future challenges facing China. In part II Maria Bondes and Ding Li present China's increasing demands for natural resources to accelerate its economic growth. China's most severe environmental problems include rapidly growing water shortages, deforestation, and air pollution. Chinese government has placed environmental issues as its top priority on domestic policy agenda, but it is also encouraging social organizations and the public to actively participate in meeting these challenges. Both authors have used extensive field research in Northwestern China's MinQuin oasis in Gansu Province—one of the most severely affected regions by desertification, water shortages, and poverty.

In chapter 6 Eduardo D. Assad, Antonio Marcio Buainain, Hilton Pinto, Miguel Rocha de Sousa, and Vanessa Duarte analyze four aspects of the impacts of climate change on the environment and its consequences for human security in Brazil. These four sectors—Agriculture, Desertification, Deforestation, and Urbanization—will be severely affected by climate change induced rainfall patterns. Agricultural production will suffer due to droughts, and migration of people from arid areas to other parts of the country.

Roopinder Oberoi and M. P. Singh in chapter 7 conceptualize linkages between food security and climate change in India. The chapter addresses the vulnerability of food security to climate change, its impact on agricultural output, and the strategies to reduce this vulnerability.

In chapter 8 Cheng Feng-Ting discusses the important role of climate change in creating natural disasters and global economic crisis as demonstrated by the Asian tsunami in 2004 and 2010 in Japan. The author analyzes the current international and national policy responses to mitigate these occurrences. The role of the United Nations International Strategy for Disaster Reductions (UNISDR), the Tokyo International Conference on African Development (TICAD), International Monetary Fund (IMF), Green Climate Fund (GCF), and International Consultations and Analysis (ICA) under UNFCCC are analyzed.

Chapter 9 Leonid Grigoryev, Igor Makarov, and Alla Salmina analyze Russia's important role in mitigation of the global climate change. They posit that Russian efforts have been hindered by many socioeconomic and political challenges it faces in its domestic sphere. Russia's diversity, its size, and its climates pose special challenges to formulate a cohesive climate change policy, however determined efforts are being made by NGOs, scientific community and policy makers in strengthening its civil society to mitigate adverse impacts of these changes.

Paul William in chapter 10 points out challenges Turkey faces in procuring and judicious usage of energy for its economic development. He points out that due to the dilemma faced by Turkey as a party to Annex I of the UN Framework Convention on Climate Change (UNFCC) and EU mandates and its developmental goals Turkey has generally shown poor substantive performance and weak administrative capacity on climate change.

In chapter 11 Natalia Eremina and Igor N. Barygin discuss the environmental debate in Russian Federation, the lack of public awareness and concern about climate change on human security due to Russian preoccupation with day to day economic challenges. The emergence of political opposition in recent years and its efforts to raise ecological issue awareness is slow. The chapter analyzes these issues in the context of cultural, political, and economic challenges in Russia.

In chapter 12 R. K. Mishra and P. S. Janaki Krishna discuss the impending disaster posed by global warming/climate change in the two most impoverished countries in South Asia, and the strategies being adopted by respective governments in these countries.

The concluding chapter 13 presents a brief synopsis of major points and observations in earlier chapters, and posits policy options and recommendations to meet the challenges posed by global climate change in the twenty-first century.

NOTES

1. The effect of changes in temperature distribution on extremes. Different changes in temperature distributions between present and future climate and their effects on extreme values of the distributions: (a) effects of a simple shift of the entire distribution toward a warmer climate; (b) effects of an increase in temperature variability with no shift in the mean; (c) effects of an altered shape of the distribution, in this example a change in asymmetry toward the hotter part of the distribution.

2. Projected annual changes in dryness assessed from two indices. Left column: Change in annual maximum number of consecutive dry days (CDD: days with precipitation < 1mm). Right column: Changes in soil moisture (soil moisture anomalies, SMA). Increased dryness is indicated with yellow to red colors; decreased dryness with green to blue. Projected changes are expressed in units of standard deviation of the interannual variability in the three twenty-year periods 1980–1999, 2046–2065, and 2081–2100. The figures show changes for two time horizions, 2046–2065 and 2081–2100, as compared to late twentieth-century values (1980–1999), based on GCM simulations under emissions scenario SRES A2 relative to corresponding simulations for the late twentieth century. Results are based on 17 (CDD) and 15 (SMA) GCMs contribution to the CMIP3. Colored shading is applied for areas where at least 66 percent (12 out of 17 for CDD, 10 out of 15 for SMA) of the models agree on the sign of the change; stippling is added for regions where at least 90 percent (16 out of 17 for CDD, 14 out of 15 for SMA) of all models agree on the sign of the change. Grey shading indicates where there is insufficient model agreement (< 66 percent).

REFERENCES

Astill, James 2012 "The Vanishing North," *The Economist*, June 16–22, 13.

Banyan 2012 "Snow Dragons," *The Economist*, September 1, 46.

Bayentin, L. Ouarda El Adlouni, P. Gosselin, et al. 2010 "Spatial Varability of Climate Effects on schemic heart disease hospitalization rates for the period 1989–2006 in Quebec, Canada" *International Journal of Heath Geographics* 9:10 1186/1476-072X-9-5.

Bohle, H. G. et al. 1994 "Climate Change and Social Vulnerability" *Global Environmental Change* 4 (1).

Brown, Lori 2006 *State of the World: A year in Review*, New York: W.W. Norton.

Brundtland, G. W. (ed.) 1987. *Our Common Future*, New York: Oxford University Press p. 9.

Bruynincx, Hans 1993 "Environmental Security: An Analyses of the Conceptual Problems Defining the Relationship between Environment and Security," Paper presented at the 34th Annual Convention of the International Studies Association, Acapulco, Mexico, March 23–27.

Calfalnoieri, U. et al. 2007 "Human Health" *Climate Change.*

Cao, J. et al. 2009 "Diurnal temperature range is a risk factor for Coronary heart disease death," *Journal of Epidemiology* 19: 328–332.

Claussen Eileen 1995 "Environmental and Security: The Challenge of Integration" *Environmental and Security Debates: An Introduction,* Washington D.C. The Woodrow Wilson Center, 40–43.

Conca, Ken 1994 "In the Name of Sustainability: Peace Studies and Environmental Discourse," Paper presented at the Conference on Environmental Change and Security Vancouver, Canada.

Danoploulos, C. P., Dhirendra Vajpeyi and Amir Bar 2004 (eds) *Civil Military Relations, Nation-Building and National Identities, Comparative Perspectives,* London: Praeger.

Deschenes, O. et al. 2009 "Extreme weather events, mortality, and migration," *The Review of Economics and Statistics* 91: 659–681.

Dessler, Andrew, and Edward A. Parson 2006 *The Science and Politics of Global Climate Change—A Guide to the Debate,* Cambridge, U.K.: Cambridge University Press.

Detraz, Nicole 2009 "Climate Change and Environmental Security: For Whom the Discourse Shifts," *International Studies Perspectives,* 10 pp. 303–320.

Deudney, Daniel 1991 "Environment and Security: Muddled Thinking" *Bulletin of the Atomic Scientist,* vol. 47 no. 3 April.

Deudney, Daniel and Richard Matthews (eds) 1995 *Contested Ground: Security and Conflict in the New Environmental Politics,* Albany: SUNY Press.

Diamond, Jared 2005 *Collapse—How Societies Choose to Fail or Succeed,* New York: Viking.

Dwivedi, O. P. 2008 "The Well-Being of Nations: Reflections on Integrating the Human and Ecosystem well-being (eds) Dhirendra Vajpeyi and Renu Khator *Globalization, Governance and Technology: Challenges and Alternatives,* New Delhi: Deep and Deep pp. 60–75.

Elmer-Dewitt, Phillips 1992 "Rich vs. Poor," *Time* June 12.

Finger, Matthias 1991 "The Military, The Nation State and The Environment," *The Ecologist,* 21 (5) pp. 220–225.

Flavin, Christopher 2005 "Redefining Global Security," in *State of the World* 2005, *Redefining Global Security* (ed.) Michael Renner et al. New York: W. W. Norton.

Gaby, Keith 2011 "Climate 411 New Report on Climate Change Says Wilder Weather is Headed Our Way" *Climate Central,* November 18.

Gobal Warming Deniers are full of Hot Air. 2011 *The Union of Concerned Scientists Fact Check* Cambridge, MA: (www.ucsusa.org).

Grubb, Michael 1995 "Seeking Fair Weather: Ethics and International Debate on Climate Change," *International Affairs,* 71 (3) 463–496.

Henk, Dan 2005 "Human Security: Relevance and Implications," *Parameters,* Summer pp. 91–106.

Henson, Robert 2011 *The Rough Guide to Climate Change,* London: Rough Guides Ltd.

Hertsgaard, Mark 2011 *Hot-Living Through the Next Fifty years on Earth*, New York: Mifflin Harcourt.

Hoffman, Stanley 2000 "Too Proud," *The New Republic*, January.

Homer-Dixon, Thomas F. 1994 "Environmental Scarcities and Violent Conflict: Evidence from Cases, *International Security* 19 (1) pp. 5–40.

Idso, C. D. and S. F. Singer 2009 *Climate Change Reconsidered: 2009 Report of the Nongovernmental International Panel on Climate Change (NIPCC)*, Chicago, IL: The Heartland Institute.

IPCC Fourth Assessment Report: Climate Change 2007 (Working Group I: The Physical Science Basis. www.nipccreport.org/about/about/html.

IPCC 2012: Summary for Policymakers' *Managing the Risks of Extreme Events and Disasters to Advance Climate Change Adaptation, A Special Report of Working Groups I and II of the IPCC Panel on Climate Change*, Cambridge, U.K.: Cambridge University Press. Pp. 1–19.

Johansson, M.A. et al. 2009 "Multiyear Climate Variability and Dengue-El Nino Southern Oscillation, Weather and Dengue incidence in Puerto Rico, Mexico, Thailand: A longitudinal data analysis," *PLOS Medicine* 6: e1000168

King, Gary and Christopher Murray 2001 "Rethinking Human Security," *Political Science Quarterly*, pp. 589–590.

Kyle, J.L. and E. Harris 2008 "Global Spread and Persistence of Dengue," *Annual Review of Microbiology* 62: 71–92.

Lacis, A. A. et al. 2010 "Atmospheric CO_2: Principal Control Knob Governing Earth's Temperature," *Science* October 356–359.

Le Prestre, Phillippe 1993 "Envitonmental Security and the Future of American Defense Polity" *Geopolitics of Environment and the New World Order: Limits, Conflicts, Insecurity*. Chartitilly France (6 January 1993).

Liverman, D. M. 1994 "Vulnerability to Global Environmental Change," in Cutter S. L. (ed.) *Environmental Risks and Hazards*, N.J., Prentice-Hall

Matthews, Jessica T. 1989 "Redefining Security," *Foreign Affairs*, Spring pp. 162–177.

Mendelsohn, Robert, Ariel Denar and Larry Williams 2006 "The Distributional Impact of Climate Change on Rich and Poor Countries" *Environmental and Development Economics* 11.

Myers, Norman 1993, *Ultimate Security: The Environmental Basis of Political Stability*, New York: W. W. Norton, p. 31.

Nef, Jorge 2008 "Political Economy of Globalization," in Vajpeyi and Khator (eds) pp. 141–169.

Nordhaus, William 2012 "In the Climate Casion: An Exchange," *The New York Review of Books*, August 16, 85.

Nordhaus, William 2012 "Why the Global Warming Skeptics are Wrong?" *The New York Review of Books*, March 22, 32–34.

O'Brien, Karen L. and Robin M. Leichenko 2000 "Double Exposure: Assessing the impacts of climate change within the context of Economic Globalization" *Global Environmental Change*, 10, 221: 232.

Paris, Roland 2001 "Human Security—A Paradigm Shift or Hot Air," *International Security*, Autumn 87–108.

Parry M. and T. Carter 1998 *Climate Impact and Adaptation Assessment* London: Earthscan Publications.

Raleigh, Clionadh, and Henrik Urdal 2007 "Climate Change, Envrionmental Degradation and Armed Conflict," *Political Geography* 26: 674–694.

Randall, David A., Richard A. Wood et al. 2007 "Climate Models and Their Evaluation" in *Climate Change 2007: The Physical Science bases. Contribution of Working Group I to the Fourth Assessment Report of the Intergovernmental Panel on Climate Change,* Cambridge: Cambridge University Press.

Reifen C. and R. Toumi 2009 "Climate Projections: Past Performance no guarantee of future skill?" *Geophysical Research Letters* 36: 10,1029GL038082.

Reuveny, Rafael 2007 "Climate Change and Induced Migration" *Political Geography,* 26, pp. 656–673.

Rice 2010. *USA Today,* June 22: 9

Rosenberg, N. J. 2010 "Climate Change, Agriculture, Water Resources: What do we tell those that need to know?" *Climate Change,* 100.

Singer, S. Fred 2012 "The Climate Contrarians," *The New York Review of Books,* August 16, 85.

Sitarz, Daniel (ed.) 1993 *Agenda 21: The Earth Summit Strategy to Save our Planet,"* Boulder, CO: Earth Press.

Soloman, Susan et al. 2009 *Proceedings of the National Academy of Sciences* 106: 1704–1769.

Specter, Michael 2012 "The Climate Fixers-Is there a Technological Solution to Global Warming," *The New Yorker,* May 14, pp. 96–103.

The Economist 2001, May 7: 14–18

———. December 10, 2010: 25–26.

———. December 18, 2010: 17–18.

The New York Times, March 15, 2012: B11

———. September 14, 2012: A16

Vajpeyi, Dhirendra 2012 *Water Resource Conflicts and International Security—A Global Perspective,* New York: Lexington Books.

Victor, David G. et al. 2009 "Geoengineering Option—A last resort against global warming?" *Foreign Affairs,* March–April 64–76.

Woolings, T. 2010 "Dynamical Influences on European Climate: An Uncertain Future," *Philosophical Transactions of the Royal Society* A 368.

2

Global Warming and Climate Extremes

The Impact of Carbon Dioxide (CO_2) on the Ozone

Clifton White and Rebecca McGuire

CLIMATE CHANGE AND CLIMATE EXTREMES

The relationship between climate change and climate extremes has gained increased attention and examination in recent years. Extreme weather events such as hurricanes, floods, and monsoons have forced the issue into the global spotlight, causing many to question if climate change is playing a role in the frequency and intensity of these events. According to the Intergovernmental Panel on Climate Change (IPCC) Fourth Assessment Report,

> Since 1950, the number of heat waves has increased and widespread increases have occurred in the numbers of warm nights. The extent of regions affected by droughts has also increased as precipitation over land has marginally decreased while evaporation has increased due to warmer conditions. Generally, numbers of heavy daily precipitation events that lead to flooding have increased, but not everywhere. Tropical storm and hurricane frequencies vary considerably from year to year, but evidence suggests substantial increases in intensity and duration since the 1970s (IPCC 2007: 107)

The IPCC also suggests that climate change, specifically surface warming in some regions, has contributed to "an increase of greenhouse gases in the atmosphere," and can lead to "increased risk of drought and increased intensity of storms, including tropical cyclones with higher wind speeds, a wetter Asian monsoon, and, possibly, more intense mid-latitude storms" (The Impact of Climate Change on Natural Disasters, *National Aeronautics and Space Administration*). According to Munich Re, the largest insurance company in the world, 2007 had the highest number of recorded natural disasters. 2010 came in second on the list, and was also one of the warmest years on record (Huber and Gulledge 2011), however July 2012, in America was the hottest month on record. While climate change could produce benefits to some regions by contributing to more mild winters, more rainfall, and an increase in crop production, it is also likely that others will suffer from heat waves, droughts, and "more erratic rainfall" (Riebeek 2010).

Although there is some evidence that climate change can play an influencing role on climate extremes and weather patterns, this connection is not straightforward or simple to establish (IPCC 2012). While many would like to attribute recent extreme weather and climate events to climate change, the actual relationship between the two proves to be ambiguous, inconsistent, and confusing. There are many factors that contribute to a change in climate extremes, and climate change is never the sole force influencing climate extremes (IPCC 2012). This chapter will review recent literature surrounding the issue to determine if, and to what extent, climate change plays a role on climate extremes.

Before examining the relationship between climate change and climate extremes, it is important to define them. This could be challenging, as there are several ways these extreme events are defined and classified. The IPCC states that *extreme weather events* and *extreme climate events* fall under the general term, *climate extremes*. "An extreme weather event is typically associated with changing weather patterns, that is, within time frames of less than a day to a few weeks" (Field et al. 2012: 117). These events can include floods, monsoons, hurricanes, and cyclones, "which do not necessarily occur every year at a given location" (Easterling et al. 2000a: 2068). Hence, natural disasters can fall into the category of weather extremes.

Extreme weather events can be influenced by extreme climate events. Extreme climate events typically take place over a longer period of time, and "can be the accumulation of several (extreme or non-extreme) weather events, for example, the accumulation of moderately below-average rainy days over a season leading to substantially below-average cumulated rainfall and drought conditions)" (Field et al. 2012: 117). Some examples of extreme climate events include "extreme daily temperatures, extreme daily rainfall amounts, [and] large areas experiencing unusually warm monthly temperatures" (Easterling, Meehl, et al. 2000b: 418).

Climate extremes are measured and classified in a variety of ways, most commonly by examining their frequency, intensity, and environmental, economic, or social impact (Beniston and Stephenson 2004). The IPCC states that an extreme weather event is generally classified as "rare within its statistical reference distribution at a particular place. Definitions of 'rare' vary, but an extreme weather event would normally be as rare as or rarer than the tenth or ninetieth percentile." (Parry 2007: 875). It is also important to note that the characteristics of climate extremes can vary based on the location (Parry 2007).

MEASURING CLIMATE EXTREMES

Climate models are the most commonly used mechanism to quantitatively research and estimate the relationship between climate change and climate extremes (Easterling et al. 2000a: 2068). The Intergovernmental Panel on Climate Change (IPCC) is a leading source of information on climate change and climate extremes. Its numerous reports assess "the scientific literature on issues that range from the relationship between climate change and extreme weather and climate events ('climate extremes') to the implications of these events for society and sustainable development." (IPCC 2012: 2).

According to the IPCC, "Confidence in observed changes in extremes depends on the quality and quantity of data and the availability of studies analyzing these data, which vary across regions and for different extremes" (IPCC 2012: 6). Although research and data on climate extremes is becoming increasingly valuable and efficient, it is important to note some of its shortcomings. One of the biggest obstacles is the fact that there is a shortage of long-term climate data needed to analyze and measure changes in climate extremes (Easterling et al. 2000b). In most countries, research utilizing "high temporal and spatial resolution observations of temperature, precipitation, humidity, winds, and atmospheric pressure" is relatively new, only going back to WWII (Easterling et al. 2000b: 418). However, the United States, Australia, Norway, and South Africa can trace their climate analysis to the beginning of the twentieth century (Easterling et al. 2000). The fact that there is a lack of long-term data, coupled with the fact that extreme events occur infrequently and randomly, makes it difficult to evaluate changes in the "frequency or intensity" of climate extremes (IPCC 2012: 6). Furthermore, the rarer the occurrence of weather events, the more difficult it is to determine "long-term changes" in those events (IPCC 2012: 6) thus increasing the difficulty to compile long term data and statistics.

PRECIPITATION EVENTS

There is general agreement among scientists that Earth's climate is changing due to "natural variability" and "increased concentrations of greenhouse gases and aerosols." (The Water Cycle and Climate Change, *National Aeronautics and Space Administration*). There is also widespread agreement "that these changes may profoundly affect atmospheric water vapor concentrations, clouds, [and] precipitation patterns . . ." (The Water Cycle and Climate Change, *National Aeronautics and Space Administration*). A warmer lower atmosphere can lead to an increase in evaporation, which produces "an increase in the amount of moisture circulating throughout the troposphere (lower atmosphere)" (The Water Cycle and Climate Change, *National Aeronautics and Space Administration*). This can lead to an increase in the occurrence of intense precipitation events, primarily over land. Warmer temperatures can also contribute to more rainfall instead of snowfall in some regions (The Water Cycle and Climate Change, *National Aeronautics and Space Administration*).

An increase in precipitation intensity and/or frequency can produce many unfavorable results. Van Aalst has outlined some of the risks suggested by the IPCC:

"Increased flood, landslide, avalanche, and mudslide damage"
"Increased soil erosion"
"Increased flood runoff could increase recharge of some floodplain aquifers"
"Increased pressure on government and private flood insurance systems and disaster relief" (van Aalst 2006:9; IPCC 2012).

More specifically, these potential outcomes of increased heavy precipitation events can contribute to crop damage, water contamination, deaths from injuries, infectious diseases, allergies, and damage to rural infrastructure and transportation (Extreme Events-Abrupt Climate Change, *The Environmental Protection Agency* 2012). Additionally, an increase in Asian monsoons can contribute to an "increase in flood and drought magnitude and damages in temperate and tropical Asia" (van Aalst 2006: 9; IPCC 2012).

On-going climate research and data generally support the claim that the amount of precipitation around the world is increasing. According to the IPCC, "there have been statistically significant trends in the number of heavy precipitation events in some regions." (IPCC 2012: 6). However, this increase is not evenly spread across every region, as some regions have not been subject to increased rainfall at all, and some may experience a decrease in precipitation (IPCC 2012; Flannery 2006: 124). While not all regions are impacted by an increase in precipitation, observations and data suggest that "north and South America, northern and central Asia, and northern Europe" have experienced an increase in the amount of precipitation between 1900–2005 (Climate Change, 2007: 7). A report by the U.S. Climate Change Science Program found that total precipitation in the continental United States increased by 7 percent over the last century, while the heaviest 1 percent of daily precipitation increased by 20 percent. (Karl 2008: 4). Additionally, a report by the Center for Climate and Energy Solutions suggests that global precipitation has increased 7 percent over the last fifty years (Huber and Gulledge 2011). The U.S. Climate Change Science Program also found that "heavy precipitation events averaged over North America have increased over the past fifty years, consistent with the observed increases in atmospheric water vapor, which have been associated with human-induced increases in greenhouse gases" (Karl 2008: 4–5). The IPCC has asserted "medium confidence" (a five-in-ten chance) that anthropogenic activities have been a factor in the increase of intense precipitation events. (IPCC 2012: 7).

This increase in precipitation intensity is expected to continue. According to the IPCC, "it is likely that the frequency of heavy precipitation or the proportion of total rainfall from heavy falls will increase in the twenty-first century over many areas of the globe" (IPCC 2012: 11). Precipitation will increase predominantly in "high altitudes and tropical regions, and in winter in the northern midlatitudes" (IPCC 2012: 11). The IPCC also asserted, "based on a range of emissions scenarios, a one-in-twenty year annual maximum daily precipitation amount is likely to become a one-in-five to one-in-fifteen year event by the end of the twenty-first century in many regions, and in most regions the higher emissions scenarios lead to a stronger projected decrease in return period" (IPCC 2012:11)

MONSOONS

According to NASA, "A monsoon is a seasonal shift in wind direction that brings alternate very wet and very dry seasons" (Cook-Anderson, Asian Summer Monsoon Stirred by Dust in the Wind, September 7, 2006, *National Aeronautic and Space Administration*). The IPCC has stated that "it is likely that warming associated with increasing greenhouse gas concentrations will cause an increase of Asian summer monsoon precipitation variability" (Houghton 2001:16). Depending on the scenario, climate change can contribute to changes in the strength and lifespan of monsoons by increasing or decreasing these features. (Houghton 2001). A report by the U.S. Climate Change Science Program stated that the Mexican monsoon season is beginning about ten days later than usual, and although there are fewer rain events in southwestern North America, the events that do occur are more intense (Karl 2008: 4).

Scientists have suggested that the "South Asian summer monsoon" could occur later in the season and be weakened as a result of rising temperatures (*USA Today* 2009: February 7). A research group at Purdue University found that climate change could contribute to less summer rainfall overall and longer intervals between "rainy periods" in the region. According to Noah Diffenbaugh, the interim director of the Purdue Climate Research Center, "almost half of the world's population lives in areas affected by these monsoons, and even slight deviations from the normal monsoon pattern can have great impact" (*USA Today* 2009: February 7). "Agricultural production, water availability, and hydroelectric power generation could be substantially affected by delayed monsoon onset and reduced surface runoff" (*USA Today* 2009: February 7). The Purdue model observed that while weakened monsoons could negatively impact many parts of this region, there will also be increases in rainfall over other areas, including Bangladesh. This could contribute to increased flooding in Bangladesh, which poses a great risk to local communities (*USA Today* 2009: February 7).

While there have been several studies regarding the relationship between monsoons and climate change, the IPCC states that "confidence is low in projections of changes in monsoons (rainfall, circulation) because there is little consensus in climate models regarding the sign of future change in the monsoons" (IPCC 2012: 13). There is also some inconsistency in projected changes in the "El Nino-Southern Oscillation variability and the frequency of El Nino episodes," and the reliability of monsoon forecasts is restricted by how efficient climate models are at "simulat[ing] the detailed seasonal evolution of the monsoons" (IPCC 2012:13–14; Houghton 2001: 16).

FLOODS

According to the IPCC, "there is limited to medium evidence available to assess climate-driven observed changes in the magnitude and frequency of floods at regional scales because the available instrumental records of floods at gauge stations are limited in space and time, and because of confounding effects of changes in land use and engineering" (IPCC 2012: 6). Because heavy floods are infrequent, it is difficult to determine the difference between 'freak events' and "real changes in the probability of a particular event" (van Aalst 2006: 11). Flood events are also influenced "by land use in the watershed area" (van Aalst 2006: 11). This makes it difficult to establish a strong link between changes in flooding and climate change (van Aalst 2006: 11).

However, the IPCC has asserted medium confidence that the predicted increases in heavy rainfall will be a contributing factor to increased "local flooding in some catchments or regions" (IPCC 2012: 13). The United States Environmental Protection Agency (EPA) has also suggested that climate change could increase or diminish coastal flooding. "Flooding from rainstorms may become worse if higher temperatures lead to increasing rainfall intensity during severe storms" (Coastal Zones and Sea Level Rise, April 14, 2011, *The Environmental Protection Agency*). A boost in tropical storms could also increase damages caused by floods. "It is likely that there has been a human influence on increasing extreme coastal high water due to an increase in mean sea level" (IPCC 2012: 7).

DRYING AND DROUGHTS

While many regions are experiencing an increase in precipitation intensity, other regions are forced to cope with a serious lack of rainfall, resulting in droughts and overall drier climates (Karl 2008; Flannery 2006). According to a report by the U.S. Climate Change Science Program, "drought is one of the most costly types of extreme events and can affect large areas for long periods of time" (Karl 2008: 5). The IPCC projects that increased drying and droughts can have many impacts on the environment, including:

A reduction in crop production
"Damage to building foundations caused by ground shrinkage"
A reduction in water supply "quantity and quality"
Increased probability of forest fires

(van Aalst 2006: 9)

According to *Climate Change 2007: The Physical Science Basis*, "increased drying linked with higher temperatures and decreased precipitation have contributed to changes in drought . . . changes in sea surface temperatures (SST), wind patterns, and decreased snowpack and snow cover have also been linked to droughts" (Climate Change 2007: 8). The IPCC expresses "medium confidence" that some regions, particularly "southern Europe and West Africa," have been affected by more extreme and prolonged droughts. (IPCC 2012: 8). Since the 1970s, droughts have been occurring in larger areas, "particularly in the tropics and subtropics" (Climate

Change 2007: 8). Dryer climates have also "been observed in the Sahel, Mediterranean, southern Africa, and parts of southern Asia" (Climate Change 2007: 7). NASA has echoed this statement with evidence from the Palmer Drought Severity Index, which shows that the Sahel region of Africa experienced harsher drought conditions from 1900 to 2002. This index also suggested opposite conditions in the south central United States and southern South America (The Water Cycle and Climate Change, *National Aeronautics and Space Administration*).

Due to overall increase of dryness and drought in Africa's Sahel region it has received a considerable amount of attention. The first sign of a decline in rainfall in the Sahel was observed in the 1960s, and since then there has been no indication "that the life-giving monsoon rains will return" (Flannery 2006: 124). Although the lack of rainfall is an enormous factor contributing to the region becoming a "dust-filled wasteland," many in the west have argued that the local populations are also to be blamed for the drought (Flannery 2006:124–125). In 2003, climatologists at the National Center for Atmospheric Research in Colorado "published a painstaking study that used computer models to simulate rainfall regimes in the region between 1930 to 2000" (Giannini et al. 2003; Flannery 2006: 125). The study found that influences due to the local population, such as land degradation caused by locals, played an insignificant role on the drought. Instead, the rising temperatures of the surface of the Indian Ocean, brought about by a collection of greenhouse gases, was a major factor responsible for the "dramatic climate shift." (Giannini 2003; Zeng 2003; Flannery 2006: 125). The study found that as the Indian Ocean warmed, the conditions that produce monsoons in Sahel deteriorated (Giannini et al. 2003; Flannery 2006). Although many different environmental and anthropogenic factors influence every weather event, many believe that climate change, specifically the warming of the Indian Ocean, has been a major contributing factor to the Sahel drought.

While droughts have increased in many regions, in other areas "droughts have become less frequent, less intense, or shorter, for example, in central North America and northwestern Australia" (IPCC 2012:8). Although droughts are decreasing in some areas, the IPCC states, "there is medium confidence that droughts will intensify in the twenty-first century in some seasons and areas, due to reduced precipitation and/or increased evapotranspiration" (IPCC 2012: 13). These regions include "southern Europe and the Mediterranean region, central Europe, central North America, Central America and Mexico, northeast Brazil, and southern Africa" (IPCC 2012: 13). The IPCC also stated that "definitional issues, lack of observational data, and the inability of models to include all the factors that influence droughts" make it difficult to have higher confidence in drought predictions (IPCC 2012: 11).

TROPICAL CYCLONES

According to van Aalst (2006: 12), "Tropical cyclones, typhoons, and hurricanes all pertain to the same phenomenon: weather systems with strong winds that circulate around low-pressure areas, with a central 'eye' that has a diameter of typically 20–150 kilometers." These weather events are usually produced over oceans with temperatures of at least 25.5°C. Tropical Cyclones can last for at least two weeks, but they quickly lose momentum as they move inland (van Aalst 2006), however they do create havoc and disarray in several ways:

1. "Damage to crops; windthrow of trees"
2. "Power outages cause disruption of public water supply"
3. "Increased risk of deaths, injuries, [and] water and food borne diseases"
4. "Disruption by flood and high winds; withdrawal of risk coverage in vulnerable areas by private insurers"

(Extreme Events—Abrupt Climate Change, March 18, 2012, *The Environmental Protection Agency*)

Cyclone events are particularly difficult to study and establish their link to climate change. It is because they occur infrequently, and there is a "high interannual and multi-decadal variability in tropical cyclone occurrence" which can be attributed to "the El Nino/Southern Oscillation (ENSO), the stratospheric quasi-biennial oscillation and multi-decadal oscillations in the North Atlantic" (van Aalst 2006: 13). Also tropical cyclones "are the product of much more complex atmospheric interactions," making it more difficult to establish a strong link to global warming (van Aalst 2006). Current research and climate models have failed to reach a consensus regarding the change in tropical cyclones; while some climate models show that climate change can contribute to a decrease in tropical cyclone frequency, others show it resulting in an increase. (IPCC Workshop on Changes in Extreme Weather and Climate Events 2002: 19).

Research on tropical cyclones is also hindered because satellite observations of tropical cyclones only date back thirty years. There is an inconsistent and uneven quality of observation across different regions, and different basins have different definitions of "tropical cyclones" (IPCC 2002). These difficulties produce a lack of consistency in the techniques for measuring tropical cyclones, necessitating more work and research to better understand the relationship between climate change and tropical cyclones (IPCC Workshop on Changes in Extreme Weather and Climate Events 2002). In addition to the inconsistency and lack of consensus regarding the change in tropical cyclones, the IPCC has "low confidence in any observed long-term (i.e., forty years or more) increases in tropical cyclone activity (i.e., intensity, frequency, duration), after accounting for past changes in observing capabilities" (IPCC 2012: 8).

Even though a strong connection between climate change and tropical storms has not been established, it is likely that global warming is increasing ocean temperatures and producing warmer, moister air, which creates a stronger energy source to fuel storms. This can produce more powerful and intense rainfall and wind velocity in tropical storms (Trenberth 2005; van Aalst 2006; The Impact of Climate Change on Natural Disasters, *National Aeronautic and Space Administration*. The maximum wind speed of tropical cyclones in some areas is projected to increase 5–10 percent, while the precipitation accompanying these storms could increase 20–30 percent (IPCC 2012; Houghton 2001). Additionally, "models based on scientists' current understanding of hurricanes suggest that if ocean temperatures increased by 2–2.5 degrees, the average intensity of hurricanes would increase by 6–10 percent" (Hurricanes: The Greatest Storms on Earth, *National Aeronautic and Space Administration*). The National Aeronautic and Space Administration (NASA) observes that the average ocean temperature has increased by half a degree since 1970, which suggests that theoretically, tropical storms could currently be one to three percent stronger than they were forty years ago (Hurricanes: The Greatest Storms on Earth, *National Aeronautic and Space Administration*).

Karl's (2008: 5) U.S. Climate Change Science Program supports the above observation "Atlantic tropical storm and hurricane destructive potential as measured by the Power Dissipation Index (which combines storm intensity, duration, and frequency) has increased." It points out that although "there have been fluctuations in the number of tropical storms and hurricanes from decade to decade and data uncertainty is larger in the early part of the record compared to the satellite era beginning in 1965," it is likely that the number of tropical storms and hurricanes have increased over the past century" (Karl 2008: 6). The intensity of hurricanes in the western north Pacific has increased since 1980, while it has decreased in the eastern Pacific, "affecting the Mexican west coast and shipping lanes" (Karl 2008: 6)

Karl's report also assessed the impact of anthropogenic forces on tropical cyclones. "It is very likely that the human-induced increase in greenhouse gases has contributed to the increase in sea surface temperatures in the hurricane formation regions" (Karl 2008: 6). Furthermore, there has been "a strong statistical connection between tropical Atlantic sea surface temperatures and Atlantic hurricane activity as measured by the Power Dissipation Index" (Karl 2008: 7). Although an anthropogenic connection to tropical cyclone activity is suggested, further research needs to be done on this topic to determine if there is a link, especially because there are many factors that affect tropical storms (Karl 2008: 7) that contribute to changes in the climate extremes. Several studies have found that climate change can play a role in altering the frequency and intensity of extreme events, and that recent changes to the earth's climate, including warming of the oceans and atmosphere, can contribute to changes in climate and weather events. These changes in climate extremes are expected to continue in the future, producing "important consequences on society and natural systems" thus impacting the quality of human life (Easterling et al. 2000a).

CARBON DIOXIDE AND THE OZONE LAYER

As discussed above there are many factors associated with climate change. With the discovery of the Ozone Hole over the Antarctic in 1985; the ozone layer became a hot button issue. Along with the ozone layer carbon dioxide (CO_2) has also attracted much attention among scientists in the ongoing debate over the past decade. It is clear that since the "first" industrial revolution starting during the mid eighteenth century the CO_2 levels in our environment have increased sharply. More recently we have seen an adverse change in weather patterns throughout the world. It has been suggested that these anomalies are due to global warming and climate change resulting in devastating storms and natural disasters.

THE OZONE LAYER

The Ozone is a gas that occurs naturally in very small amounts in Earth's upper atmosphere that protects Earth and its inhabitants from the Sun's radiation (*Ozone Hole Watch: What is Ozone*, March 15, 2012). The Ozone itself is most commonly measured in Dobson Units, which can be a complicated measurement. Simply put, "One Dobson Unit is the number of molecules of ozone that would be required to create a layer of pure ozone 0.01 millimeters thick at a temperature of 0 degrees Celsius and a pressure of 1 atmosphere (the air pressure at the surface of the Earth)" (*Ozone Hole Watch: What is a Dobson Unit*, March 15, 2012). The average Ozone Layer is around 3 millimeters thick or 300 Dobson Units, while the Ozone Hole (or depleted area) averages around one millimeter thick or 100 Dobson Units (*Ozone Hole Watch: What is a Dobson Unit*, March 15, 2012). To put that in perspective the width of a penny is about 1.5 millimeters thick; so the Ozone Layer that protects us from the sun is essentially two pennies thick. Thus any substantial reduction in the ozone layer could potentially pose serious health risks not only to humans but also to plant and animal life on earth.

The ozone layer didn't become a topic of serious debate until a group of scientists in the mid 1980s discovered the depletion of the ozone over Antarctica. They found that it was largely due to the use of chlorofluorocarbons (CFCs) which were heavily used in aerosol cans. Because when chlorine is released from the CFCs it combines with the ozone layer and slowly destroys it (*Ozone Hole Watch: What is the Ozone Hole,* March 15, 2012). With this scientific discovery it was realized that serious steps were needed to remedy the situation to avert future calamities. In September 1987 leaders from around the world signed the Montreal Protocol. The Protocol was designed to eliminate the use of ozone-depleting substances (ODS) to curb and eventually mend the ozone hole. Within the Protocol there were specific deadlines that countries were to meet when it came to phasing out ODSs. By 2030 the United States is supposed to completely phase out the use and production of ODSs. The deadline for developing countries to meet their "phase out" is by 2040.

Since the Protocol's implementation, The United States has successfully "phased out nearly ninety-seven of one-hundred ozone-depleting chemicals," (*The New York Times,* Nov. 8, 2010). Though the first "wave" of deadlines has passed the United States is still in the process of phasing out the rest of the prohibited chemicals under the Montreal Protocol's guidelines, latest by 2030. Progress made under the Montreal Protocol has been integral to the survival of the ozone and it is no wonder that former UN Secretary General Kofi Annan called it, "perhaps the most successful international agreement to date" (*The New York Times,* Nov. 8, 2010).

Table 2.1. United States Production of First Generation ODS Phase Out Schedule

Chemical Group	*Production Phase-out Dates*	*Deadline Met*
Halons	January 1, 1994	YES
Chlorofluorocarbons (CFCs)	January 1, 1996	YES
Carbon tetrachloride	January 1, 1996	YES
Hydrobromofluorcarbons (HBFCs)	January 1, 1996	YES
Methyl chloroform	January 1, 1996	YES
Chlorobromethane	August 18, 2003	YES
Methyl Bromide	January 1, 2005	YES

Source: "Montreal Protocol," Environmental Protection Agency, 2007, pg. 2 (http://www.epa.gov/ozone/downloads/MP20_Backgrounder.pdf)

The European Union has taken a hard line approach to reduce their use of ODSs. The Montreal Protocol had stipulated the European Union to phase out all ODSs by the year 2020. The European Union met its deadline ten years ahead of schedule in 2010. Not only did it complete its obligation ten years earlier it has also regulated chemicals that are "not considered as consumption under the Montreal Protocol" ("Climate Action," March 11, 2011, The European Commission). The success of the European Union shows that the task at hand is clearly feasible and can be achieved by all countries.

Though there has been much effort and much progress to reduce the depletion of the ozone there are still instances of recent depletion. In 2011 the first ever significant Arctic ozone hole was reported. While this was considered temporary it just shows us that the past damage that was done to the ozone cannot be fixed over night. According to Santee, a planetary scientist at NASAs Jet Propulsion Lab, "The root cause is the residual products from the CFCs that were released throughout the twentieth century. But they are very long-lived, and it will take a few decades for them to be cleansed from the atmosphere" (*The New York Times,* Oct. 3, 2011). The evidence points out that the Montreal Protocol (1987) has done an excellent job of regulating the use of ozone depleting substances; yet the emissions from these chemicals remain close to 25 percent higher than when the Protocol was enacted. Depletion of the ozone is still taking place reminding us "how sudden anomalies can occur as a result of human activity that occurred years ago" (*The New York Times,* Oct. 3, 2011).

One of the most disturbing challenges to ozone depletion efforts is the new scientific data on chemicals such as nitrous oxide as an ODSs impacting the Ozone Layer. According to the National Oceanic and Atmospheric Administration the overall health of the ozone has improved since the implementation of the Montreal Protocol, however "nitrous oxide looms large today as an artificial destroyer of the ozone layer" (*New York Times,* Aug. 27, 2009). The extent of damage to ozone by nitrous oxide is further complicated due to scientific data's uncertainty as to "how much nitrous oxide comes from where, and that the uncertainties are significant," however, it is quite clear that it does contribute to the depletion of our ozone (*New York Times,* Aug. 27, 2009).

CARBON DIOXIDE AND CLIMATE CHANGE

Carbon Dioxide levels are another aspect of global warming that have been becoming more prevalent in the recent years. CO_2 gases can be emitted into

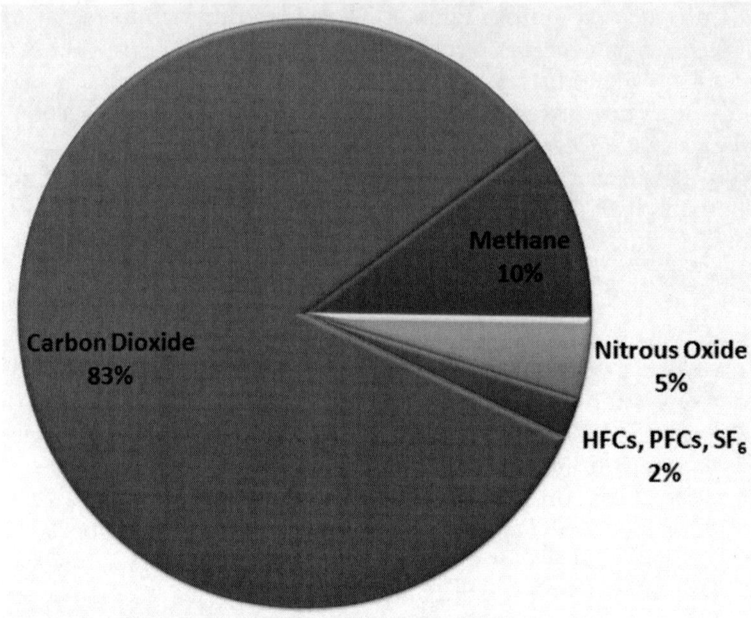

Figure 2.1. U.S. Greenhouse Gas Emissions 2009
Source: "U.S. Greenhouse Gas Emissions by Gas," *Center for Climate and Energy Solutions* (http://www.c2es.org/facts-figures/us-emissions/gas)

the atmosphere through either human activities or the carbon cycle which is a natural process (Carbon Dioxide, *The Environmental Protection Agency*, April 14, 2011). The Industrial Revolution (mid eighteenth century) brought about immense political, economic, and social changes. The vast natural resources were made available for human consumption, and industries run by coal emitted vast amounts of CO_2 in the atmosphere. As a result in 2005, "global atmospheric concentrations of CO_2 were 35 percent higher than they were before the Industrial Revolution" (Carbon Dioxide, *The Environmental Protection Agency*, April 14, 2011). CO_2 is the largest contributor of greenhouse gases to Earth's atmosphere. In the United States CO_2 contributes 83 percent of its greenhouse gas emissions, which is higher than any other country. (Figures 2.1 and 2.2)

Though CO_2 and greenhouse gas emissions in general are seen as a negative, it is important to note that they do occur naturally within our environment. The problem lies when they're buttressed by external human activities such as production and consumption processes. While CO_2 is naturally emitted into the environment, it is also naturally removed as well through "sinks." According to the EPA the primary ways that CO_2 gases are emitted and removed naturally from our atmosphere are:

1. Plant and animal respiration—"oxygen and nutrients are converted into CO_2 and energy, and plant photosynthesis by which CO_2 is removed from the atmosphere and stored as carbon in plant biomass."
2. Ocean-atmosphere exchange—This is where "oceans absorb and release CO_2 at the sea surface."
3. Volcanic eruptions also release a small amount of carbon

(Natural Sources and Sinks of Carbon Dioxide, April 14, 2011, *The Environmental Protection Agency*).

The data shows that these natural sinks are overworked and they can't keep up and remove the amount that humans emit. The major human generated causes of CO_2 consist of:

1. Combustion of fossil fuels
2. Industrial production processes and product uses—This consists of "mineral production, metal production, and the use of petroleum-based products."
3. Deforestation

(Human-Related Sources and Sinks of Carbon Dioxide, April 14, 2011, *The Environmental Protection Agency*).

Carbon can also be removed by humans through capture techniques that capture the gases prior to their release into the atmosphere and then injected underground (geoengineering). Growing and planting new trees can also contribute to the creation of more carbon sinks (Human-Related

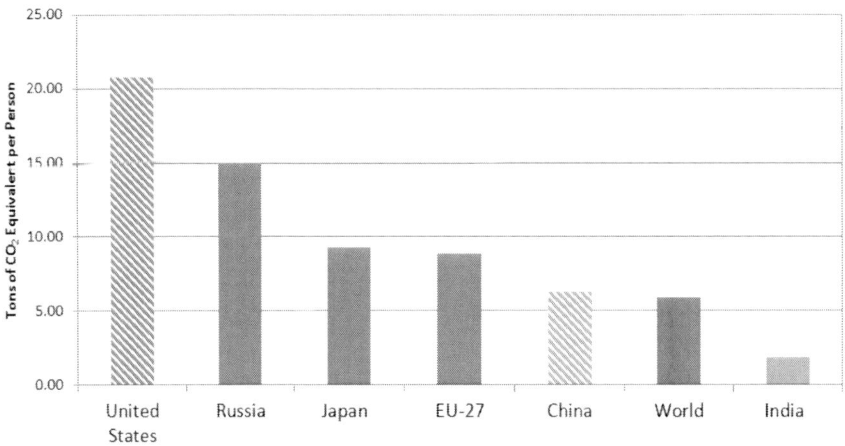

Figure 2.2. Per Capita GHG Emissions 2009

Source: "Per Capita GHG Emissions," *Center for Climate and Energy Solutions* (http://www.c2es.org/facts-figures/international-emissions/capita).

Sources and Sinks of Carbon Dioxide, April 14, 2011 *The Environmental Protection Agency*). But this is difficult because many countries don't have the expansive and untested carbon capture technology and just planting trees won't solve the problem; especially due to worldwide deforestation (Vajpeyi 2001). Another problem is related to the use of carbon credits. In the United States, for example industries are allowed to sell their unused "credits" to industries which exceed the regulated emissions level. This essentially allows companies to pollute more without being punished.

Though there are many sources of CO_2 emissions the largest one by far is due to fossil fuels. When we burn fossil fuels to create energy, the emissions byproduct that is produced is nearly entirely CO_2 (Human-Related Sources and Sinks of Carbon Dioxide, *The Environmental Protection Agency*, April 14, 2011). Since our entire industrial world infrastructure is based on fossil fuels usage the emission of CO_2 is a difficult issue to tackle. The dire need to adapt alternative energy has not met with much success despite an increased use of solar energy and other alternative measures.

Throughout the earth's history CO_2 levels have risen and declined, but over the past fifty years these levels have skyrocketed to new levels as never before seen (Figure 2.3). Due to this very real and measurable change the clear consensus among most scientists is that our climate is warming and greenhouse gases are major factors contributing to the problem. The Intergovernmental Panel on Climate Change stated, "Scientific evidence for warming of the climate system is unequivocal" (Global Climate Change, *National Aeronautics and Space Administration*).

With this increase in CO_2 levels there have also been clear and measurable changes to our environment. Over the past century the global

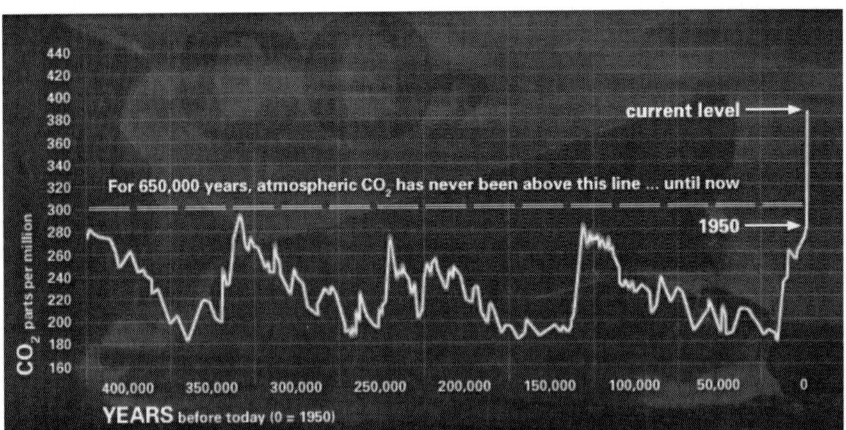

Figure 2.3. Atmospheric CO_2 Levels

Source: "Global Climate Change: Vital Signs of the Planet," National Aeronautics and Space Administration (http://climate.nasa.gov/evidence/).

sea level has increased by close to 17 centimeters; but within the last decade, this rate has nearly doubled (Global Climate Change, *National Aeronautics and Space Administration*). With this continued increase in sea levels some coastal areas and islands could potentially be reclaimed by the sea. Another measurable aspect of climate change is the global temperature, and "all three major global surface temperature reconstructions show that Earth has warmed since 1880" (Global Climate Change, *National Aeronautics and Space Administration*). Not only has earth clearly warmed in the past century, the majority of the temperature increases have occurred since 1980; and the ten warmest years on record have occurred within the past fifteen years (Global Climate Change, *National Aeronautics and Space Administration*). With temperatures rising at such a rapid rate it has adverse effects on other environmental areas as well; particularly the ice sheets. Between the years 2002 and 2006 Greenland's ice sheets were depleted by 36 to 60 cubic miles each year, and measurements clearly show that the "Arctic sea ice has rapidly declined over the past several decades Global Climate Change, *National Aeronautics and Space Administration*)." Studies such as the one done by Hellmer's point out that warming water is actually melting the ice from below the surface, and the rising sea level could potentially be much more rapid than expected (*The Chicago Tribune*, May 9, 2012).

Oceans have also been greatly affected by CO_2 emissions. The waters of the oceans have increased in acidity by close to 30 percent (Global Climate Change, *National Aeronautics and Space Administration*). Oceans are carbon sinks, and due to increased emissions they have been absorbing more carbon causing the acidity of the ocean to increase. Currently the upper portion of the oceans absorption of CO_2 is increasing by around 2 billion tons per year (Global Climate Change, *National Aeronautics and Space Administration*). With the increased acidity levels of the oceans the potential effect on plant and animal life is quite diverse. One recent study has linked increased ocean acidification to oyster die-offs. In 2006 oyster hatcheries in the Pacific Northwest in the United States began to notice catastrophic die-offs in the oyster population directly related to the rising CO_2 that cause ocean acidity (*The New York Times*, April 12, 2012). Miyoko Sakashita, Oceans Director for the Center for Biological Diversity, stated "Oyster die-offs are an unmistakable warning that our oceans are in trouble" (*The New York Times*, April 12, 2012).

Recent research by Shakun indicates that increased CO_2 levels have caused an end to the ice age, which they feel potentially solves the debate as to what came first: temperature changes or increase carbon dioxide levels (*The Chicago Tribune*, April 5, 2012). This topic has been debated heavily in the past years especially in the political community after the release of former Vice President Al Gore's movie "An Inconvenient Truth."

Skeptics such as Texas Representative Joe Barton question the extent and nature of these developments, "The temperature appears to drive CO_2, not vice versa. On this point, Mr. Vice President, you're not just off a little. You're totally wrong" (*The Chicago Tribune,* April 5, 2012).

Though there is an overwhelming consensus amongst the scientific community that global warming is occurring and CO_2 is increasing the Earth's temperature, there are still many deniers that dispute this fact. William Happer a professor of Physics at Princeton is one of the outspoken challengers of global warming and the damage it is doing. Happer disputes the fact that CO_2 is a problem and in fact says it is good for the environment and not in fact a pollutant. He states:

> CO_2 is not a pollutant. Life on earth flourished for hundreds of millions of years at much higher CO_2 levels than we see today. Increasing CO_2 levels will be a net benefit because cultivated plants grow better and are more resistant to drought at higher CO_2 levels, and because warming and other supposedly harmful effects of CO_2 have been greatly exaggerated. (*The Wall Street Journal,* March 27, 2012)

When looking at the data, it is clear that we have seen a substantial increase in our land surface and ocean temperature in the past century which by most accounts is likely due to the increasing CO_2 levels in our atmosphere (Figure 2.4).

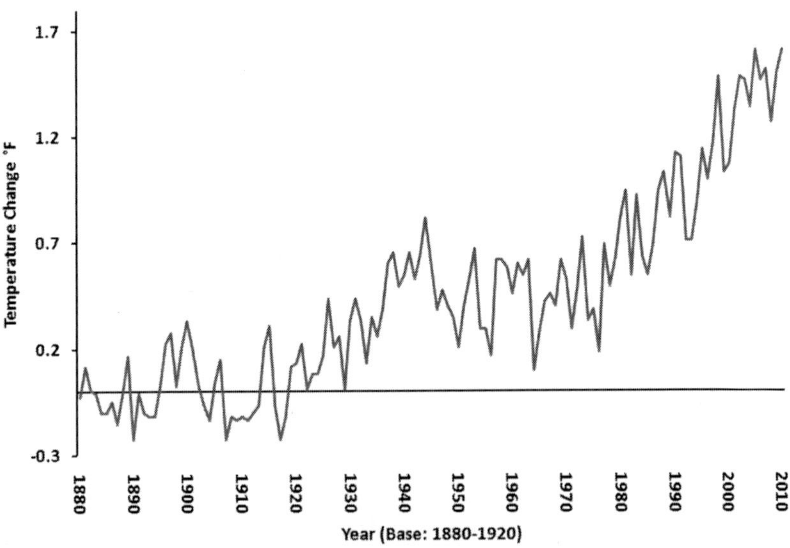

Figure 2.4. Global Surface Temperature Trends

Source: *Center for Climate and Energy Solutions* (http://www.c2es.org/facts-figures/trends/surface-temp).

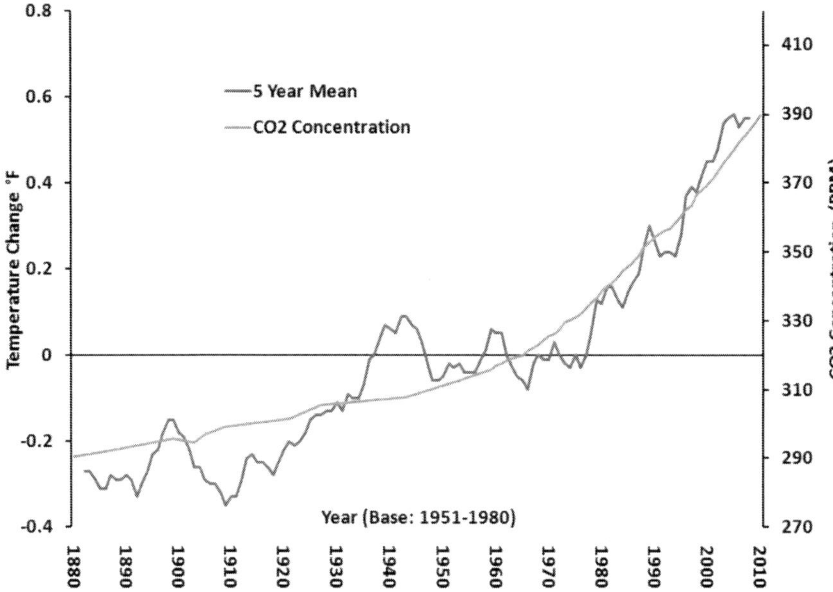

Figure 2.5. Atmospheric Carbon Dioxide and Global Surface Temperature Trends
Source: *Center for Climate and Energy Solutions* (http://www.c2es.org/facts-figures/trends/co2-temp).

Though temperatures have fluctuated from year to year, it is clear that they have risen over 1 degree Fahrenheit since 1880, with the sharpest increase occurring within the past thirty years (Figure 2.5).

The CO_2 and temperature levels are related and fluctuate together. As CO_2 concentration has increased over the past 130 years so has a steady rise in temperature.

Despite the Montreal Protocol's success in regulating ODSs there has been little progress made since then on to climate change as a whole. Post Montreal Protocol has witnessed several international commitments to meet the global warming challenges (chapter 1). The Kyoto Protocol calls for reduction of emissions by 5.2 percent below 1990 levels by 2012 (International Negotiations, *Center for Climate And Energy Solutions*). Though the Kyoto Protocol was specifically meant to curb the emissions from industrialized countries; the United States still has yet to officially ratify the Protocol.

Climate change is a major challenge facing humanity and the well being of the planet and can be solved only by global efforts by all respective political ideologies and geographical locations. Finger pointing and pursuit of short range narrow national interest as demonstrated at Copenhagen in 2009 are definitely going to affect human security and the quality of human life and sustainable development. ". . . that for the planet to survive in some habitable form, the world has to live within a fixed carbon budget of about 750 gigatons of CO_2 emissions between now and 2050 . . . The cold, hard fact is that a drastic reduction in aggregate emissions is required if we are to achieve a reasonable probability of keeping temperatures at a livable levels" (Mattoo and Subramanian, 2013: 7)

REFERENCES

"Atmospheric Carbon Dioxide and Global Surface Temperature Trends," *Center for Climate and Energy Solutions*, www.c2es.org/facts-figures/trends/co2-temp.

Barringer, Felicity. "A Significant Ozone Hole is Reported Over the Arctic," *The New York Times*: October 3, 2011, www.nytimes.com/2011/10/04/science/earth/04ozone.html?_r=3.

Beniston, Martin, and David B. Stephenson. 2004. "Extreme Climatic Events and Their Evolution Under Changing Climatic Conditions." *Global and Planetary Change.* 44, no. 1: 1. g01.cnitc.cn/uploadfile/file/□□/□□□□□/1995-2005/2012010620521617.pdf.

Broder, John M. "A Novel Tactic in Climate Fight Gains some Traction," *The New York Times*: November 8, 2010, www.nytimes.com/2010/11/09/science/earth/09montreal.html?_r=1.

"Carbon Dioxide," April 14, 2011, *The Environmental Protection Agency*, www.epa.gov/climatechange/emissions/co2.html.

"Climate Action," March 11, 2011, *The European Commission*, ec.europa.eu/clima/policies/ozone/index_en.htm.

Climate Change 2007: The Physical Science Basis-Summary for Policymakers. Contribution of Working Group I to the Fourth Assessment Report of the Intergovernmental Panel on Climate Change. 2007. www.ipcc.ch/pdf/assessment-report/ar4/wg1/ar4-wg1-spm.pdf.

Coastal Zones and Sea Level Rise. April 14, 2011. *The Environmental Protection Agency*. www.epa.gov/climatechange/effects/coastal/index.html.

Cook-Anderson, Gretchen. "Asian Summer Monsoon Stirred by Dust in the Wind," September 7, 2006. *National Aeronautic and Space Administration*. www.nasa.gov/vision/earth/environment/asian_rain.html.

Dean, Cornelia. "New Culprit Seen in Ozone Depletion," *The New York Times*: August 27, 2009, www.nytimes.com/2009/08/28/science/earth/28nox.html?_r=1.

Easterling D. R., G. A. Meehl, C. Parmesan, S. A. Changnon, T. R. Karl, and L. O. Mearns. 2000 (a). "Climate Extremes: Observations, Modeling, and Impacts." *Science* (New York, N.Y.). 289, no. 5487: 2068–74. clivac.eri.ucsb.edu/classes/Tropical_meteorology/classes_ucsb/climate_change/articles_climate_change_tropics/Easterling_et_al_2000%20extreme%20events.pdf.

Easterling, D. R., J. L. Evans, P. Ya Groisman, T. R. Karl, K. E. Kunkel, and P. Ambenje. 2000 (b). "Observed Variability and Trends in Extreme Climate Events: A Brief Review." *Bulletin of the American Meteorological Society.* 81, no. 3. g01.cnitc.cn/uploadfile/file/□□/□□□□□/2000/2012010714282449.pdf.

Extreme Events—Abrupt Climate Change. March 28, 2012. *The Environmental Protection Agency*. www.epa.gov/climatechange/effects/extreme.html.
Field, Christopher, Barros, Vicente, Stocker, Thomas F., and Dahe, Qin. Managing the Risks of Extreme Events and Disasters to Advance Climate Change Adaptation Special Report of the Intergovernmental Panel on Climate Change. Cambridge University Press, 2012.
Flannery, Timothy Fridtjof. The Weather Makers: The History and Future Impact of Climate Change. London: Allen Lane, 2006.
Giannini, A., Saravanan, R., and Chang, P. 2003. Oceanic Forcing of Sahel Rainfall on Interannual to Interdecadal Timescales. *Science* 302, pp. 1027–30. pangea.stanford.edu/research/Oceans/GES205/SahelianPrecip.pdf.
Gillis, Justin and John Broder, "With Carbon Dioxide Emissions at Record High, Worries on How to Slow Warming," *The New York Times*, December 3, 2012.
"Global Anthropogenic GHG Emissions by Gas," Center for Climate and Energy Solutions, www.c2es.org/facts-figures/international-emissions/gas.
"Global Climate Change: Vital Signs of the Planet," *National Aeronautics and Space Administration*, climate.nasa.gov/evidence/.
"Global Greenhouse Gas Data," April 14, 2011, The Environmental Protection Agency, www.epa.gov/climatechange/emissions/globalghg.html.
"Global Surface Temperature Trends," *Center for Climate and Energy Solutions*, www.c2es.org/facts-figures/trends/surface-temp.
Harper, William, "Global Warming Models are Wrong Again," *The Wall Street Journal*, March 27, 2012, online.wsj.com/article/SB10001424052702304636404577291352882984274.html.
Houghton, J. T. Climate Change 2001: The Scientific Basis: Contribution of Working Group I to the Third Assessment Report of the Intergovernmental Panel on Climate Change. Cambridge: Cambridge University Press, 2001. www.acrim.com/%5C/Reference%20Files/CLIMATECHANGE%202001%20-%20The%20Scientific%20Basis.pdf.
Huber, Daniel G. and Gulledge, Jay. Extreme Weather & Climate Change: Understanding the Link and Managing the Risk. *Center for Climate and Energy Solutions*, December 2011. www.c2es.orgwww.c2es.org/docUploads/white-paper-extreme-weather-climate-change-understanding-link-managing-risk.pdf.
"Human-Related Sources and Sinks of Carbon Dioxide," April 14, 2011, *The Environmental Protection Agency*, www.epa.gov/climatechange/emissions/co2_human.html.
Hurricanes: The Greatest Storms on Earth. *National Aeronautics and Space Administration*. earthobservatory.nasa.gov/Features/Hurricanes/hurricanes_3.php.
"International Negotiations," *Center for Climate and Energy Solutions*," www.c2es.org/international/negotiations.
IPCC, 2007: Climate Change 2007: The Physical Science Basis. Contribution of Working Group I to the Fourth Assessment Report of the Intergovernmental Panel on Climate Change-Frequently Asked Questions. [Solomon, S., D. Qin, M. Manning, Z. Chen, M. Marquis, K. B. Averyt, M. Tignor and H. L. Miller (eds.)]. Cambridge University Press, Cambridge, United Kingdom and New York, N.Y., USA. www.ipcc.ch/pdf/assessment-report/ar4/wg1/ar4-wg1-faqs.pdf.

IPCC, 2012: Summary for Policymakers. In: Managing the Risks of Extreme Events and Disasters to Advance Climate Change Adaption [Field, C. B., V. Barros, T. F. Stocker, D. Qin, D. J. Dokken, K. L. Ebi, M. D. Mastrandrea, K. J. Mach, G. K. Plattner, S. K. Allen, M. Tignor, and P.M. Midgley (eds.)]. A Special Report of Working Groups I and II of the Intergovernmental Panel on Climate Change. Cambridge University Press, Cambridge, UK, and New York, N.Y., USA, pp. 1–19. ipcc-wg2.gov/SREX/images/uploads/SREX-SPM_FINAL.pdf.

IPCC Workshop on Changes in Extreme Weather and Climate Events. IPCC Workshop on Changes in Extreme Weather and Climate Events, Beijing, China, 11–13 June 2002: Workshop Report. Beijing, China: The Workshop, 2002. www.ipcc.ch/pdf/supporting-material/ipcc-workshop-2002-06.pdf.

Karl, Thomas. Weather and Climate Extremes in a Changing Climate Regions of Focus: North America, Hawaii, Caribbean, and U.S. Pacific Islands. Washington, D.C.: U.S. *Climate Change Science Program*, 2008. purl.access.gpo.gov/GPO/LPS110676.

Kaufman, Leslie. "Study Links Raised Carbon Dioxide Levels to Oyster Die-Offs," *The New York Times:* April 12, 2012, green.blogs.nytimes.com/2012/04/12/study-links-raised-carbon-dioxide-levels-to-oyster-die-offs/.

Koebler, Jason. "Study: Carbon Dioxide Increase Caused End of Ice Age," *The Chicago Tribune*, April 5, 2012, (http://articles.chicagotribune.com/2012-04-05/news/sns-201204051128usnewsusnwr201204040404warmingapr05_1_global-carbon-dioxide-co2).

Mattoo, Aaditya and Arvind Subramanian. 2013. *Greenprint-A new Approach to Cooperation on Climate Change*, Washington DC Center for Global Development.

"Montreal Protocol," *Environmental Protection Agency*, 2007. www.epa.gov/ozone/downloads/MP20_Backgrounder.pdf.

"NASA Ozone Watch: What is a Dobson Unit," March 15, 2012, *National Aeronautics and Space Administration*, ozonewatch.gsfc.nasa.gov/facts/dobson_SH.html.

"NASA Ozone Watch: What is Ozone," March 15, 2012, *National Aeronautics and Space Administration*, ozonewatch.gsfc.nasa.gov/facts/SH.html.

"NASA Ozone Watch: What is the Ozone Hole," March 15, 2012, *National Aeronautics and Space Administration*, ozonewatch.gsfc.nasa.gov/facts/hole_SH.html.

"Natural Sources and Sinks of Carbon Dioxide," April 14, 2011, *The Environmental Protection Agency*, www.epa.gov/climatechange/emissions/co2_natural.html.

Parry, M. L. Climate Change 2007: Impacts, Adaptation and Vulnerability: Contribution of Working Group II to the Fourth Assessment Report of the Intergovernmental Panel on Climate Change-Glossary. Cambridge: Cambridge University Press, 2007. www.ipcc.ch/pdf/assessment-report/ar4/wg2/ar4-wg2-app.pdf.

Riebeek, H. July 6, 2010. Why is global warming a problem? *National Aeronautics and Space Administration*. earthobservatory.nasa.gov/blogs/climateqa/why-is-global-warming-a-problem/.

Study: Climate change to affect monsoon in South Asia. *USA Today*: February 27, 2009. www.usatoday.com/weather/climate/globalwarming/2009-02-27-climate-change-asian-monsoon_N.htm.

The Impact of Climate Change on Natural Disasters. *National Aeronautics and Space Administration.* earthobservatory.nasa.gov / Features / RisingCost / rising_cost5 .php.

The Water Cycle and Climate Change. *National Aeronautics and Space Administration.* earthobservatory.nasa.gov / Features / Water / page3.php.

Trenberth, K. (2005). Uncertainty in Hurricanes and Global Warming. *Science.* 308. Pp. 1753–1754. yly-mac.gps.caltech.edu / reprintsyly / A_RecentPapers / li%20 jiang_h2o / zjunk / early%20versions / ref / Trenberth_Hurricanes_Science2010%20 copy.pdf.

"U.S. Greenhouse Gas Emissions by Gas," *Center for Climate and Energy Solutions,* www.c2es.org / facts-figures / us-emissions / gas.

Vajpeyi, Dhirendra, 2001, *Deforestation, Environment and Sustainable Development,* London: Praeger.

Van Aalst, Maarten K. 2006. "The Impacts of Climate Change on the Risk of Natural Disasters." *Disasters.* 30, no. 1: 5–18. onlinelibrary.wiley.com.proxy.lib.uni.edu / store / 10.1111 / j.1467-9523.2006.00303.x / asset / j.1467-9523.2006.00303.x.pdf ?v=1&t=h29j55oe&s=cc973d62daaa4263381486c5adb3a5fa538973f2.

Wickham, Chris. "New Antarctic Ice Shelf Threatened by Warming," The Chicago Tribune, May 9, 2012, www.chicagotribune.com / business / sns-rt-us-antarctica -global-warmingbre84811e-20120509,0,4563187.story.

Zeng, N. 2003. "Drought in the Sahel." *Science* 302, pp. 999–1000. www.sciencemag .org / content / 302 / 5647 / 999.short.

3

Health, Human Security, and Climate Change

Laura L. Janik

In 2011, comic strip artist Joel Pett published a humorous, yet very telling, cartoon in *USA Today*. The setting for the cartoon is an environmental conference where one conference presenter is demonstrating, via PowerPoint, the effects of embracing more eco-friendly policies. These include: energy independence, rainforest preservation, sustainability, green jobs, livable cities, renewables, clean water, clean air, and healthy children. Not fully believing that climate change is a real or dangerous thing, one conference attendee asks, "what if it's a big hoax and we create a better world for nothing?"

The climate change debate pits believers and skeptics against one another and is the source of much disgust, anger, and outrage for both camps. More specifically, modernists (also known as skeptics) tend to argue against the severity of certain climate change predictions and firmly believe that technology will enable us to overcome resource shortages and create new, and perhaps even cleaner, forms of energy. In contrast, neotraditionalists feel that the world is quickly reaching its limits to growth and argue that we must act now to address the threats posed by climate change or face the dire consequences of inaction (Shimko 2009: 322–342). The number of climate change disbelievers has, decreased over the past several decades; moreover, as the Pett (2011) comic strip suggests, regardless of where you might fall in the climate change debate, embracing greener policies is a universal good thing and can produce positive externalities for many different issue areas regardless of how fast or slow climate change and environmental degradation is occurring.

This chapter does not seek to further engage the modernist and traditionalist debate, as this task has already been ably accomplished by other contributors to this edited volume and elsewhere. Rather, in line with Pett's rather in-your-face revelation, this investigation assumes that changing our current approach to managing the global environmental commons can "create a better world" regardless of how fast or how much global temperatures are increasing. As such, the goal of this chapter is to highlight the connection between environmental degradation and health and suggest ways to more effectively tackle environmental and health problems at the global level. I will begin by discussing the broader, multidimensional concept of health and then move into the specific health consequences that result from environmental degradation. Next, I will highlight various international efforts, both successful and unsuccessful, that have been undertaken to address environmental degradation and the resulting health consequences of such degradation, with a specific emphasis placed on the Montreal Protocol (1987), Kyoto Protocol (1997), and newly formed International Renewable Energy Agency (IRENA) (2009). In doing so, this chapter aligns itself with neoliberal institutionalist and constructivist theory because of its emphasis on the positive externalities that multilateral cooperation in the form of treaties and institutions can generate. Treaties and institutions, however, can and do fail; therefore, this investigation is bent on examining how to properly structure such entities so that they can achieve their stated goals.

DEFINING HEALTH

Health is a multifaceted concept that can mean different things in different contexts. In other words, my understanding of what it means to be healthy, and unhealthy, may be radically different from your understanding; further, our conceptualizations of what it means to be healthy are ultimately shaped by our economic standings, geopolitical locations, able-bodiedness, age, and sex. In the broadest sense possible then, health implies the absence of disease and well-being of the body *and* mind. The preamble to the constitution of the World Health Organization (1946), charged with promoting good health and solving global health problems, formally defines health as, "a state of complete physical, mental and social well-being and not merely the absence of disease or infirmity." From the WHO perspective, therefore, being healthy implies more than simply not being sick. It means well-being of the physiological and psychological. As such, enabling people to become and remain healthy in body and mind can be an onerous and expensive operation that extends beyond the state level, particularly in the global South where states face severe resource and financial shortcomings.

While key intergovernmental organizations such as the WHO and United Nations Children Fund (UNICEF) embrace an all-encompassing definition of health, until recently the majority of health studies conducted by social scientists tended to focus on the prevention and treatment of communicable diseases, likely because infectious threats such as tuberculosis (TB), hepatitis A, B, and C, human immunodeficiency virus (HIV), measles, malaria, and polio, are easily transmitted from person to person or from species to person and can seamlessly cross borders in the absence of structured cross-border health regulations (see for example: Archibugi and Bizzare 2005: 33–51; Fidler 2004: 799–804; Janik 2011: 1–18; Kirton 2009: xv–xxxi; Price-Smith 2009: 11–32). In fact, even with health-based border regulations in place, sickness can be masked and certain diseases can lay dormant for months, or even years. Therefore the control of communicable pathogens is not only difficult but often requires states and non-state actors to work together.

More recently, the study of non-communicable diseases, such as diabetes, high blood pressure, bipolarity, heart disease, schizophrenia, and cancer have gained the attention of the global health community. Whereas non-communicable diseases may appear less threatening given that they cannot be transmitted from person to person or from species to person, studies have demonstrated that, particularly in the global North, the increase in non-communicable diseases may devastate health care systems in the future given how costly it is to treat and cure these long-term ailments (see for example: Beaglehole and Yach 2003: 903–908; Collin, Lee, and Bissell 2002: 265–282; DeLaet and DeLaet 2012: 69–83). While non-communicable disease threats exist everywhere, the high fat, high calorie diet that many people in the global North indulge in, at present, makes ailments such as high blood pressure and heart disease more prevalent in wealthier states. This study intends to demonstrate that climate change and environmental degradation can have negative implications for human health and can impact the prevalence of both communicable *and* non-communicable diseases.

In addition to our understanding of various types of health issues that plague the global community, increasingly social scientists are approaching health issues using different issue lenses including, but not limited to: security, economics, human rights, and development. Beginning with the security lens, there is a growing recognition and concern with the security implications of unhealthy populations, which can threaten peace within the state, drain state resources, and ultimately make the state itself vulnerable, weak, and unprepared. In *The Health of Nations*, for example, Price-Smith (2002: 171–172) argues that

> The increasing prevalence of infectious disease will increase human morbidity and mortality, resulting in gradual erosion of state capacity and in increasing poverty. Because of this relationship, pathogen induced economic decline will increase the demands of the population on the state for the provision of basic services, even as the ability of the state to provide for those goods and to govern effectively declines. This combination of declining state capacity and increasing deprivation may contribute to increasing governance problems and development problems in affected states and may in extreme cases contribute to political destabilization. This confluence of negative trends may compromise the ability of transitional states (e.g., Russia and South Africa) to consolidate democratic and effective systems of governance.

The security implications of uncontrolled pathogens run deep. Perhaps one of the biggest testaments as to the security repercussions of unhealthy populations was the passage of United Nations Security Council Resolution 1308, which addresses the severity of the HIV/AIDS crisis, most notably in Africa, and exposes the negative repercussions that HIV can have on peacekeeping missions and ultimately international peace and stability (S/Res/1308/2000; Elbe 2006: 119–144).

Health can also be viewed through an economic lens. Whiteside (2002: 313–332) has ably documented the causal relationship between poverty and disease arguing that disease not only contributes to poverty at the individual and community level, but also at the state level. Further, Whiteside (2002: 313–332) notes that the causal relationship between poverty and disease is bidirectional meaning that poverty also increases people's likelihood of contracting disease and ultimately impacts a state's ability to grow and prosper.

Additionally, the human rights lens has enabled social scientists to not only push health as a human right that states need to more fully realize and engage, but has also exposed various human rights violations committed by states and non-state actors that refuse to promote health or who implicitly and explicitly work to undermine people's health and well-being. For example, human rights activists have revealed that many people die not because they are sick, but because they cannot afford certain preventative medicines or treatments due to intellectual property laws that drive up the cost of health resources (Cohen and Illingworth 2003: 27–48; Lanoszka 2003: 181–197). Powerful international financial institutions, such as the World Trade Organization (WTO), force member states to adopt strict intellectual property laws and have catalyzed some social scientists to claim that free trade and the neoliberal economic agenda trump respect for health and human rights in the international community.

Taken as a whole, the field of global health studies is complex and multilayered. Social scientists have successfully expanded the global health agenda to include both a communicable and non-communicable disease focus; further, there are multiple lenses and theoretical perspectives from which to view health issues of the communicable and non-communicable sort. To add to this complexity, the global health agenda must expand further to more fully tackle the health issues that climate change is catalyzing. This requires not only cross-disciplinary dialogue but also cross-field conversations between the social sciences and natural sciences. This is a difficult, but not insurmountable, task.

CLIMATE CHANGE AND ITS IMPACT OF HUMAN HEALTH AND SECURITY

The relationship between the environment and health is complex and multifaceted. Price-Smith (2002: 141) argues that, "accelerating global environmental change will have significant negative effects on the health and prosperity of human societies which in turn will increase stress on state capacity and increase human deprivation." Climate change and its connection to health problems, therefore, can and should be approached from multiple levels, beginning with the micro, that of the individual, and extending to the macro, that is, the global level. Whereas the costs of altering the human activities and dependencies that contribute to climate change and environmental degradation may be expensive, the health consequences of failing to do so are likely much larger. For example, the WHO (2012: para. 3) estimates that the "direct damage costs to health" of climate change could be 2–4 billion dollars per year by 2030 if we fail to more fully address climate change issues. The WHO (2012: para. 2) mortality statistics are even more startling and suggest that "the modest warming that has occurred since the 1970s was already causing over 140,000 excess deaths annually by the year 2004."

As Vajpeyi notes in the introductory chapter of this book, climate change should not be solely associated with increasing temperatures, which are only a part of the broader climate change phenomenon. Instead, climate change can induce erratic weather patterns that spur tornadoes, hurricanes, violent storms, flooding in coastal areas, as well as both temperature increases and decreases, all of which have direct and indirect implications for human health and can increase the prevalence of both communicable and non-communicable disease threats.

Vector-borne Disease Transmission and Rising Temperatures

Beginning with rising temperatures, scholars such as Craig (1999: 105–111), Epstein, Farber, and Sachs (2011: 1–5) McMichael et al. (2004: 1543–1650), Platt (1996: 31–41), Price-Smith (2002: 141–170), and Shope

(1992: 171–174) have demonstrated that vector-borne disease transmission—viruses, bacteria, and protozoa transmitted via insects and other intermediate hosts—can accelerate with warmer temperatures. Temperature increases have amplified the prevalence of diseases including, but not limited to: trypanosomiasis, cholera, malaria, dengue, schistosomiasis, Ebola, and hanta virus. The reasons for this acceleration are manifold. Price-Smith (2002: 141–170) suggests that warmer temperatures can catalyze replication of pathogens. The WHO (2012: para 1–6) notes that rising temperatures may extend warm seasons thus giving certain vector-borne illnesses a prolonged period of time to subsist, multiply, and infect. Increased prevalence might also elicit new forms of drug resistant infections that health care systems are not prepared to tackle given that drug resistant infections are very difficult, if not impossible, to cure. Additionally, rising temperatures might contribute to the reemergence of certain parasitic and zoonotic diseases in non-endemic, but previously affected, areas as well as presenting themselves for the first time in certain populations. As Price-Smith (2002: 145) notes, "owing to climate change, vector-borne disease may also return to immunologically naïve populations that have been unaffected by such parasites for a very long time, through the dynamics of recrudescence."

The increased prevalence of vector-borne diseases further strains health care systems given the increased number of people that will seek out medical care and require medical treatment, particularly in the global South where vector-borne infections are more common and where the ability of the health care system to reach affected persons is weak. Moreover, with more vector-borne infections, this will require health care systems to increase their staff and medical personnel, purchase more and newer medical treatments, as well as acquire new diagnostic and testing equipment, all of which are expensive. At the individual and household levels, the sickness or death of a family member can be expensive due to increased medical bills, the loss of a bread winner, and funeral costs.

Climate Change, Flooding, and Health

Environmental degradation related to urbanization and modernization processes that depend on harmful energy sources is also associated with increased flooding, particularly in coastal areas. Flooding can have indirect consequences for human health. It can damage crops and farmland leading to deceased productivity and malnutrition in poorer areas (Postel 1996: 13–25). It also displaces people from their homes for extended periods of time. This type of dislocation can have serious psychological repercussions as well as physical repercussions for the displaced. For example, the major floods in Pakistan in 2010 left more than 20 million people homeless, some of whom still remain displaced (Chamberlain and Shah 2010: para. 1–4). Not only do individuals get separated from their families, but the basic living conditions in camps for displaced persons and the common issue of overcrowding can make such camps breeding grounds for disease. Such issues are not limited to developing countries. For example, recent "super storms" in the New England area of the United States, including Hurricane Irene and Hurricane Sandy, further demonstrate the health damages that extreme weather can exert on affected populations. While the death toll from Hurricane Sandy is much smaller when compared to the floods in Pakistan (it is estimated that 130 people were killed in the Sandy storm compared to 1,500 in Pakistan), tens of thousands of people remain displaced due to this major weather event. Additionally, Sandy relief efforts could cost the U.S. federal government up to sixty billion dollars (Levs and Watkins 2013: para. 1–4) in addition to billions more in state level funding and individual losses.

The more direct health consequences from flooding and storm surges come from water contamination such as, drinking water, sewage, and human waste intermingle, leading to an increase in water-borne illnesses such as cholera, polio, and even the plague (Platt 1996: 42–51). The 2010 floods in Pakistan not only led to the emergence of cholera outbreaks but forced health officials to preemptively treat 36,000 people "as if they were suffering from cholera" (Chamberlain and Shah 2010: para. 13). Just as concerning as the outbreak of cholera was the spread of poliomyelitis. We have almost achieved the full eradication of polio, but the Global Polio Eradication Initiative still faces major challenges in the remaining endemic areas of Pakistan, Afghanistan, and Nigeria (Global Polio Eradication Initiative 2010: para. 1–5; Janik 2012: 1–2). The floods in Pakistan not only led to an increase in polio prevalence rates in the country, but also put neighboring non-endemic countries at risk for reinfection as some individuals who fled from their homes simultaneously crossed internationally recognized borders in search of safety. Flooding created additional obstacles for the eradication initiative, as displacement interfered with vaccinator attempts to conduct vaccination campaigns and keep accurate records of those immunized and missed (Tuscano 2010: para. 1–8).

Unfortunately, it does not appear that flooding trends will reverse anytime soon without the deliberate and concerted actions of states and non-state actors; for example, as the Centers for Disease Control and Prevention (2010: para. 2) states, "a 40 cm rise in sea level is expected to increase the average annual numbers of people affected by coastal storm surges from less than 50 million at present to nearly 250 million by 2080." As such, related human health concerns will continue to present themselves as weather shifts contribute to rising sea levels, leading to flooding and displacement in areas surrounded by water.

Non-communicable Disease and Global Warming

Climate change can also have a detrimental impact on non-communicable health issues including cataracts and skin cancer. Barrett (1999: 192–219) has demonstrated that the depletion of the ozone layer caused by an increase in chlorofluorocarbons (CFCs), attributed to common aerosol sprays and household items such as air conditioners, refrigerators, and other coolants, can contribute to eye problems, namely cataracts, and more generally an increase in skin cancer due to increased exposure to the sun. Further, the Centers for Disease Control and Prevention (2010: para. 1) has documented the impact of climate change may have on air quality "including production and allergenicity of aeroallergens such as pollen and mold spores and increases in regional ambient concentrations of ozone, fine particles, and dust." Price-Smith (2002: 160–161) notes that increased exposure to ultraviolet B radiation may contribute to immunosuppression in humans and make us more susceptible to both communicable and non-communicable disease threats, including cancer. Whereas allergies, asthma, and respiratory problems are not always life-threatening, they can decrease the quality of one's life and also place an added burden on an already overwhelmed health care system as citizens seek out medical care to deal with these problems. Further, many non-communicable health problems, such as cataracts and asthma, are long term ailments, making them even more troubling and burdensome to both the person afflicted and the health care provider he/she seeks out.

Loss of Biodiversity, Famine, and Drought

Yet another health consequence of climate change is the loss of biological diversity as certain plant and animal species are killed off via the loss of habitat, loss of food supply for the flora and fauna species, or the introduction of new, invasive and predatory species during the modernization and development process. There is a growing recognition that biodiversity and the diversity of microorganisms can generate positive externalities for the human health condition in the form of food and material well-being, economic livelihood and work, and scientific and pharmacological breakthroughs leading to the generation of new and better medicines (Millennium Ecosystem Assessment 2005: 1–16). International efforts have been made to increase awareness about the need to promote and protect biological diversity, such as the current UN Decade on Biodiversity, the UN Convention on Biodiversity, and the Convention on International Trade in Endangered Species of the Wild Fauna and Flora. These multilateral cooperation attempts positively contribute to the protection of biodiversity, but they are not sufficient by themselves.

As the Millennium Ecosystem Assessment (2005: 6) notes, "the costs and risks associated with biodiversity loss are expected to increase, and to fall disproportionately on the poor."

Finally, just as climate change has the ability to generate extreme weather patterns leading to floods, tornadoes, and hurricanes, it can also contribute to dry conditions leading to famine and malnourishment. One of the natural consequences of drought, or a deficiency in the water supply, is the inability of crops to naturally flourish or for people to fully intervene in the absence of natural rainfall. Particularly in underdeveloped rural areas, drought can damage the local population's ability to feed itself, leading to malnourishment, sickness, and sometimes death. While the global community has created institutions to aid drought victims and combat famine, people still die from hunger and famine-related diseases. In 2011, a drought in East Africa led to the death of more than ten thousand individuals in Ethiopia, Djibouti, Kenya, and Somalia, and left nearly nine million people in need of food aid and assistance (UN Office for the Coordination of Humanitarian Affairs 2011: 1–11; Woolridge 2011: para. 1–10). Drought not only contributes to human morbidity and mortality via malnourishment and hunger, but it can also encourage crop pests that can damage the human condition. The Centers for Disease Control and Prevention (2010: para. 1) notes that drought has generated bouts of "aphids, locusts, and whiteflies, as well as the spread of the mold *Aspergillus flavus* flavus that produces aflatoxin, a substance that may contribute to the development of liver cancer in people who eat contaminated corn and nuts." In the absence of stronger international efforts to combat environmental degradation and climate change, these trends are only likely to increase, depriving poor areas of the necessary resources to become and remain productive and further drain the international community of resources to ease these trends.

THE PUBLIC GOODS DIMENSIONS OF HEALTH AND WELLBEING

A healthy environment and a healthy population are public goods that produce positive externalities that all can benefit from. Public goods possess two defining features: non-rivalry and non-excludability (Boyer and Butler 2006: 75–91; Kaul and Mendoza 2003: 78–111; Sandler 1992: 23–50; Sandler and Arce 2002: 195–222; Samuelson 1954: 387–389; Touffut 2006: 1–12). Non-rivalry implies that one person's consumption of a good detracts none whatsoever from others' abilities to equally consume that good. For example, decreasing greenhouse gas emissions can generate positive externalities such as more stable global temperatures, decreases in acid rainfall, and fewer erratic weather outbursts. My enjoyment of these positive spillover effects detracts none whatsoever from your ability to equally and fully enjoy them because these spillovers are equally shared, that is joint, and thus non-rivalrous. Next, non-excludability means that no one can be barred from consuming the benefits of a good, such as decreased greenhouse gas emissions, because these benefits are global and cannot be confined to one area or group of persons. Because of the all-inclusive nature of these entities, public goods tend to go under-provided in the absence of larger entities capable and willing to provide them. In other words, I may find a clean global environment both desirable and good, but I alone do not have the resources or know-how to create it; further, because everyone would benefit from my hypothetical actions, it is likely that I will downplay my preference for a cleaner environment hoping that someone else will step up to the task, allowing me to "free-ride" on their goodwill and effort. Public goods are frequently referred to as market failures because of their non-rivalrous and non-excludable characteristics, meaning that they cannot be adequately provided in the free market; by contrast, persons are much more willing to express their preferences for private goods that are excludable and rivalrous making the free market an appropriate venue to buy and sell such entities.

Whereas public goods analyses, which are a part of the broader liberal theoretical tradition, were once confined to an understanding of non-rival and non-excludable goods at the national and subnational levels (Musgrave 1939: Part II; Samuelson 1954: 387–389), however more recently the public goods agenda has evolved and expanded to include the provision of goods at the regional and global levels (Barrett 1999: 192–219; Jayaraman and Kanbur 1999: 418–435; Chen, Evans and Cash 1999: 284–305; Sandler 1998: 221–247; 2006: 5–25). Providing global public goods, of the health and environment type, is extremely challenging given that: 1) states alone cannot or will not provide such goods, thus non-state actors, or a combination of public and private actors, may be necessary; 2) there are so many desirable public goods that crowd the global agenda and ultimately require difficult decisions to be made; and 3) the provision of such goods is costly and may infringe on state sovereignty in uncomfortable ways.

Additionally, whereas all states may agree and desire certain goods (such as, the eradication of polio), most states would rather not ante up money, personnel, and other resources to accomplish this task, but instead would rather free-ride and see that other states provide these entities. In other words, because of the all-inclusive and non-rival characteristics of global public goods, the classic free rider and collective action dilemmas (Olson 1965: 1–52) come into play when referencing the betterment of the global environment and global health. As Kaul, Grunberg, and Stern (1999: 6) claim, "[public goods] elicit patterns of behavior that, from the individual agent's viewpoint, are quite rational. Yet from a collective viewpoint—such as that of a local community, nation or humanity as a whole—the result is suboptimal and can be disastrous." By contrast, national public goods, such as national security and education, are arguably easier to generate given that the state not only gains from providing these goods, but can limit the persons that benefit from its provision of certain goods (namely citizens benefit).

Given the anarchical nature of the international system, that is, a system that lacks a higher authority in the form of a world government, it is not possible to force states to ante up and provide global public goods. This investigation intends to reveal, however, is that it is possible to craft multilateral treaties and institutions that can tackle environmental degradation as well as the human health consequences resulting from this degradation. An examination of the Montreal (1987) and Kyoto (1997) Protocols reveal that multilateral cooperation can reduce environmental bads, but the failure to structure treaties properly, as with the Kyoto Protocol, can be disastrous. By contrast, the common-but-differentiated carrots and sticks approach of the Montreal Protocol makes it a model treaty that future multilateral agreements would be wise to emulate. Finally, this investigation will explore the International Renewable Energy Agency (IRENA) and demonstrate how states and energy corporations can work together to better the global environment and thus the human health condition. The IRENA investigation suggests that environmental institutions can be effective in the fight against climate change, but must appeal to the self-interest of appropriate parties in order to succeed. Again, the structure of international institutions is quite important.

The Do's and Don'ts of Multilateral Treaties

The Montreal Protocol (1987)

The Montreal Protocol (MP), which was adopted in 1987 and entered into force in 1989, is a multilateral agreement intended to reduce the production and consumption of ozone depleting substances (ODS) such as chlorofluorocarbons and halons. More than a decade before the enforcement of the Montreal Protocol, atmospheric scientists became acutely aware of the negative consequences that ODS can have on the thin layer of protective ozone that limits the amount of ultraviolet radiation people are exposed to (Barrett 1999: 192–219). The depletion of the ozone layer has both environmental and health consequences. As the preamble of the MP recognizes, "worldwide emissions of certain substances can significantly deplete and otherwise modify the ozone layer in a manner that is likely to result in adverse effects on human health and the environment" (UNEP 1987: preamble).

The relationship between ozone depletion and human health is multifaceted and ozone depletion can damage the human condition in a plethora of ways. Ozone depletion further exposes people to sunlight that can increase the prevalence of skin cancer, cataracts, and genetic mutations, as well as damaging fishery and agricultural efficiency and output. Beginning with cataracts, this eye problem leads to a clouding of the eye's "naturally clear lens" which leads to vision impairment and even blindness (U.S. Environmental Protection Agency 2010: iii). It is possible to treat patients suffering from cataracts by removing the affected, cloudy lens and replacing it with an artificial intraocular lens, but this procedure can be very expensive. The United States Environmental Protection Agency (2010: iii) documents that "Prevent Blindness America estimated in its 2007 *Economic Impact of Vision Problems* report that the direct medical cost of cataract treatment for Americans over the age of forty totaled $6.8 billion annually."

Skin cancer can also result from increased exposure to ultraviolet radiation as a result of ozone depletion. Only two years after the Montreal Protocol went into effect, research on ozone depletion and cancer incidence suggested that "it is estimated that a sustained 10 percent reduction in ozone would result in a 26 percent increase in non-melanoma skin cancer (NMSC) worldwide. Using very conservative assumptions . . . it can be estimated that such an increase would be equivalent to more than 300,000 additional cases of NMSC and 4,500 cases of melanoma. This may be an extremely conservative estimate, possibly off by a factor of two or more" (Longstreth, de Gruijl, Takizawa, van der Leun 1991, para. 4). The costs of treating skin cancer are enormous for the individual and his/her health care provider. Further, skin cancer treatment, such as radiation, may generate new health issues for the patient given that radiation can kill off unhealthy *and* healthy cells.

Whereas cataracts and skin cancer are the most frequently cited health effects of ozone depletion, the WHO (2013: para. 1–9) suggests that the impact of ozone depletion runs much deeper and can also cause photodermatoses, photokeritis, pterygium, acute solar retinopathy, and cancer of the cornea. Further, as previously stated, ozone depletion and increased exposure to sunlight may impact an individual's immune system, and accelerate the progression of certain viruses such as HIV/AIDS (Price-Smith 2002: 141–170). As such, multilateral cooperation to address ozone depletion not only has positive environmental impacts, but can save healthcare systems billions of dollars in out-patient and in-patient treatment in addition to catalyzing healthier and happier people.

Multilateral meetings convened by the United Nations Environment Program lead to the creation of the Vienna Convention for the Protection of the Ozone Layer in 1985. The Vienna Convention was intended to "take appropriate measures . . . to protect human health and the environment against adverse effects resulting or likely to result from human activities which modify or are likely to modify the ozone layer" (UNEP 1985: Article 2); however the convention lacked some degree of specificity given that parties to it did not commit to specific reduction requirements. Instead, the framework was a commitment on behalf of states to "co-operate in the formulation of agreed measures, procedures and standards for the implementation of this Convention, with a view to the adoption of protocols and annexes" (UNEP 1985: Article 2). What this meant was that individual states could decide for themselves the amount of CFC emission reduction to strive for and how quickly.

Almost immediately after the creation of the Vienna Convention (1985), scientific evidence suggested that the initial predictions regarding ozone depletion were understated, prompting state parties to reconsider their commitments under the Vienna Convention and establish a more binding, rigorous, and specific agreement involving as many state parties as possible. The final product of these negotiations was the Montreal Protocol on Substances that Deplete the Ozone Layer. The MP committed parties to binding and specific reduction targets for certain CFCs and halons. One of the more remarkable factors regarding the MP is that it has been amended several times: the London Amendment (1990), the Copenhagen Amendment (1992), the Montreal Amendment (1997), and the Beijing Amendment (1999). The recognition that science and technology change constantly catalyzed the MP negotiators to incorporate an adjustment mechanism that could "accelerate and increase the stringency of controls on previously agreed ozone-depleting substances by simple majority" (UNEP 2007: 5). Each amendment, therefore, expands the scale and scope of the MP in terms of the number of ozone depleting substances that are targeted, the number of countries involved, as well as the time frame for CFC reduction. In sum, the MP incorporates a degree of flexibility that allows for its modification and amendment.

In what is known as common-but-differentiated responsibilities, the multilateral agreement expects all state parties, regardless of economic standing, to actively engage in reducing CFCs, but allows developing and transition economies a "grace period" (UNEP 2007: 5). This means that certain economies are provided a longer period of time to incorporate MP mandates. The Montreal Protocol also offers developing countries financial support for participating in the MP in what is known as the "Multilateral Fund for the Implementation of the Montreal Protocol" (UNEP 2007: 9). As Barrett (1999: 195) notes, the London Amendment "offered to pay developing country parties for the incremental costs of complying with the agreement."

Next, the MP and its additional protocols have enacted punishments for noncompliance with the objective of the protocol. The MP limits and bans trade in ozone depleting substances with parties and nonparties alike. In dong so, it attacks what Barrett (1999: 215) refers to as "leakage" problems in that it does not provide holdouts with a comparative advantage given that the MP bans and limits trade in ODS products. There are 197 state parties to the MP, although there are a few holdouts that have yet to ratify some of the additional protocols.

As a result of the MP, developed countries have reduced the production and consumption of more than 99 percent of ODS controlled by the MP and developing countries have achieved a 72 percent reduction (UNEP 2007: 11). As a result of these changes, most scientists, policy makers, and academics have hailed the MP as a successful environmental treaty with positive health spillover effects. The lessons that emerge from the MP are significant.

The Kyoto Protocol (1997)

The Kyoto Protocol (KP), which was adopted in 1997 and came into force in 2005, was designed to reduce the emission of green house gasses (GHGs) that can contribute to climate change and, in turn, reduce disease transmission of the communicable and non-communicable diseases. While the KP was implemented less than a decade ago, the task of creating the KP was both arduous and intensive. The science surrounding climate change has existed for decades, and organizations such as the United Nations Environment Program, International Energy Agency, Club of Rome, and World Meteorological Organization continue to advocate changes in how states produce and consume energy because current trends are not sustainable and are quite damaging to both the environment and people's health.

The connection between climate change and health is multifaceted and there are a multiplicity of health issues that rising temperatures, erratic weather patterns, extended warm seasons, flooding, and in some cases temperature decreases can generate. These include:

An increase in vector-borne disease transmission including dengue, Ebola, trypanosomiasis, malaria, schistosomiasis, and hanta virus due to extended warm seasons and the migration of such vectors to new, warm habitats.

An increase in water-borne diseases, such as cholera and polio, due to flooding and the intermingling of drinking water, waste, and sewage.

Human displacement that contributes to psychological damage and disease (from displacement camp overcrowding) as a result of flooding, hurricanes, tornadoes, and other erratic weather patterns.
Hunger and famine resulting from drought and crop devastation due to increasing temperatures.
Increased prevalence of respiratory problems and cardiovascular issues due to increased pollution and air quality issues.

As if the environmental damages caused by climate change are not enough, when combined with its impact on human health, climate change is extremely costly and necessitates more concerted action from states and nonstate actors. A recent study conducted by researchers from the Natural Resources Defense Council, the University of California, Berkeley, and the University of California, San Francisco, examined the morbidity and mortality data resulting from six recent weather-related events in the United States and the findings were alarming. Knowlton et al. (2011: 2167–2176) established that six recent extreme weather events—including the 2004 Florida hurricane season and 2006 California heat wave—had significant and negative impacts on individual health and seriously overburdened already strained health care system. Knowlton et al. (2011: 2167) found "that the health costs exceeded $14 billion, with 95 percent due to the value of lives lost prematurely. Actual health care costs were an estimated $740 million. This reflects more than 760,000 encounters with the health care system." If industrialized, Northern states, such as the United States, struggle to deal with the resultant health issues generated by climate change, then the impact on developing, Southern states can only be that much larger and more damaging. As such, amplified multilateral cooperation to address climate change is a necessity.

The first serious international push to better understand the impact of GHG emission on the global environment came in 1988 when the United Nations Environment Programme and World Meteorological Organization jointly established the Intergovernmental Panel on Climate Change (IPCC). The IPCC is a scientific body, and the leading global authoritative voice, that aims to assess climate change trends and patterns. The IPCC does not conduct research on climate change; rather, it draws together some of the best and brightest climate scientists that do and that have voluntarily come together to assess climate change patterns and objectivity in the climate change research being conducted around the world (IPCC 2012: para. 1–5). Under United Nations General Assembly Resolution 43/53, the first major task of the IPCC was to provide a comprehensive and objective assessment of climate change and its impact and also to craft potential mechanisms that could counteract these anthropogenic trends. The *IPCC Scientific Assessment* (1990), the first ever assessment report by the newly formed scientific body, unwaveringly stated,

> We know, with certainty, that the concentrations of naturally occurring greenhouse gases in the atmosphere have varied on palaeo time-scales. For thousands of years prior to the industrial revolution the abundance of these gases were relatively constant. However, as the world's population increased, emissions of greenhouse gases such as carbon dioxide, methane, chlorofluorocarbons, nitrous oxide and troposphere ozone have increased substantially due to industrialization and changes in agriculture and land use. (Houghton, Jenkins, Ephaums 1990: XXXVII)

The IPCC assessment report played a critical role in the creation of the United Nations Framework Convention on Climate Change (UNFCCC), a multilateral treaty aimed at reducing GHG emissions. The framework convention attracted more than 150 parties demonstrating, once again, that climate change posed serious consequences for humans and their surrounding environment. As with the 1985 Vienna Convention targeting CFC and halon reduction, however, the UNFCCC lacked a degree of specificity and did not commit state parties to binding reduction targets. Instead, as Article 3.1 of the UNFCCC states, "Parties should protect the climate system for the benefit of present and future generations of humankind, on the basis of equity and in accordance with their common but differentiated responsibilities and respective capabilities. Accordingly, the developed country Parties should take the lead in combating climate change and the adverse effects thereof." Further, while the UNFCCC encouraged states to reduce GHG emissions, specific targets were left up to countries to determine for themselves. To remedy this lack of specificity problem, the first Conference of Parties (COP1) to the Framework Convention on Climate Change conceded that specific reduction targets, most notably for industrialized countries, should be established and that a climate change treaty similar to the Montreal Protocol was desirable (Barrett 1999: 192–219).

The second Conference of Parties (COP2), held in Kyoto, Japan, led to the establishment of the Kyoto Protocol, which operationalizes the UNFCCC by committing industrialized countries, also referred to as Annex I countries, to binding GHG reduction targets. The KP officially entered force in 2005 and the five-year framework for reducing GHG emissions under the KP extended from 2008–2012. The KP did not achieve its overarching goal of reducing global GHG emissions by 5 percent against 1990 levels (Paris 2012: para. 1–6); however, calling the KP an outright failure, which many have done, is also a mistake (Verweij 2006: 31–60). Under the KP, some Annex I countries have significantly reduced GHG emissions. For example, Latvia decreased GHG emissions by 148 percent between

1990–2010 and Lithuania by 77 percent (UNFCCC 2012: para. 1). However, other countries have significantly increased GHG emissions; for example, Turkey saw a GHG increase of 147 percent between 1990–2010 and New Zealand a 59 percent increase (UNFCCC 2012: para. 1).

There are many reasons that the KP did not achieve its ultimate GHG reduction goals and much of the blame lies in the structure of the protocol itself, which is quite different from the aforementioned, successful Montreal Protocol. To start, the KP divides states into three main categories: Annex I, Annex II, and non-Annex I parties. Annex I parties include industrialized states and economies in transition. Annex II are industrialized countries (a part of the Organization for Economic Cooperation and Development) sans economies in transition. Non-Annex I parties are developing countries. Binding reduction targets have only been set for Annex I parties. Non-Annex I countries that have ratified the treaty have signaled their intent to reduce emissions, but by no set amount and while simultaneously pursuing development processes that may be environmentally damaging.

Next, the KP's carrots (incentives) and sticks (punishments) are inappropriate. It was not until 2010, at the sixteenth session of the KP Conference of Parties, that the Green Climate Fund (GCF) was established (GCF 2013: para. 1–4). The purpose of the GCF is to enable countries to combat climate change in line with the UNFCCC objectives (GCF 2013a: para. 1–5). In other words, it provides developing countries with funds to pursue more eco-friendly development processes. While the GCF is certainly an incentive that non-Annex I parties will benefit from, its establishment so late in the game is disappointing.

The KP sticks approach to punish parties that do not comply with the terms of the treaty have wavered from one extreme to the other. To expand, parties to the KP agree to estimate and monitor anthropogenic emissions and submit progress reports for assessment. Under article 6 of the protocol, non-compliant countries are not able to acquire emission reduction units (ERUs), which are a part of the larger KP emissions trading process whereby countries with emission units to spare can sell units to countries that have exceeded their targets (United Nations 1998: Article 6). In sum, the KP encompasses three flexibility mechanisms that are intended to aid Annex I countries as they attempt to achieve emissions targets. These are known as Joint Implementation, Emissions Trading, and Clean Development Mechanism. At best, non-compliance was, at least initially, met with a weak punishment whereby states could not engage in part of the KP emissions trading process, a process that some find controversial given that states can buy and sell pollution credits.

Later on in, the KP established the Compliance Committee under the Marrakech Accords at the Seventh Conference of Parties in 2001. The

Compliance Committee became fully operational in 2006 (when it submitted its first annual report) and could be seen as a welcome addition that can push countries to meet their reduction commitments by forcing noncompliant states to "make up the difference between its emissions and its assigned amount during the second commitment period, plus an additional deduction of 30 percent" (UNFCCC 2012a: para. 7). By many accounts, however, this punishment is perhaps too extreme for a protocol that previously lacked harsh sticks and may encourage states to withdraw from the protocol if forced to undertake such drastic measures (Hovi and Kallbekken 2004: 1–16).

Finally, emission limits under the KP are not permanent; rather they extend from 2008–2012. The most recent Conference of Parties (COP18) that took place in December 2012 has been heralded as a success given that many state parties agreed to extend the KP by another eight years with the intent of creating a more comprehensive climate change agreement covering all states by 2015 (to be implemented in 2020). However, some Annex I parties to the original KP, such as Canada, Japan, Russia, and New Zealand refused to commit to the extended version of the climate change treaty and seventeen of the twenty-five largest GHG emitters have still not committed to specific, binding reduction targets (Ritter 2012: para. 1–7; Yang 2012: para. 1–10).

Lessons Learned

In the anarchical world in which we live, multilateral treaties can enable states to achieve common goals by working together towards a common objective. Some treaties are more effective than others, however, and an examination of the Montreal Protocol and Kyoto Protocol reveals that the structure a treaty takes is critical. The MP and KP vary greatly in certain respects and this is a major reason that the MP has produced better results.

First, the MP abides by the concept of common-but-differentiated responsibilities in that developing states are provided a longer time period to enact the MP reduction targets. However, all states, at the end of the day, have agreed to completely phase out specific chlorofluorocarbons in the near future. By contrast, while the KP claims to abide by the common-but-differentiated responsibility mechanism, it only places binding reduction targets on Annex I parties, allowing non-Annex I parties the ability to determine how much and how quickly to reduce GHG emissions. Beneath the surface, the KP really only embraces uneven responsibilities; parties to the KP may encompass a similar vision, but responsibilities lie squarely with Annex I parties. Further, because non-Annex I parties do not encompass binding targets, Barrett (1999: 192–219) suggests that this may lead to leakage issues whereby polluting activities shift to non-parties or, in the case of the KP, non-Annex I states. This is because non-parties have no responsibilities vis-à-vis the KP and non-Annex I parties do not have to meet specific reduction targets, making them even more attractive to businesses that generate pollutants that degrade the environment.

Second, at present both environmental treaties encompass carrots in the form of financing for developing countries seeking to offset the cost of CFC and GHG reduction; The Multilateral Fund for the Implementation of the Montreal Protocol (the financing arm of the MP) was established in 1990 to enable member states to the MP, a treaty with *permanent* emissions limits, to deal with the cost of phasing out ODS. By contrast, the KP financing arm, known as the Global Climate Fund, was founded just one year before the terms of the KP were to expire. That the KP now encompasses this type of funding mechanism is promising and will likely catalyze developing countries to agree to binding reduction targets in the future; one is left to wonder if the KP would likely have been more successful had the GCF been established much earlier.

Third, the MP encompasses effective sticks/punishments in that it can enact trade sanctions against MP violators and bans and limits the trade in ozone depleting substances from both parties and nonparties. The KP sticks approach is much different whereby noncompliance was initially met with the threat of cutting off a state's eligibility for emission reduction units (a fairly weak punishment). With the addition of the Compliance Committee, the KP can now demand that noncompliant states, "make up the difference between its emissions and its assigned amount during the second commitment period, plus an additional deduction of 30 percent" (UNFCCC 2012a: para. 7). For many analysts, this punishment may be too extreme and encourage states to withdraw and still does not impact nonparties (Barrett 2002: 529–547; Hovi and Kallbekken: 1–16).

In sum, the Montreal Protocol has been much more successful in its attempts to embed carrots and sticks in its overall framework when compared to the Kyoto Protocol, which partially accounts for the MP success and KP problems. Combined with the MPs flexibility, as previously discussed, the MP is simply a more robust and well-structured environmental and health treaty.

A final point is worth noting in regard to the success of the MP and questionable outcome of the first KP implementation round (2008–2012). Barrett (1999: 192–219) notes that the benefits of implementing the MP, measured in terms of avoidance of cancer cases, significantly outweigh the costs, measured in terms of costs for cancer treatment and deaths from cancer and cancer-related illness. In short, the benefits of implementing the MP outweigh the costs by a significant margin. Additionally, Barrett (1999: 192–219) notes that the cost of shifting from ODS products to non-ODS products were minimal. By contrast, cost-benefit models designed for the KP reveal a much different picture leading some to conclude that, "for climate change the incentives to reduce emissions unilaterally are much more muted. And while a case can be made that full cooperation requires substantial abatement, the opposite conclusion can also be supported by the data" (Barrett 1999: 206).[1] If we assume that states are self-interested, rational actors, then we can also accept that they are more likely to commit to policy changes that elicit more benefits in relation to costs. Even when a specific policy change has significant and positive intergenerational impacts, such as a switch to renewables or a reduction in GHG emission, states' decisions are usually guided by cost-benefit analysis. With this in mind, perhaps it should not be so surprising that a country such as the United States fully committed to the MP and resisted the KP. Cost played a serious role in this policy choice. As such, in order for any future climate change treaty to attract more participants (especially large polluters), cost considerations become even more pertinent. Two suggestions come to mind. The first relates to creating cheaper renewables that are just as, if not more, cost competitive with traditional fossil fuels such as oil and coal. This would create a competitive market whereby renewables and fossil fuels compete on a level playing field. The second suggestion relates to the scope of the KP itself which is grand. Future KP negotiators should consider creating smaller climate change agreements focusing on select greenhouse gasses. Doing so might convince nonplayers to begin becoming active in certain GHG areas, particularly areas where they might have a niche or where cost-benefit analysis reveals the benefits of reducing *specific* GHGs. This suggestion will likely be criticized by traditionalists who feel that the extent of climate change will be so devastating that only a grand effort targeting all GHGs and all countries can possibly save the environment. However, approaching climate change practically and incrementally may be the only hope for getting all polluters on board and willing to make changes.

INSTITUTIONS, CLIMATE CHANGE, AND HEALTH

The above discussion has clearly posted that climate change is a transborder problem given that environmental pollutants can seamlessly cross borders and do not respect sovereign national boundaries. Further, no one state can solve the climate change problem, and resultant health issues that result from it, regardless of the actions it takes at the domestic level (Kaul, Grunberg, and Stern 1999: 1–19). Instead, as a global problem, climate change, and environmental degradation broadly speaking, requires that states and non-state actors work together for the betterment of the human condition.

International institutions are multi-actor entities that have the ability to solve transborder problems and provide public goods such as a cleaner environment and healthier people. In almost every issue area, there are now institutions that exist to grapple with collective action dilemmas: economy, environment, health, security, and human rights. When adequately funded and efficiently administered, institutions can alleviate transborder trends from pollution and communicable and non-communicable diseases.

However, international support for such institutions, at present, is in crisis. For example, there is growing frustration with the United Nations and its inability to more fully address the human rights abuses taking place in Syria (Karam 2012: para. 15). Financial woes in the European Union have catalyzed disgust and antipathy for the International Monetary Fund (IMF) and other international financial institutions (Koutantou 2011: para. 2–3). Climate change concerns and the unknown post-Kyoto era have critics slamming the UN sponsored climate change talks and the Intergovernmental Panel on Climate Change (Harris 2011: para. 2), just as drone strikes and continued insecurity along the Pakistan/Afghanistan border have made some analysts increasingly skeptical of security alliances such as the North Atlantic Treaty Organization (Foust 2012: para 1–12).

Because international institutions have the potential to provide public goods and assuage public bads, this growing skepticism is problematic. In our anarchical international system that lacks a hierarchical world order with effective enforcement and punishment mechanisms, institutions are, at times, one of the few mechanisms that states can utilize to solve disputes, work together, mitigate uncertainty concerns, and develop bonds of trust (Janik and Sterling-Folker 2011). Yet, amidst all of this cynicism and wariness, one international institution that has the ability to simultaneously solve certain economic, environmental, and health issues has not only emerged as a strong international presence, but has also gained the support of critical players, such as the United States, despite that the agency requires members to contribute *mandatory* funds to its budget (Press 2011, Sach 2011, Schöpe 2011).

The International Renewable Energy Agency (IRENA) was founded in Bonn, Germany in January 2009 and its Statute officially came into force in July 2010. At present, IRENA's one hundred and four members have signed and ratified the IRENA Statute, which is an impressive number given that the agency has existed for only three years. Located in Abu Dhabi, UAE, IRENA is both an economic and environmental agency that seeks to promote renewable energy, and consequentially the economic development and the health of persons globally. It does so by connecting countries in search of energy diversification with companies that can aid them in this process (Press 2011). In other words, IRENA promotes the use of renewable energy sources that can combat climate change and the negative health issues that climate change produces. With this in mind, the remainder of this chapter is dedicated to understanding the success of the IRENA institution. In doing so, this brief investigation aligns itself with a variety of theoretical perspectives, such as constructivism and neoliberal institutionalism, that find value in institutions; it also stresses the importance of how institutions are structured and argues that institutional structure is a critical factor that dictates, at least in part, an institution's success or failure.

Given that IRENA emphasizes renewable energy and energy diversification, it is an institution that has the ability to simultaneously address economic development dilemmas, environmental woes, and the multiplicity of health issues resulting from climate change discussed above and elsewhere in this edited collection. Put differently, the agency's intentions are focused on the economy and the environment; however, as IRENA enables states to enter into and expand their presence in the renewables market, it is simultaneously contributing to the human condition. By pushing states to adopt renewables, IRENA can address communicable disease threats resulting from the increased presence of vector-borne disease (due to extended warm seasons and vector movement and reemergence), waterborne diseases that result from flooding and displacement, as well as non-communicable disease threats resulting from air pollution and ozone depletion.

IRENA is a young agency, having been created only three years ago; hence assessing its success in terms of output and its ability to meet its goals is too early. IRENA's ultimate goal is to act as "a facilitator and catalyst, providing experience for practical applications and policies, offering support on all matters relating to renewable energy and helping countries to benefit from the efficient development and transfer of knowledge and technology" (IRENA/FC/Statute Article IV A 2009). IRENA is only now moving out of its startup phase, in mind therefore its success is measured via its ability to attract more than one hundred members in fewer than three years. Membership in IRENA requires mandatory contributions from member states. As Elizabeth Press, a Senior Program Officer for IRENA, (2011) notes, "it is not a voluntary thing—something that 90 percent of other IGOs have to deal with. In this economic climate, the fact that countries are signing up to these mandatory contributions is another miracle." Its membership includes some of the world's largest polluters, including Australia, India, and the United States, in addition to nearly all EU states and most Middle Eastern states. Also IRENA was founded in the midst of a global financial meltdown and requires members to contribute mandatory funds, its structure, mandate, and vision are even more pertinent to understand.

Perhaps the most important factor that has contributed to IRENA's growth spurt can be located in its mandate. IRENA emphasizes renewables as a source of energy diversification and as a path towards development. The formation of IRENA is an acknowledgement that renewables can address a number of issues. Some see the renewable energy field as a source of jobs and new opportunities (Press 2011). Others see renewable energy as an answer to dwindling petroleum resources. Further, nuclear energy, a potentially cleaner source of energy when compared to fossil fuels, has recently lost some of its attraction since the disaster in Fukushima, Japan. The Fukushima incident demonstrated that there are serious downsides to reliance on such a potentially harmful energy source, especially when compared to other renewables (Sach 2011). By contrast, the renewable energy market, as many IRENA officials stress, has the potential to simultaneously address issues related to economic growth and development, energy diversification, energy access, and energy security. As such, participation in IRENA is attractive to countries in both the global North and the global South. As Rafael Conde de Saro (2011), a Spanish Ambassador, notes, "the creation of IRENA is the natural consequence of an across-the-board recognition, on an international level, of the importance of promoting the use of renewable energies worldwide."

Most representatives of IRENA declare that the agency is both an environmental *and* an economic organization and some go so far as to claim that IRENA is primarily an economic agency that produces environmental and health goods (Press 2011). As such, for developing economies whose primary concerns are growth and development, and not necessarily eco-friendly development processes, IRENA membership is attractive because the agency is viewed as a mechanism that can generate jobs, attract foreign direct investment, increase energy independence and security, and generally speaking, help economies grow while simultaneously promoting the health and well-being of citizens. As Press (2011) claims, "access for developing countries is such a big thing and renewables can do for energy what mobile telephones did for communication avoiding this huge infrastructure and financial liability."

For developed economies, IRENA presents their business sectors with novel opportunities for overseas investment and business expansion. As such, countries such as the United States that are not traditionally seen as promoters of eco-friendly policies view IRENA to bolster economic growth for their energy-based corporations. The North, generally speaking, also would like to see more eco-friendly development policies in the South where the industrialization process, for some, has just begun. And finally, for "greener" countries such as Germany, Spain, and Denmark, the trio that was initially very supportive of the renewables agency, IRENA has the potential to combat environmental bads such as climate change and pollution, and membership in IRENA is viewed as a way to catalyze green development policies. Put differently, IRENA has the potential to provide much needed global public goods by appealing to the self-interest of public and private entities. People are becoming increasingly weary of price and supply issues, especially because of overwhelming dependence on the Middle East and North Africa, two regions with a substantial amount of instability. (Press 2011). Renewables not only address climate change problems but also address a multiplicity of other equally important issues related to energy security, communicable and non-communicable disease threats, energy access, and economic development.

IRENA's ability to mesh economic need with environmental and health concerns has made the agency palatable to developing and developed countries alike. It has allowed the agency to overcome resistance from developing voices in the global South that reject Northern calls for more environmentally friendly development policies, while simultaneously attracting Northern countries that see in IRENA economic opportunity and the potential for good business investments. In other words, IRENA caters to the *self-interest* of state and non-state actors. While some states may sign up to diversify energy sources and reduce dependence on finite and expensive fossil fuels, others are doing so because of the benefit IRENA generates for domestic corporations or simply because IRENA's end goal is to promote a clean and healthy environment for all. Non-state actors also see in IRENA an agency that can help them expand their businesses abroad and thus generate profit. And finally, United Nations intergovernmental organizations, as well as many nongovernmental organizations, view IRENA as an added contribution to the global community that can help promote eco-friendly policies and a healthier global citizenry.

IRENA's ability to generate a plethora of public goods of the environmental and health sort should not blind us from acknowledging that states have ratified the IRENA statute for different self-interested reasons. In other words, as we move forward and look to clean up the global environment and generate public health goods, we must remember that most actors in the international system will not act unless there is something in it for them. Treaties and institutions that fail to acknowledge this will likely not be successful.

IRENA's success, measured in terms of its ability to attract so many members and in such a short timeframe, is not solely confined to its appeal to self-interest. Interviews with IRENA staff and country representatives indicate that there are other factors to consider, such as the choice to locate IRENA in the UAE, a Middle Eastern state in a strategic geopolitical location for major polluters such as the United States (Janik and Carter 2011: 12–16). Another important factor to consider was that IRENA attempted to reach out to potential competitors in the UN system, such as the UN Environment Programme and International Environment Agency, in order to win over competitors and thus ensure that they did not undermine the new agency (Janik and Carter 2011: 12–16). Perhaps the most important factor, however, was in the agency's ability to cater to self-interest and provide different types of services to a multiplicity of players and at the same time.

CONCLUSION

This chapter has attempted to demonstrate that environmental degradation has negative implications for the human condition, and more specifically, human health. Our overreliance on finite fossil fuels to meet the energy needs of a growing population in pursuit of development and industrialization not only damages the environment we rely on to live, but also generates health bads in the form of communicable and non-communicable disease threats. In short, the interconnections between the environment and health run deep.

No individual actor in present international system can singlehandedly clean up the environment, and thus, positively contribute to people's health. Instead, this enormous task requires the effort of state and non-state actors alike including intergovernmental organizations, nongovernmental organizations, and private entities. And just as no single entity can solve global health and environmental problems, there is no single solution to these complex issues. In the above discussion I have attempted to put forward two potential solutions that can simultaneously help us generate a cleaner environment and healthier people. At the core of this argument is an acknowledgement that multilateral cooperation can generate global public goods of the health and environment sort. More specifically, the discussion has attempted to demonstrate that international treaties and international institutions can aid the global community in meeting the health and environmental needs of present and future generations.

However, neither treaties nor institutions are foolproof, and there are unfortunately many of these entities in existence that have failed to meet their mandate. This is problematic in and of itself and also because failed attempts at multilateral cooperation can generate mistrust for such entities and sour individual and state attitudes regarding future attempts. An examination of the of the Montreal Protocol on Substances that Deplete the Ozone Layer and the Kyoto Protocol indicates that treaty structure is of utmost importance. The Montreal Protocol's permanent emissions limits, effective carrots and sticks approach, attempts to prevent leakage, and common-but-differentiated responsibility mechanisms have made the MP a relative success by lessening the use of ozone depleting substances and thus the related health problems, such as cataracts and skin cancer, that ODS produce. In other words, the MP is a model environmental treaty that future treaties should emulate. By contrast, the KP's failure to fully incorporate effective incentives in a timely manner, extreme approach to sticks/punishments, inability to prevent leakage, and its limited emissions time frame (2008–2012/2012–2020), have restricted the effectiveness of the KP. As KP negotiators look forward to renegotiating the Kyoto Protocol in 2015 (for a 2020 implementation date), they would do well to reexamine the lessons learned from Montreal.

International institutions can also aid states and nonstate actors seeking to invest in a cleaner environment and the positive health externalities that this process generates. A brief investigation into the newly formed International Renewable Energy Agency demonstrates that institutional structure is a critical factor that can dictate the success or failure of an institution. IRENA's success, as measured by its ability to attract one hundred and four members (with fifty-five additional states in the ratification process) in fewer than three years, can be attributed to a number of factors. The most important factor, however, rests in IRENA's ability to cater to the self-interest of state and nonstate actors that seek to diversify their energy sources and increase foreign direct investment. By connecting states seeking to diversify their energy dependencies with companies that can assist them in this process, IRENA simultaneously meets the needs of states in the global North and global South as well as energy corporations and other UN intergovernmental organizations. In contrast to realist claims that institutions either do not matter or are merely puppets of the great powers, this investigation suggests that institutions can enable the global community to meet the environmental and health needs of developed and developing states when properly structured.

The complex relationship between health and the environment will require new and innovative thinking if we are to provide for the needs of current and future generations. With this in mind, one final suggestion is warranted. The complex United Nations system, with myriad intergovernmental programs and specialized agencies, has recognized the intricate link between health and the environment. UN entities including the World Health Organization and UN Environment Program continue to stress the interconnection between health and the environment, but now more than ever, we need to see greater interagency cooperation between these like entities to generate novel, pioneering, and practical solutions to climate change and the resultant health issues that stem from a hotter, erratic, and less predictable global environment. A failure to do so would be very unhealthy.

NOTE

1. The author specifically uses cost/benefit data for the United States.

REFERENCES

Archibugi, Danielle, and Kim Bizzari. 2005. The Global Governance of Communicable Diseases: The Case for Vaccine R. and D. *Law and Policy* 27: 33–51.

Barrett, Scott. 1999. Montreal versus Kyoto: International Cooperation and the Global Environment. In Inge Kaul, Isabelle Grunberg, and Marc Stern, eds. *Global Public Goods: International Cooperation in the 21st Century.* New York: United Nations Development Programme and Oxford University Press.

Barrett, Scott. 2002. Consensus Treaties. *Journal of Institutional and Theoretical Economics* 158: 529–547.

Beaglehole, Robert and Derek Yach. 2003. Globalization and the Prevention and Control of Non-Communicable Disease: The Neglected Diseases of Adults. *Lancet* 32: 903–908.

Boyer, Mark A. and Michael Butler. 2006 Public Goods Liberalism: The Problem of Collective Action. In Jennifer Sterling-Folker, ed., *Making Sense of International Relations Theory.* Boulder: Lynne Rienner.

Centers for Disease Control and Prevention. 2010. Asthma, Respiratory Allergies, and Airway Diseases. www.cdc.gov/climateandhealth/effects/airway_diseases.htm Accessed 14 December 2012.

Centers for Disease Control and Prevention. 2010. Waterborne Diseases. www.cdc.gov/climateandhealth/effects/waterborne.htm Accessed 14 December 2012.

Centers for Disease Control and Prevention. 2011. Cholera in Haiti: One Year Later. www.cdc.gov/haiticholera/haiti_cholera.htm Accessed 14 December 2012.

Chamberlain, Gethin and Saeed Shah. 2010. Pakistan Floods: Aid Trickles in for Victims as Cholera Spreads in Pakistan's Worst-Ever Floods. *The Observer.* www.guardian.co.uk/world/2010/aug/15/pakistan-floods-international-aid-slow Accessed 14 December 2012.

Chen, Lincoln C., Tim G. Evans, and Richard A. Cash. 1999. Health as a Global Public Good. In Inge Kaul, Isabelle Grunberg, and Marc Stern, eds., *Global Public Goods: International Cooperation in the 21st Century.* New York: United Nations Development Programme and Oxford University Press.

Cohen, Jillian Clare and Patricia Illingworth. 2003. The Dilemma of Intellectual Property Rights for Pharmaceuticals: The Tension between Ensuring Access of the Poor to Medicines and Committing to International Agreements. *Developing World Bioethics* 3: 27–48.

Collin, J., K. Lee and K. Bissell. 2002. The Framework Convention on Tobacco Control: The Politics of Global Health Governance. *Third World Quarterly* 23 (2): 265–282.

Conde de Saro, Rafael, Spanish Ambassador. 2011. E-mail Correspondence. 5 October.

Craig, M. 1999. A Climate-Based Distribution Model of Malaria Transmission in Sub-Saharan Africa. *Parasitology Today* 15: 105–111.

DeLaet, Debra L. and David E. DeLaet. 2012. *Global Health in the 21st Century: The Globalization of Disease and Wellness.* Boulder: Paradigm Publishers.

Elbe, Stefan. 2006. Should HIV / AIDS be Securitized? The Ethical Dimensions of Linking HIV / AIDS and Security. *International Studies Quarterly* 50: 119–144.

Epstein, Paul R., Dan Farber, and Jeffrey Sachs. 2011. *Changing Planet, Changing Health: How the Climate Change Crisis Threatens our Health and What We Can Do about It.* Berkley: University of California Press.

Federal Emergency Management Agency (FEMA). 2012. Hurricane Sandy. www.fema.gov / sandy. Accessed 03 January 2013.

Fidler, David P. 2004. Germs, Governance, and Global Public Health in the Wake of SARS. *The Journal of Clinical Investigation* 113: 799–804.

Ford, J. *The Role of Typansomiases in African Ecology: A Study of Tsete Fly Problem.* Claredon Publishers.

Foust, Joshua. 2012. U.S. Drones Make Peace With Pakistan Less Likely. *The Atlantic.* 12 July: www.theatlantic.com / international / archive / 2012 / 07 / us-drones-make-peace-with-pakistan-less-likely / 259756 /.

Global Polio Eradication Initiative. 2010. Polio Eradication Staff Mobilized to Support Pakistan Flood Relief. www.polioeradication.org / tabid / 167 / iid / 66 / Default.aspx. Accessed 14 December 2012.

Green Climate Fund (GCF). 2013. Background. gcfund.net / about-the-fund / background.html Accessed 04 January 2013.

Green Climate Fund (GCF). 2013a. Mandate and Governance. gcfund.net / about-the-fund / mandate-and-governance.html Accessed 04 January 2013.

Harris, Richard. 2011. At Climate Talks, Frustration And Interruptions. *NPR.* 8 December: www.northcountrypublicradio.org / news / npr / 143379789 / at-climate-talks-frustration-and-interruptions.

Houghton, J.T., G.J. Jenkins, J.J. Ephaums, eds. 1990. 1990. *Climate Change: The IPCC Scientific Assessment.* UK: Cambridge University Press.

Hovi, Jon and Steffen Kallbekken. 2004. The Price of Non-Compliance with the Kyoto Protocol: The Remarkable Case of Norway. *CICERO Working Paper.* http: / / www.cicero.uio.no / media / 2773.pdf Accessed 14 January 2013.

Intergovernmental Panel on Climate Change (IPCC). 2012. Organization. ipcc.ch / organization.shtml. Accessed 03 January 2013.

IRENA / FC / Statute Article IV, A. 2009. Conference on the Establishment of the International Renewable Energy Agency www.irena.org / documents / uploadDocuments / Statute / IRENA_FC_Statute_signed_in_Bonn_26_01_2009_incl_declaration_on_further_authentic_versions.pdf Accessed 27 July 2011.

Jayaraman, Rajshri and Ravi Kanbur. 1999. International Public Goods and the Case for Foreign Aid. In Inge Kaul, Isabelle Grunberg, and Marc Stern, eds., *Global Public Goods: International Cooperation in the 21st Century.* New York: United Nations Development Programme and Oxford University Press.

Janik, Laura. 2011. Understanding Hepatitis B and the Global Failure to Act. *Global Health Governance* IV (2).

Janik, Laura L. and Marci Carter. 2011. International Environmental Innovation: When Does it Work? Working Paper, Draft presented at 2011 ISA-Midwest Conference in St. Louis, MO.

Janik, Laura and Jennifer Sterling-Folker. 2011. Neoliberal Institutionalism. In *IPSA Encyclopedia of Political Science*. Edited by Dirk Berg-Schlosser, Bertrand Badie, and Leonardo Morlino. SAGE Publications.

Karam, Zeina. 2012. Ambassador: United States 'Disgusted' by Russian, Chinese Veto of UN Resolution on Syria. *CnsNews*. 5 February: cnsnews.com/news/article/ambassador-us-disgusted-russian-chinese-veto-un-resolution-syria.

Kaul, Inge, and Ronald U. Mendoza. 2003. Advancing the Concept of Public Goods. In Inge Kaul, Pedro Conceicao, Katell de Goulven, and Ronald Mendoza (eds.) *Providing Global Public Goods: Managing Globalization*. New York: United Nations Development Programme and Oxford University Press.

Keohane, Robert and Joseph Nye. 1977. *Power and Interdependence*. Longman Publishing.

Kirton, John J, editor. 2009. *Global Health*. Burlington, VT: Ashgate Publishing.

Koremenos, Barbara, Charles Lipson, Duncan Snidal. 2001. The Rational Design of Institutions. *International Organization* 55 (4): 761–799.

Koutantou, Angeliki. 2011. EU, IMF Expected to Judge Greece as Protests Swell. *Reuters*. 29 May: www.reuters.com/article/2011/05/29/us-greece-idUS TRE74S2NJ20110529.

Knowlton, Kim, Miriam Rotkin-Ellman, Linda Geballe, Wendy Max, and Gina M. Solomon. 2011. Six Climate Change–Related Events In The United States Accounted For About $14 Billion In Lost Lives And Health Costs. *Health Affairs* 30 (11): 2167–2176.

Lanoszka, Anna. 2003. The Global Politics of Intellectual property Rights and Pharmaceutical Drug Policies in Developing Countries. *International Political Science Review* 24 (2): 181–197.

Levs, Josh, and Tom Watkins. 2013. Limited Sandy Vote to Occur on Friday. *CNN* 03 January: www.cnn.com/2013/01/03/politics/house-sandy-bill-vote/index.html?iref=allsearch Accessed 03 January 2013.

Longstreth, J.D., F.R. de Gruijl, Y. Takizawa, and J.C. van der Leun. 1991. Human Health. In *Environmental Effects of Ozone Depletion: 1991 Update*. United Nations Environment Programme.

McMichael, Anthony, Diarmid Campbell-Lendrum, Sari Kovats, Sally Edwards, Paul Wilkinson, Theresa Wilson, Robert Nicholas, Simon Hales, Frank Tanser, David Le Sueur, Michael Schlesinger, and Natasha Andronova. 2004. Global Climate Change. In WHO ed., *Comparative Quantification of Health Risks: Global and Regional Burden of Disease Attributable to Selected Major Risk Factors*. Geneva: WHO.

Millennium Ecosystem Assessment, 2005. *Ecosystems and Human Well-being: Biodiversity Synthesis*. World Resources Institute, Washington, DC. www.unep.org/maweb/documents/document.354.aspx.pdf. Accessed 03 January 2013.

Musgrave, Richard Abel. 1939. The Voluntary Exchange of Public Economy. *The Quarterly Journal of Economics*. 53: 213–237.

Olson, Mancur. 1965. *The Logic of Collective Action: Public Goods and the Theory of Groups*. Harvard University Press.

Paris, Max. 2012. Kyoto Climate Change Conference Sputters to a Sorry End. *CBC News* 12 December. www.cbc.ca/news/canada/story/2012/12/20/pol-kyoto-protocol-part-one-ends.html?cmp=rss Accessed 03 January 2013.

Pett, Joel. 2011. i.usatoday.net/news/opinion/cartoons/2009/December/e091207_pett.jpg.

Platt, Anne E. 1996. *Infecting Ourselves: How Environmental and Social Disruptions Trigger Disease.* World Watch Institute.

Press, Elizabeth, Senior Programme Officer, IRENA. 2011. Phone Interview 12 September.

Press, Elizabeth, Senior Programme Officer, IRENA. 2011. Email Correspondence 4 March.

Price-Smith, Andrew. 2002. *The Health of Nations: Infectious Disease, Environmental Change, and Their Effects on national Security and Development*. Boston: The MIT Press.

Price-Smith. 2009. *Chaos and Contagion: Disease, Ecology and National Security in the Era of Globalization*. Boston, MS: The MIT Press.

Ritter, Karl. 2012. UN Climate Conference: Kyoto Protocol Extended At Doha, Qatar Talks. Huffington Post 08 December. www.huffingtonpost.com/2012/12/08/un-climate-conference-kyoto-doha-qatar_n_2262371.html Accessed 14 January 2013.

Sach, Karsten, Deputy Director-General for International Cooperation in the Federal Ministry for Environment, Nature Conservation and Nuclear Safety of Germany. 2011. Phone Interview. 26 September.

Sach, Karsten, Deputy Director-General for International Cooperation in the Federal Ministry for Environment, Nature Conservation and Nuclear Safety of Germany. 2012. Written Correspondence. 5 October.

Samuelson, Paul A. 1954 The Pure Theory of Public Expenditure. *Review of Economics and Statistics* 36(4): 387–389.

Sandler, Todd. 1992. *Global Challenges: An Approach to Environmental, Political, and Economic Problems*. Cambridge, UK: Cambridge University Press.

Sandler, Todd. 1998. Global and Regional Public Goods: A Prognosis for Collective Action," *Fiscal Studies* 19 (3): 221–247.

Sandler, Todd and Daniel G. Acre M. 2002. A Conceptual Framework for Understanding Global and Transnational Public Goods for Health. *Fiscal Studies* 23(2): 195–222.

Sandler, Todd. 2006. Regional Public Goods and International Organizations," *Review of International Organizations* 1 (1): 5–25.

Schöpe, Martin, Division Head for the International and European Affairs of Renewable Energy, German Federal Ministry for the Environment, Nature Conservation and Nuclear Safety in Berlin. 2011. Phone Interview. 26 September.

Schöpe, Martin, Division Head for the International and European Affairs of Renewable Energy, German Federal Ministry for the Environment, Nature Conservation and Nuclear Safety in Berlin. 2012. Phone Interview. 9 October.

Shimko, Keith. 2009. *International Relations: Perspectives and Controversies*, 3rd edition. Boston: Wadsworth Publishing.

Shope, R. 1992. Global Climate Change and Infectious Diseases. *Environmental Health Perspectives* 96: 171–174.

Touffut, Jean-Philippe. 2006. *Advancing Public Goods.* Northampton, MA: Edward Elgar.

Tuscano, Antionette. 2010. Rotarians in Pakistan Boost Polio Eradication, Flood Relief. www.rotary.org/en/MediaAndNews/News/Pages/101004_news_pakfloodspolio.aspx Accessed 14 December 2012.

United Nations. 1998. Kyoto Protocol to the United Nations Framework Convention on Climate Change. unfccc.int/resource/docs/convkp/kpeng.pdf Accessed 15 December 2012.

United Nations Environment Programme (UNEP). 1985. The Vienna Convention for the Protection of the Ozone Layer. ozone.unep.org/new_site/en/Treaties/treaty_text.php?treatyID=1 Accessed 15 December 2012.

United Nations Development Programme (UNDP). 1994. *New Dimensions of Human Security.* New York: Oxford University Press.

United Nations Environment Programme (UNEP). 1987. The Montreal Protocol on Substances that Deplete the Ozone Layer. ozone.unep.org/new_site/en/Treaties/treaty_text.php?treatyID=3&secID=94 Accessed 15 December 2012.

United Nations Environment Programme (UNEP). 2007. A Success in the Making: The Montreal Protocol on Substances that Deplete the Ozone Layer. ozone.unep.org/Publications/MP_A_Success_in_the_making-E.pdf Accessed 15 December 2012.

United Nations Framework Convention on Climate Change (UNFCCC). 2012. Global Map Annex I. http://maps.unfccc.int/di/map/. Accessed 03 January 2013.

United Nations Framework Convention on Climate Change (UNFCCC). 2012a. An Introduction to the Kyoto Protocol Compliance Mechanism. unfccc.int/kyoto_protocol/compliance/items/3024.php Accessed 14 January 2013.

United Nations Security Council Resolution 1308. S/Res/1308/(2000). www.un.org/ga/search/view_doc.asp?symbol=S/RES/1308 percent282000 percent29

United Nations General Assembly. 1988. A/RES/43/53. www.un.org/documents/ga/res/43/a43r053.htm.

United Nations Office for the Coordination of Humanitarian Affairs. 2011. Eastern Africa Drought Humanitarian Report No. 3. reliefweb.int/report/burundi/eastern-africa-drought-humanitarian-report-no-3. Accessed 14 December 2012.

United States Environmental Protection Agency. 2010. Protecting the Ozone Layer Protects Eyesight: A Report on Cataract Incidence in the United States Using the Atmospheric and Health Effects Framework Model. www.epa.gov/Ozone/science/effects/AHEFCataractReport.pdf. Accessed 18 January 2013.

Verweij, Marco. 2006. Is the Kyoto Protocol Merely Irrelevant, or Positively Harmfu, for the Efforts to Curb Climate Change? In Marco Verweij and Michael Thompson, eds. *Clumsy Solutions for a Complex World: Governance, Politics and Plural Perceptions.* New York, NY: Palgrave Macmillan.

Whiteside, Alan. 2002. Poverty and HIV/AIDS in Africa. *Third World Quarterly* 23 (2): 313–332.

World Health Organization (WHO). 1946. Constitution of the World Health Organization. apps.who.int/gb/bd/PDF/bd47/EN/constitution-en.pdf.

World Health Organization (WHO). 2008. Climate Change will Erode Foundations of Health. www.who.int/mediacentre/news/releases/2008/pr11/en/index.html. Accessed 14 December 2012.

World Health Organization (WHO). 2009. Global Health Risks: Mortality and Burden of Disease Attributable to Selected Major Risks. Geneva: WHO.

World Health Organization (WHO). 2012. Climate Change and Health: Fact Sheet N 266. www.who.int/mediacentre/factsheets/fs266/en/index.html. Accessed 14 December 2012.

World Health Organization. 2013. Climate Change and Human Health—Risks and Responses. Summary. www.who.int/globalchange/summary/en/index7.html Accessed 18 January 2013.

Woolridge. Horn of Africa Tested by Severe Drought. BBC News 4 July www.bbc.co.uk/news/world-africa-14023160. Accessed 14 December 2011.

Yang, Tseming. 2012. Quicker Fix to Climate Change. CNN 26 December www.cnn.com/2012/12/26/opinion/yang-climate-change/index.html?iref=allsearch Accessed 04 January 2013.

4

Accommodating "Climate Refugees"

Models of Sovereignty and Security in the International Climate Regime

Craig A. Johnson

A number of recent high-profile discussions have raised the concern that climate change may soon displace large numbers of poor people in the developing world.[1] In 2009, UN Secretary General Ban Ki Moon warned that

> . . . migration is a likely consequence of climate change impacts. Populations will relocate due to more extreme weather including prolonged droughts, intensive storms and wildfire. In some cases, as with small island nations, whole countries are under threat. Protecting vulnerable communities must be a priority in both national and international adaptation efforts.[2]

In its 2007 report, "the Human Tide," the British Aid charity Christian Aid (2007) called climate-induced migration "the most urgent threat facing poor people in the developing world. The time for action is now." More recently, the UK government's "Foresight Report" on migration and global environmental change concluded that:

> The challenges of migration in the context of environmental change require a new strategic approach to policy. Policy makers will need to take action to reduce the impact of environmental change on communities yet must simultaneously plan for migration. (Foresight 2011: 7)

At the heart of this discourse is the idea that climate change poses both a "human security threat" to populations facing the risk of forced displacement and a broader "strategic security threat" to the sovereignty and stability of individual nation states.[3] Within the strategic security field, particular emphasis has been placed on the idea that forced international migration will overwhelm the ability of sovereign nation states to accommodate and absorb new flows of people displaced by climate change. The German Advisory Council (WBGU, 2008: 1), for instance, maintains that:

> . . . without resolute counteraction . . . climate change could exacerbate existing environmental crises such as drought, water scarcity and soil degradation, intensify land-use conflicts and trigger further environmentally induced migration. Rising global temperatures will jeopardize the bases of many people's livelihoods, especially in the developing regions, increase vulnerability to poverty and social deprivation, and thus put human security at risk. Particularly in weak and fragile states with poorly performing institutions and systems of government, climate change is also likely to overwhelm local capacities to adapt to changing environmental conditions and will thus reinforce the trend towards general instability that already exists in many societies and regions.

Similarly, and more squarely directed towards the American security regime, Werz and Conley (2012: 7–8) warn that "Environmentally induced migration, resource conflicts, and unstable states will not only have an impact upon the nations where they occur, but also on the United States and the broader international community."

Notwithstanding substantial improvements in adaptive capacity, projections have been made that climate change will entail large-scale displacement of populations affected by the disruption of food supplies, health systems, human settlements and major livelihoods, especially ones deriving from agriculture, forestry and fishing (Biermann and Boas 2007; 2008a; IPCC WG-II, 2007; Brown et al., 2007; Reuveny, 2007; Christian Aid, 2007; WBGU, 2008). In a widely cited article, Norman Myers observes (1993) that climate change may displace as many as 250 million people by the year 2100. Drawing upon existing trends and future scenarios, Warner et al., (2009) project that 250 million people could be displaced as a result of climate change (Bogardi and Warner, 2009; cf. Biermann and Boas, 2008a; WBGU, 2008). Others put the figure at 1 billion (Christian Aid, 2007).[4]

Reflecting upon the limitations of reforming the existing refugee regime, many commentators are now calling upon the international community to create a new protection regime that would accommodate the special needs and circumstances of populations displaced as a result of climate change (Biermann and Boas 2007; 2008a; Risse, 2009a; 2009b; Foresight, 2011). Others contend that the international community needs to be more proactive, using international financing for adaptation and development (including the newly-created Green Fund) to support relocation and resettlement in advance of dangerous climate change (Biermann and Boas 2007; 2008a; Risse 2009a; 2009b; Barnett and Webber 2010; Johnson and Krishnamurthy 2010; Foresight 2011; Johnson 2012).

When it is planned and supported through public policy, migration can provide an important means of diversifying and strengthening poor people's assets, livelihoods and income sources, thereby reducing vulnerability to climate change (Biermann and Boas 2008a; Johnson and Krishnamurthy 2010; de Haan 1999; Deshingkar 2005; McLeman and Smit 2006; Perch-Nielson et al., 2008; de Sherbinin et al., 2008; 2011b Bogardi and Warner 2009; Barnett and Webber 2010). However, governments often actively discourage migration, using labour codes, land use restrictions and other policy instruments to control the movement and settlement of itinerant populations (cf. Bakewell 2008). Moreover, national and international debates about immigration and asylum policy have become increasingly hostile towards the rights of migrants, "asylum seekers" and refugees (Joppke 1999; Neumayer 2006; Zetter 2007).

In short, there appears to be a significant disconnect between the laudable norms being espoused by the international community and the ground realities that are now shaping the contemporary politics of migration, immigration and asylum. This chapter explores the challenge of supporting migration through the "international climate regime," a term I use to describe the rules, norms, and procedures that have been established through the UN Framework Convention on Climate Change (hereafter UNFCCC).[5] By examining the priorities outlined in the UNFCCC-sponsored (NAPAs), I make the case that (1) migration has been characterized almost universally in terms of the threats and vulnerabilities it creates for societies, ecological systems and migrants and (2) that the UNFCCC has effectively upheld the principles of state power and national sovereignty at the expense of human security.

The chapter proceeds as follows. The following section first outlines the risks of climate-induced displacement, contesting the notion that climate change will automatically or necessarily lead to future floods of "climate refugees." Section 3 then explores the ways in which migration has been characterized in relation to the priorities outlined in the NAPAs and the UNFCCC. Section 4 situates the discourse on climate change and displacement within a wider discourse on migration, immigration and asylum. Finally, section 5 concludes the chapter by discussing the implications of these findings and insights for human security, national security and the international climate regime.

A "COMING TIDE"? FORCED DISPLACEMENT AND CLIMATE CHANGE

There is now a growing consensus that human emissions of greenhouse gases are leading to unprecedented transformations in the earth's climate, creating new forms of vulnerability to climate-induced disasters and long-term environmental change.[6] In its fourth assessment report, the IPCC predicts that "hot extremes, heat waves, and heavy precipitation will continue to become more frequent," during the next century (IPCC WG-I, 2007: 15), projecting that the effects of climate change will be worst in Africa and Asia, where agriculture, public health systems, food supplies, and human settlements are least able to adapt to extreme climate events, such as flooding, windstorms, disease, and drought (IPCC WG-II, 2007). In its latest assessment on disasters and extreme events the Intergovernmental Panel on Climate Change (IPCC, 2012) reports that it is "virtually certain" that daily temperature extremes will become increasingly frequent and intense over the next century.

Such projections have serious implications for human security and vulnerability. One is the prospect that extreme weather events will become increasingly frequent and intense, exacerbating human exposure to climate-related disasters, such as windstorms and floods. A second implies longer-term transformations in which changes in rainfall, temperature, and seasonality become increasingly inconsistent with the strategies farmers, resource managers, and policy makers have used to sustain livelihoods in environmentally sensitive regions and sectors, such as farming, forestry, and fishing.

To conceptualize the range of policies that may be used to accommodate populations displaced as a result of climate change, Johnson and Krishnamurthy (2010) distinguish between displacement occurring as a result of longer-term *processes* and displacement due to relatively discrete environmental *events*. Broadly speaking, *processes* refer to cumulative environmental changes that lead to the destruction or degradation of livelihoods over a relatively long period of time. *Events*, on the other hand, refer to sudden meteorological hazards such as tropical cyclones and floods, which can destroy lives and livelihoods, forcing people to resettle either temporarily or permanently. Depending on the nature and scale of the stress, and on the assets of a household, such strategies may involve new household organizational structure or even migration and abandonment of their place of origin.

Whether displacement is due to processes or events has important implications for the kinds of migration and adaptation strategies affected populations may be able to pursue. First, longer-term processes (e.g., desertification, soil erosion, sea level rise) can lead to the permanent degradation of livelihoods and ecological services. Events, in contrast, may affect livelihoods temporarily, suggesting that migration may be used as a short-term strategy. Second, the duration of the impact on livelihoods has implications for the scale and proximity of displacement. In the case of process-induced migration, resettlement may occur over larger distances, potentially over national borders. Finally, process-related migration may open new economic opportunities, highlighting the challenge of distinguishing between migration that occurs as a result of climatic factors as opposed to migration aimed at securing better incomes and livelihoods (Johnson and Krishnamurthy 2010; Gemenne 2011; and below).

On this basis, conceptual distinctions can be made between *distress migration*, in which household decisions are largely *ad hoc* responses to external environmental processes and events; and *economic migration*, which suggests the existence of forward and strategic planning on the part of the household. Finally, "forced displacement" implies a form of distress migration that obviates all but the most desperate of coping strategies (cf. de Haan 1999; Deshingkar 2005; McLeman and Smit 2006; Perch-Nielson et al., 2008; de Sherbinin et al., 2008; Boano et al., 2008; Raleigh et al., 2008; IOM, 2009).

Arguably the most immediate form of displacement will come as a result of sea level rise. The IPCC (Nicholls et al., 2007) projects with "very high confidence" that coastal areas will become increasingly vulnerable to "an accelerated rise" in eustatic sea levels, ranging from 0.2 to 0.6 metres or more by the year 2100. Others (e.g., Young and Pilkey 2010) contend that if the Greenland and West Antarctic ice sheets are factored into the equation, global sea levels could rise by a range of 4 to 6 metres. According to a study commissioned by the US National Science Foundation, an estimated 600 million people (or 10 percent of the Earth's population) now live in areas that are within 100 kilometres of the coast and at elevations less than 100 metres above sea level (McGranahan et al., 2007). Of these, 360 million people live in urban areas.

The IPCC (WGI 2007) identifies three "key hot spots of societal vulnerability," in which a combination of land use and population pressure may exacerbate exposure to windstorms, erosion and floods: river deltas, especially the seven Asian "mega-deltas;" low elevation coastal urban areas, especially ones prone to subsidence; and small islands, especially coral atolls. Drawing primarily upon IPCC data, the German Advisory Council (WBGU 2008) identifies five regional "hotspots" in which major forms of climate-induced displacement and conflict are expected to occur:

The Caribbean and the Gulf of Mexico
North Africa, especially the Nile Delta
The Sahel zone
South Asia, especially the Ganges-Brahmaputra Basin; and
Eastern China

With the exception of the Sahel, all of these are coastal areas and river basins, suggesting that the most immediate forms of climate-induced displacement will be the result of flooding, erosion, windstorms, and rising sea levels.

Finally, it is worth noting that planned efforts to mitigate and adapt to the impacts of climate change may themselves become drivers of displacement. De Sherbinin and colleagues (2011b), for instance, raise the concern that international efforts to fund mitigation and adaptation will likely come in the form of large-scale projects (e.g., seawalls, coastal defences, biofuel projects) that will displace local populations either as a result of planned resettlement or outright eviction. Such insights highlight an important distinction between the anticipated damages that may be expected to occur as a result of climate change and the various threats posed by planned efforts to mitigate and adapt to climate change.

To assess the ways in which—and extent to which—migration has been defined and supported at the international level, the following section first describes the main sources of funding for adaptation through the UNFCCC and the Kyoto Protocol. Next it explores in more detail the ways in which migration has been characterized in relation to the priorities outlined in the NAPAs.[7]

DEALING WITH DISPLACEMENT: IS "ADAPTATION" UP TO THE TASK?

Any effort to understand the ability of the international climate regime to support migration as an adaptation option faces two methodological problems: first, adaptation financing is still in its infancy; second, official definitions of adaptation tend to vary.

In theory, the UNFCCC provides some scope for mobilizing international action to reduce the risk of climate-induced displacement. In its preparatory document for the Copenhagen Conference of the Parties in December 2009, the Ad Hoc Working Group on Long Term Cooperative Action (AWG-LCA, 2010: p.17, Article 4 [g]) invited parties to the Convention to undertake measures "to enhance understanding and cooperation related to national, regional and international climate change induced displacement, migration and planned relocation, where appropriate." Looking forward, the Fifth Assessment Report—scheduled for 2014—will contain a number of cross cutting themes that address the challenge of migration and forced displacement.

The UNFCCC has also created a number of new financial mechanisms aimed at supporting adaptation (and potentially migration) in the developing world. Prior to the Cancun Conference of the Parties (COP) in 2010, the principal UNFCCC funding mechanisms were the Least Developed

Country Fund (LDCF), which is intended to help with the identification of adaptation priorities in the NAPAs; the Special Climate Challenge Fund (SCCF), which is designed to support adaptation, as well as mitigation and technology transfer; and the Adaptation Fund, which is supported under the Kyoto protocol and aimed at supporting "concrete adaptation" actions. Following the Cancun meetings, a Green Climate Fund was established in conjunction with a commitment on the part of developed countries to provide $US 30 billion of "fast-start finance" in 2010—2012 and US$100 billion per year by 2020.

However, neither these documents nor the UNFCCC provides explicit mechanisms for accommodating large-scale movements of populations displaced by climate change, highlighting wider normative questions about whether migration constitutes a viable adaptation strategy. Chapter 17 of the IPCC's Fourth Assessment Report (Adger et al., 2007: 736), for instance, suggests that migration leading to the permanent abandonment of land and livelihood is not a desirable adaptation option. Moreover, as noted earlier, national governments and international institutions are often strongly predisposed against the idea of accommodating new populations displaced (either permanently or temporarily) as a result of environmental and non-environmental factors.

Moreover, there is still a lack of consensus about what adaptation means, and how it should be applied to existing decisions about future vulnerability and risk (cf. Burton 2004 [2009]; Huq and Reid 2004 [2009]; Moench 2007 [2009]; Peskett et al., 2009; Ayers and Huq, 2009; Brooks et al., 2009; Boyd et al., 2009).[8] Although the UNFCCC provides a number of Articles (4.1, 4.4, 4.8 and 4.9) that identify the ways in which parties to the Convention may be expected to support adaptation, the Convention lacks a precise definition that can be operationalized through the UNFCCC (Burton, 2004 [2009]). Similarly, although the GEF has responsibility for assisting developing countries in the preparation of NAPA documents, neither its website nor its brochure "Linking Adaptation to Development" (GEF 2007) provides a precise definition of what adaptation may entail.

To understand the ways in which migration has been defined and supported through the UNFCCC, a content analysis of the forty-seven NAPAs currently listed on the UNFCCC website was used to establish:

whether specific reference is made to migration in the NAPA documents;
whether migration is considered a form of adaptation;
whether it is associated with "positive" or "negative" policy outcomes, including for instance, urbanization, over-population and resource degradation; and
whether there is a specific policy statement in the NAPA aiming to support migration.

Table 4.1 presents the results of the survey, which reveal that of the forty-seven NAPAs currently eligible for LDC funding, only two (Sao Tome and Principe and Tanzania) describe migration in positive terms. Twelve NAPAs (including, surprisingly, the Maldives) make absolutely no reference to migration; the remaining thirty-three describe migration in terms of the threat it poses to resource scarcity, urban crowding, agrarian conflicts (involving primarily herders and farmers) and social stability.

Among many of the NAPAs, a common concern is that climate change will exacerbate resource scarcities and population pressures, thereby fueling a vicious cycle in which migration leads to additional resource degradation, chronic poverty, and ultimately some form of social breakdown. The following quotation from Uganda's NAPA (2007: 37) captures the essence of this concern:

> If affected communities have no option for coping with climate-induced stress, especially in drought-prone areas then, victims migrate to urban areas or resource-endowed neighbourhoods . . . In the protected areas such as the national parks and game reserves, these negative aspects of the strategy are more pronounced . . . In the pastoral communities where livestock is the major source of food, migration of the men (family leaders) with the livestock herds in search of water and pasture often leaves the family behind more vulnerable to famine.

Similarly, Sudan (2007: 45) highlights the problems that can arise when

> . . . local farmers are unable to harvest during the rainy season but equally unable to harvest during summers where climate variability produces drought, many are forced to migrate to find more suitable land, resulting in tribal confrontation over land resources and internal displacement.[9]

In almost every instance migration is being conceptualized in terms of the threats and vulnerabilities it poses to societies, ecosystems, and migrants themselves. In almost no instances is it characterized as a means of diversifying or improving livelihoods and income sources, understating dramatically the very large literature that now exists on migration, remittances, and poverty reduction (cf. de Haan 1999; Perch-Nielson et al., 2008; de Sherbinin et al., 2008; Barnett and Webber 2010).

The only exceptions to this general pattern can be found in the NAPAs prepared by Sao Tome and Principe, where migration is listed as a possible form of adaptation to coastal erosion and Tanzania, where human resettlement and a "good land tenure system" are listed among a number of measures that may be taken to reduce the risks of forced displacement or to facilitate migration when it becomes inevitable. Mali and Senegal have specific policies on migration, although these are essentially aimed at using

Table 4.1. Support for Migration in the National Adaptation Programs of Action

Country	Reference to migration?	Migration considered adaptation?	Negative reference?	Specific policy on migration?
1. Afghanistan	✓	X	✓	X
2. Angola	✓	X	✓	X
3. Bangladesh	✓	X	✓	X
4. Benin	X	X	X	X
5. Bhutan	✓	X	✓	X
6. Burkina Faso	✓	X	✓	X
7. Burundi	✓	X	✓	X
8. Cambodia	✓	X	✓	X
9. Cape Verde	✓	X	✓	X
10. Central African Republic	✓	X	✓	X
11. Chad	✓	X	✓	X
12. Comoros	✓	X	✓	X
13. Democratic Republic of Congo	X	X	X	X
14. Dijbouti	✓	X	✓	X
15. Eritrea	✓	X	✓	X
16. Ethiopia	✓	X	✓	X
17. Gambia	✓	X	✓	X
18. Guinea	✓	X	✓	X
19. Guinea-Bissau	✓	X	✓	X
20. Haiti	✓	X	✓	X
21. Kiribati	✓	X	✓	X

(continued)

Table 4.1. (*continued*)

22. Lao PDR	X	X	X	X
23. Lesotho	✓	X	✓	X
24. Liberia	X	X	X	X
25. Madagascar	✓	X	✓	X
26. Malawi	X	X	X	X
27. Maldives	X	X	X	X
28. Mali	✓	X	✓	✓
29. Mauritania	✓	X	✓	X
30. Mozambique	✓	X	✓	X
31. Nepal	X	X	X	X
32. Niger	X	X	X	X
33. Rwanda	✓	X	✓	X
34. Samoa	✓	X	✓	X
35. Sao Tome + Principe	✓	✓	X	✓
36. Senegal	✓	X	✓	✓
37. Sierre Leone	X	X	X	X
38. Solomon Islands	X	X	X	X
39. Sudan	√	X	√	√
40. Tanzania	✓	✓	✓	✓
41. Timor-Leste	✓	X	√	X
42. Togo	✓	X	√	X
43. Tuvalu	✓	X	X	X
44. Uganda	✓	X	✓	X
45. Vanuatu	X	X	X	X
46. Yemen	✓	X	✓	X
47. Zambia	X	X	X	X

Source: http://unfccc.int/cooperation_support/least_developed_countries_portal/submitted_napas/items/4585.php Last accessed 18 December 2012 zoning restrictions and development programming to prevent poor and primarily rural people from migrating into urban and other rural areas.[10]

TO WHAT EXTENT DO THE PRIORITIES OUTLINED IN THE NAPA DOCUMENTS REFLECT THE PRIORITIES OF THE ORGANIZATIONS ADMINISTERING AND FUNDING THE NAPAS?

As noted earlier, the GEF does not provide a precise definition of what adaptation may entail. That said, the UNFCCC does provide a number of policy documents that have bearing on the ways in which the GEF and national governments interpret the meaning of adaptation. According to its report (UNFCCC 2007: 10) on "impacts, vulnerabilities, and adaptation in developing countries,"

> Adaptation is a process through which societies make themselves better able to cope with an uncertain future. Adapting to climate change entails taking the right measures to reduce the negative effects of climate change (or exploit the positive ones) by making the appropriate adjustments and changes. There are many options and opportunities to adapt. These range from technological options such as increased sea defenses or flood-proof houses on stilts, to behaviour change at the individual level, such as reducing water use in times of drought and using insecticide-sprayed mosquito nets. Other strategies include early warning systems for extreme events, better water management, improved risk management, various insurance options and biodiversity conservation.

Like many of the NAPA documents, the Secretariat's take on questions concerning migration and environmental degradation is also largely negative:

> Competition for scarce resources, such as fresh water, land or fishing grounds, brought about by changes in climate, has the added potential to cause conflict over resources with impacts on the achievement of the Millennium Development Goals, and on human migration. For example, in Africa increased pressure on resources related to food and water insecurity can deepen tensions between communities and ethnic groups resulting in violence and war. (UNFCCC 2007: 42)

In short, the NAPAs, the UNFCCC Secretariat and the GEF all appear to be strongly predisposed against the idea of either recognizing or supporting migration as a form of adaptation. Rather, most are framed in a way that equates migration with overpopulation, environmental degradation and social breakdown.

Although there is of course every good reason to be concerned about forced displacement (on both strategic and human security grounds), the principal policy issue here is that migration is being framed only or primarily in Malthusian terms, underplaying the idea that migration also provides an important means of diversifying incomes, improving livelihoods and reducing vulnerability to environmental shocks and stresses (cf. de Haan 1999; Deshingkar 2005; McLeman and Smit 2006; Perch-Nielson et al., 2008; de Sherbinin 2008; 2011a; 2011b; Barnett and Webber 2010; Johnson and Krishnamurthy 2010; Gemenne 2011).

To understand the bias that appears to exist within the international climate regime, the following section situates the discourse on migration and immigration within a wider discourse on national sovereignty and international security.

COMPETING MODELS OF SOVEREIGNTY AND SECURITY

The above discussion suggests an apparent dichotomy between the national security concerns that have been raised by national governments within the NAPA documents and the human security concerns that could conceivably be addressed by supporting migration. This section now situates this dichotomy within a wider debate about the real and ideal relationship between migrant populations and the nation state. It does so by characterizing a number of dominant policy narratives that have been used to frame the rights and perceived threats of populations displaced as a result of climate change.

Work within the Existing Refugee Regime

One narrative that has informed some of the legal-normative writing about the rights of "climate refugees" is the idea of using the existing international refugee regime to protect populations displaced as a result of climate change. According to the UN High Commisssion for Refugees, a refugee is defined as a person who *"owing to a well-founded fear of being persecuted for reasons of race, religion, nationality, membership of a particular social group, or political opinion, is outside the country of his nationality and is unable or, owing to such fear, is unwilling to avail himself of the protection of that country."* (CSR, Art. 1.A.2).[11] Kolmannskog (2008) and Westra (2009) identify a number of conditions under which environmental factors may be used as grounds for protection under the Convention. One is the principle of non-refoulement, which holds that people cannot be returned to places where their lives or freedoms are under threat. In theory, returning populations to homelands that have been rendered effectively uninhabitable would violate non-refoulement, thereby creating obligations on the part of national governments (Kolmannskog 2008). A second concerns the nature of persecution, where it can be established that environmental degradation (e.g., draining of marshlands, etc.) is being intentionally used to harm populations on the basis of "race, religion, nationality, membership of a particular social group, or political opinion," such groups may conceivably claim protection on the grounds of political persecution (Kolmannskog 2008; Westra 2009).

The Convention therefore provides some scope for protecting populations whose homelands have been rendered permanently uninhabitable, but these provisions apply only to individuals seeking refugee status, which effectively precludes efforts to encourage voluntary resettlement in advance of displacement (cf. Biermann and Boas 2008a). Moreover, the idea of revising the Convention to include new categories of refugee raises difficult ethical and political questions about the rights of populations currently recognized under the 1951 Convention (i.e., those facing persecution) and about the terms on which national governments may be expected to grant new rights of citizenship and asylum (Kolmannnskog 2008; Biermann and Boas 2008a).

Create a New Regime

Recognizing the limitations of working through the existing refugee regime, a number of analysts have argued in favour of long-term strategies aimed at supporting economic (as opposed to distress) migration (Risse 2009a; 2009b; Johnson and Krishnamurthy 2010; McAdam and Saul 2010; Barnett and Webber 2010; de Sherbinin et al. 2011a; 2011b; Johnson 2012). In their background report to the 2010 World Development Report on climate change and development (World Bank, 2010), Barnett and Webber (2010: 2), for instance, made the following recommendations:

> Policy responses to minimize the risks associated with migration in response to climate change, and to maximize migration's contribution to adaptive capacity include: ensuring that migrants have the same rights and opportunities as host communities; reducing the costs of moving money and people between areas of origin and destination; facilitating mutual understanding among migrants and host communities; clarifying property rights where they are contested; ensuring that efforts to assist migrants include host communities; and strengthening regional and international emergency response systems.

More ambitiously, Biermann and Boas (2008a) have argued in favour of creating an entirely new protection regime that would be defined primarily on the rights and entitlements of populations facing the immediate and long-term threat of displacement dues to climate change. Their proposal has five components. One is a commitment on the part of the international community to encourage "planned and voluntary resettlement and reintegration of affected populations over periods of many years and decades, as opposed to mere emergency response and disaster relief," (Biermann and Boas 2008a: 3). The second is that refugees whose homes and communities have been rendered permanently uninhabitable should be considered "permanent immigrants to the regions or countries that accept them," (Biermann and Boas 2008a: 3). A third is that the new regime be "tailored not to the needs of individually persecuted people (as in the current UN refugee regime), but of entire groups of people, such as populations of villages, cities, provinces, or even entire nations," (Biermann and Boas 2008a: 3). The fourth is that protection be targeted less toward populations displaced "outside their states" and more toward "the support of governments, local communities, and national agencies to protect people within their territories," (Biermann and Boas 2008a: 3). Finally, the protection of climate refugees must be the responsibility of the international community, particularly "the wealthy industrialized countries," which have "caused most past and present greenhouse gas emissions," (Biermann and Boas 2008a: 3).

In principle, the idea of acting early to support voluntary resettlement could provide an important means of diversifying and strengthening poor people's assets, livelihoods, and income sources, thereby reducing vulnerability to rapid onset disasters and long-term environmental change. In practice, it faces a number of logistical and ethical challenges. First, notwithstanding safeguards that would ensure the rights of individuals, the emphasis on groups may exclude important subgroups, such as widows, landless laborers, and people with disabilities, raising questions about the terms on which groups would be recognized and defined for the purposes of international climate refugee protection. Second, the idea of vesting responsibility in the hands of national governments puts considerable faith in states whose capacity to govern cannot be taken for granted, especially in areas plagued by chronic poverty, environmental degradation, and social breakdown (Brown et al., 2008; WBGU 2008). Third, it raises questions about the specific ways in which and terms on which national governments and international institutions may support the needs of populations displaced as a result of climate change (e.g., Hulme 2008; Johnson and Krishnamurthy 2010). Finally, it highlights the methodological challenge of establishing whether displaced populations were in fact displaced by climatic factors (cf. Black 2001; Boano et al., 2008; Hulme 2008; Hartmann 2010; Gemenne 2011).

In their crudest forms, environmental migration models appear to suggest that climate phenomena are the only or primary factors motivating migration when in fact a host of factors (including seasonality, wage rates, persecution, life cycles, *institutions*) may have equal or more bearing on the decision or not to migrate (Black 2001; Castles 2002; Boano et al., 2008; Hulme 2008; IOM 2009). At the very least, the lines of causality are more complex than they are sometimes construed in the literature (Reuveney 2007), highlighting the need to scrutinize far more carefully the theories and assumptions being used to construct these seemingly dire scenarios (cf. Barnett and Adger 2007; Nordas and Gleditsch 2007; Hartmann 2010; Gemenne 2011).

Lifeboat Ethics

A third narrative that has direct bearing on the political viability of either reforming or working within the existing system are the political discourses that are now shaping the politics of immigration and asylum in the West. If we accept the idea that the wealthier industrialized economies have a moral responsibility to protect populations displaced by climate change (cf. Biermann and Boas 2008a; Risse 2009a; 2009b; Johnson (2012), what are the challenges of achieving this in practice? Space restrictions preclude an extended analysis of the factors affecting contemporary immigration policy, but suffice to say that public discourses surrounding the rights of refugees and immigrants in the western industrialized economies have narrowed considerably on a wide range of issues, including the rights of refugees (cf. Joppke 1999; Neumayer 2006; Zetter 2007; Bauder 2008). Roger Zetter (2007), for instance, has argued that the labels being used to describe migrants and refugees have shifted in ways that now question and effectively criminalize the legitimacy of transnational migrants and refugees. Take, for instance, the reaction of one conservative commentator—Ezra Levant—to the arrival of a group of Sri Lankan Tamils seeking refugee status in Canada in 2010:

> How bad is life back in Sri Lanka for Tamil refugees? Are they tortured? Do they have a well-founded fear of persecution? Are things so bloody bad over there that we have to let a boatload of them into Canada, just because they showed up? (Levant 2010)

Similarly, the former Minister of Immigration, Monte Solberg had the following to say about the issue:

> Four hundred ninety Tamils have paid human smugglers large amounts of money to bring them to Canada on a rickety boat because the world knows Canada is soft on illegal immigration. I should know. I served as Canada's minister of immigration for a little less than 10 months back in 2006. (Solberg 2010)

Whether these statements point to a narrowing of political tolerance, they certainly highlight the contested ways in which refugees and immigrants more generally have been framed in the public discourse.

Returning to the idea of using immigration policy to protect populations facing the threat of displacement due to climate change, it is useful to consider one final case of a proposal floated by Michael Clemens of the Center for Global Development to use expanded immigration quotas as a way of assisting the victims of Haiti's 2010 earthquake. Writing in the Washington Post, Clemens (2010) argued in favour of using a "golden door visa, to be issued in limited numbers to people from the poorest countries, such as Haiti." His principal justification was that the new economic opportunities and remittances provided by an expanded immigration quota would outweigh the benefits of traditional humanitarian relief. More important than the logic Clemens was using to justify the proposal was the reaction it generated in the general public. Here, for instance, is one anonymous e-mail Clemens received after publishing the article:

> *You America hating scum. Millions of people out of work and you want more uneducated, unskilled people to flood America. . . . We have given billions of dollars to Haiti. Poverty is not to blame. The corrupt government is to blame. . . . Is Mao one of your favorite people also?* Anonymous email sent to CGD Fellow Michael Clemens responding to his Washington Post op-ed proposing a "golden door visa" that would give immigration priority to the world's poor.[12]

The point here is not that views of this kind are part of a wider pattern or trend. Rather it is that proposals aimed at using immigration and asylum policies to accommodate the needs of populations displaced by climate change will operate in a reality that is often decidedly hostile to the idea of expanding national immigration quotas. Moreover, and this is the key point, the political undercurrents that are driving the politics of immigration and asylum are often overlooked in the policy literature on climate change and displacement.

CONCLUSIONS, LIMITATIONS AND IMPLICATIONS FOR ANALYSIS

This chapter has explored the possibilities of using adaptation to support the rights and capabilities of populations displaced as a result of climate change. By examining the priorities outlined in the NAPAs, it makes the case that migration has been characterized almost universally in terms of the threats and vulnerabilities it creates for societies, ecological systems, and migrants themselves. By documenting (some of) the discourses that are currently shaping the politics of immigration and asylum in the West, it highlights the challenges of constructing a regime that would extend new rights and entitlements to populations displaced as a result of climate change.

The preceding analysis therefore suggests that the UNFCCC and the NAPAs have effectively upheld the principles of state power and national sovereignty at the expense of human security. Within countries, the bias against migration reiterates "old" debates about the industrialization of agriculture and manufacturing, about the perceived "urban bias" in development policy (e.g., Lipton 1983; Corbridge 1983) and about the more general tendency on the part of states and state bureaucracies to define and therefore control their populations in relation to fixed social categories and physical boundaries (Scott 1998; Bakewell 2008). Internationally, the discourses that have emerged in the American and Canadian contexts suggest a political narrowing that questions both the legitimacy of *potential* refugee claimants and of immigration more generally.

The challenge of accommodating populations rendered homeless and stateless by climate change therefore entails a difficult paradox of taking action to reduce the risks of climate-induced displacement (conceivably through support for migration) while at the same time having to work within a system of sovereign nation-states. The paradox has at least two dimensions. First, the idea of allowing or supporting large-scale movements of people challenges the territorial sovereignty of nation states, whose governments are typically reluctant to expand the terms on which displaced populations may seek asylum in their own countries and jurisdictions (Biermann and Boas 2007; 2008a; 2008b; Boano et al., 2008; Kolmannskog 2008; Raleigh et al. 2008). Second, it contradicts the very strong tendency on the part of development programs and policies to target and control populations whose rights of citizenship, permanent settlement and entitlement are strong (Bakewell 2008).

That said, the preceding discussion also raises a number of questions about the sources and methodologies being used in this analysis. First, questions can be raised about the extent to which the NAPAs are a reliable indication of the ways in which national governments are framing the rights and threats of populations displaced as a result of climate change. As noted earlier, the NAPAs are a product of a UNFCCC process aimed at establishing national priorities for climate change adaptation. In other words, they are basically official statements about the ways in which national governments perceive their own adaptation needs and priorities. Moreover, the NAPAs are exceedingly top-down and technocratic, promoting plans that arguably reflect the priorities of donors, international agencies (especially the GEF) and national governments, as opposed to vulnerable populations themselves (Ayers and Huq 2009).

At the same time, the NAPAs provide an important indication of the ways in which national governments are articulating their needs and priorities to the international community, whose actors, interests and potential funding constitute an important part of the international climate regime. On this basis, we can conclude that the NAPAs are at the very least reiterating a model that appears to be strongly predisposed against the idea of either recognizing or supporting migration as a form of adaptation. Like the international negotiations that gave them force, the UNFCCC and the Kyoto Protocol are products of an international system whose principal aim is to uphold the jurisdictional and territorial sovereignty of individual nation states. Correspondingly most of the adaptation programming that has so far been supported under the Kyoto Protocol and the UNFCCC has transpired on a country-by-country basis (cf. Huq and Reid 2004 [2009]).

Second, the preceding raises questions about the reliability of using selected quotes and events to construct the "political realities" that are currently shaping the politics of immigration and asylum in the West. As noted earlier, the various American and Canadian perspectives on immigration are by no means intended to represent a wider pattern or trend. On the contrary, they are only meant to provide a glimpse into the kinds of political rhetoric that have been used to frame the rights and perceived threats of international migrants and refugees. However, as Zetter (2007) has argued, the labels being used to describe refugees, migrants and "asylum seekers" can reveal a great deal about the ways in which official and unofficial policy discourses are shaping the politics of immigration. And here it is worth emphasizing that some of the most hostile positions being presented were ones that were posted *anonymously* on the Internet (what we might call the "online version" of bathroom graffiti).[13]

Third, the preceding discussion has focused primarily on the threats posed by international migrants to states and societies in the West. However, according to the 2009 Human Development Report on human mobility (UNDP 2009), the vast majority of migration—roughly three-quarters—has been happening domestically. Moreover, of the 200 million people who are estimated to have migrated internationally in 2008, the vast majority were migrating to neighboring countries.

Recognizing the costs of engaging in long-distance migration (and assuming that disaster-induced displacement will exacerbate these costs substantially), it seems reasonable to conclude that the principal threats to national sovereignty and international security will manifest themselves in close proximity to the countries and regions that are most vulnerable (and least resilient) to the effects of climate change. As Barnett and Webber (2010) point out, future patterns of climate-induced migration are likely to exacerbate the patterns that are already established in contemporary migration.

That said, Western discourses about immigration have important bearing in at least two ways. The Western countries are still by far the largest contributors to the UNFCCC. Notwithstanding a dramatic reallocation of costs and responsibilities (e.g., towards China, India, etc.), it seems likely that their needs and perspectives (including ones on immigration) will continue to have a disproportionate influence on the UNFCCC. Second, the "internal" factors that are now shaping the politics of immigration and asylum in the West have important bearing on the perceived legitimacy of the international refugee regime. As Zetter (2007) has argued, resettlement policies, refugee board decisions and op-ed articles are all part of a wider discourse that affect the ways in which governments (and migrants themselves) perceive the legitimacy of the international refugee regime. So here, a narrowing of political space in the West could very well undermine the legitimacy of the international refugee regime, thereby undermining its ability to provide new forms of protection to populations displaced as a result of climate change.

To the extent that they will have to work with these fairly well-entrenched institutions and practices, adaptation and development programming will need to develop new and creative ways of dealing with populations displaced by the effects of climate change (cf. Biermann and Boas 2008a; Johnson and Krishnamurthy 2010; de Sherbinin et al., 2011a; 2011b; Johnson 2012). Although policy makers are right to worry about large-scale displacement, Malthusian fears about over-population and urban crowding provide what is at best an impoverished understanding of the ways in which migration and support for migration may reduce vulnerability to rapid onset disasters and long-term environmental change.

If we assume that resilient systems are ones that have the capacity and the flexibility to diversify, to pool risk and to learn from past events (Berkes and Folke 1998; Moench 2007 [2009: 257]; Nelson et al., 2007), the preceding suggests that the international climate regime is in its current form woefully inadequate for dealing with the new vulnerabilities created as a result of climate change. Where climate change requires transnational cooperation and flexibility, the international system appears mired in a post-Westphalian discourse about the apparent sanctity of international boundaries and nationally-defined citizenship. Whether it can develop new and creative ways of pooling and spreading risk and whether it can learn from past experiences and events will be crucial.

"Power-based realist theories" contend that international regimes reflect the power and resources that states are able to exert on international institutions (e.g., world trade) and the inter-state system (Viotti and Kaupi 2010: 132). At the heart of this formulation is the realist assumption that states have identifiable interests (including especially hegemonic stability) that they pursue through multiple international channels. "Knowledge-based cognitive theories," on the other hand, contend that individual national interests "are not given, but created" (Viotti and Kaupi 2010: 132) by ideas whose narratives can change the norms and behaviour of international regimes and nation states (cf. Finnemore and Sikkink 1998).

The preceding suggests that the norms and ideas that are currently shaping the international climate regime are a reflection of the national interests and discourses that (in the official NAPAs at least) are heavily predisposed against the idea of accommodating populations displaced as a result of climate change. Correspondingly, the ability of NGOs (like Christian Aid) to challenge the state-based bias of the UNFCCC is constrained by the actors and interests that are currently shaping national and international discourses about migration, immigration and asylum (cf. Risse 2009a; 2009b; Johnson 2012).

Finally, the chapter raises important questions about the politics and ethics of international burden-sharing (cf. Caney 2005; Risse 2009a; 2009b; Johnson 2012). At the heart of recent climate policy debates is the idea that climate change constitutes a form of wilful negligence, whose negative consequences now justify future claims for financial compensation on the part of the developing world (cf. ActionAid et al., 2012). Similarly, debates about the rights of populations displaced as a result of climate change have articulated the idea that the burden of accommodating future "climate refugees" be borne by countries whose greenhouse gas emissions have most contributed to global warming (Byravan and Rajan 2006; Biermann and Boas 2008a). Beyond the ethical implications of deciding future immigration policies on the basis of historical emissions (cf. Risse 2009a; 2009b; Johnson 2012), the idea of compensating affected populations through expanded immigration underplays significantly the discourses that are currently driving—and narrowing—the politics of immigration and asylum in the West.

NOTES

1. Important recent contributions in the field include Byravan and Rajan (2006), Biermann and Boas (2007; 2008a; 2008b; 2010), Brown et al., 2007; Reuveny, 2007; Burton (2008), Raleigh et al. (2008), the German Advisory Council (WBGU, 2008), the Norwegian Refugee Council (Kolmannskog, 2008), Risse (2009a; 2009b); Barnett and Webber (2010); McAdam and Saul (2010); de Sherbinin et al. (2011a; 2011b); Johnson (2012) as well as the UK Government's "Foresight" Report on

Migration and Global Environmental Change, the collection of essays in Couldrey and Herson (2008) and research carried out by the Institute for Environment and Human Security at the United Nations University in Bonn (summarized in Bogardi and Warner, 2009). Counter-arguments can be found in Hulme (2008), Boano et al., (2008); Hartmann (2010) and Gemenne (2011). Recent international conferences on these themes include the Threatened Island Nations Conference held from 23–25 May 2011 at the Columbia University Center for Climate Change Law (http://www.law.columbia.edu/centers/climatechange/resources/threatened-island-nations) and the Nansen Conference on Climate Change and Displacement held from 6–7 June 2011 in Oslo, Norway. www.nansenconference.no/.

2. Downloaded 17 December 2012 from www.un.org/esa/population/migration/Opening_remarks_SG_Athens.pdf

3. For the purposes of this analysis, "human security" implies "the survival and dignity of human beings through freedom from fear and freedom from want," (Khagram et al., 2003: 292). "Strategic" or state security on the other hand implies the survival and territorial integrity of individual nation states facing real and constructed threats to their national sovereignty.

4. Note that these and other projections have been challenged for extrapolating estimates about future patterns of migration and displacement on the basis of existing vulnerability assessments. See for instance, Gemenne (2011); Hartmann (2010); Hulme (2008); Castles (2001); Black (2001). For the purposes of this chapter, the credibility of the projections matters less than the ways in which they are being used to construct a policy discourse (cf. Finnemore and Sikkink, 1998).

5. International regimes can be usefully defined as "rules agreed to by states . . . concerning their conduct in specific issue areas (trade, monetary exchange, navigation on the high seas or in the air, non proliferation . . . , etc.) and often associated with international and non-governmental organizations linked to these regimes," (Viotti and Kaupi, 2010: 131).

6. Assuming that post-Kyoto negotiations are able to achieve stabilization targets (at 450 ppm CO_2 equivalents), there is a good chance that global warming can be maintained at 2 degrees Celsius above pre-industrial levels (IPCC, WGI, 2007). However, existing concentrations are already believed to be leading to a 0.4–0.6 degree increase by 2030, irrespective of emissions scenarios (IPCC, cited in WBGU, 2008: 55). In the absence of "effective climate protection," it is projected that future climate scenarios will entail a further rise in temperature of 2 to seven degrees by the end of the next century.

7. Last accessed 18 December 2012 from unfccc.int/cooperation_support/least_developed_countries_portal/submitted_napas/items/4585.php

8. According to the official IPCC definition, adaptation involves a wide range of "initiatives and measures to reduce the vulnerability of natural and human systems against actual or expected climate change effects." Downloaded 17 December 2012 at www.ipcc.ch/publications_and_data/ar4/syr/en/annexess glossary-a-d.html

9. In this case, the NAPA was produced before the bifurcation of North and South Sudan in 2011. A close reading of the NAPA document reveals underlying concerns about the ways in which resource shortages and forced displacement

may exacerbate pre-existing ethnic and tribal tensions. At the time of writing, the Republic of South Sudan had yet to release a NAPA document of its own.

10. The Government of Mali (2007: 53), for instance, is quite explicit in its aim of promoting la "sédentarisation des population et réduction des migrations."

11. (CSR, Art. 1.A.2) downloaded 5 July 2011 from www.unhcr.org.ua/main.php?article_id=5&view=full

12. Source: blogs.cgdev.org/globaldevelopment/2010/01/reactions-to-my-proposal-for-a-new-visa-to-the-united-states.php

13. The implication being that anonymous Internet postings and bathroom graffiti both provide an effective means by which people can say what they really feel without the risk of being identified personally.

REFERENCES

AWG-LCA (Ad Hoc Working Group on Long-Term Cooperative Action under the Convention) 2010 Report of the Ad Hoc Working Group on Long-term Cooperative Action under the Convention on its eighth session, held in Copenhagen from 7 to 15 December 2009 FCCC/AWGLCA/2009/17, at unfccc.int/resource/docs/2009/awglca8/eng/17.pdf (12.05.2010).

ActionAid, CARE and WWF 2012 *Tackling the Limits to Adaptation: An International Framework to Address 'Loss and Damage' from Climate Change Impacts* at www.actionaid.org/publications/tackling-limits-adaptation-international-framework-address-loss-and-damage-climate-chan (14.12.2012).

Adger, N. et al., 2007 "Assessment of adaptation practices, options, constraints and capacity," in (M.L. Parry et al., Eds.) *Climate Change 2007: Impacts, Adaptation and Vulnerability: Contribution of Working Group II to the Fourth Assessment Report of the Intergovernmental Panel on Climate Change* Cambridge: Cambridge University Press, 717–743.

Ayers, J. and Huq, S. 2009 "Supporting Adaptation to Climate Change: What Role for Development Assistance?" *Development Policy Review* (27): 6, 657–92.

Bakewell, O. 2008 "'Keeping them in their place:' The ambivalent relationship between development and migration in Africa," *Third World Quarterly* (29): 7, 1341–58.

Barnett, J. and Adger, W. N. 2007 "Climate change, human security and violent conflict," *Political Geography* (26): 6, 639–655.

Barnett, J. and Webber, M. 2010 "Accommodating Migration to Promote Adaptation to Climate Change," *Background Paper to the 2010 World Development Report. World Bank Policy Research Working Paper 5270*, World Bank: Washington DC.

Bates, D. C. 2002 "Environmental Refugees? Classifying Human Migrations Caused by Environmental Change" *Population and Environment* (23): 5, 465–77.

Bauder, H. 2008 "Media Discourse and the New German Immigration Law" *Journal of Ethnic and Migration Studies* 34(1): 95–112.

Beck, U. 1992 Risk Society: Towards a New Modernity London: Sage.

Bell, D. 2004 "Environmental Refugees: What Rights? Which Duties?" *Res Publica* (10): 135–52.

Berkes, F. and Folke, C. 1998 'Linking social and ecological systems for resilience and sustainability,' In: Berkes, F., Folke, C. (Eds.), *Linking Social and Ecological Systems* Cambridge: Cambridge University Press, 1–25.

Biermann, F. and I. Boas 2007 "Preparing for a Warmer World: Towards a Global Governance System to Protect Environmental Refugees," Global Governance Working Paper No 33–November 2007 at glogov.org, (13.06.2013)

Biermann, F. and I. Boas 2008a "Protecting climate refugees: The case for a global protocol," *Environment* 50: (6), 8–16.

Biermann, F. and I. Boas 2008b "Response to Hulme," *Environment* 50, (6) at In: www.environmentmagazine.org/Archives/Back%20Issues/November-December%202008/hulme-full.html (13.06.2013).

Black, R. 2001. "Environmental Refugees: Myth or Reality?" at www.unhcr.ch/refworld/pubs/pubon.htm (13.06.2013).

Boano, C., R. Zetter and T. Morris (2008) "Environmentally displaced people: Understanding the linkages between environmental change, livelihoods and forced migration" Forced Migration Policy Briefing 1 Refugees Studies Centre, Oxford University www.rsc.ox.ac.uk/publications/policy-briefings/RSCPB1-Environment.pdf (13.06.2013).

Bogardi, J. and K. Warner 2009 "Here comes the flood," *Nature Reports Climate Change* (3), 9–11.

Boyd, E., N. Grist, S. Juhola and V. Nelson 2009 "Exploring Development Futures in a Changing Climate: Frontiers for Development Policy and Practice," *Development Policy Review* (27): 6, 659–74.

Brooks, N., N. Grist and K. Brown 2009 "Development Futures in the Context of Climate Change: Challenging the Present and Learning from the Past," *Development Policy Review* (27): 6, 741–65.

Brown, Oli, Anne Hammill and Robert McLeman 2007 "Climate change as the 'new' security threat: implications for Africa," *International Affairs* 83: 6, 1141–1154.

Burton, I. 2008 "Beyond borders: The need for strategic global adaptation," IIED Opinion December 2008 pubs.iied.org/17046IIED.html (13.06.2013).

Burton, I. 2004 [2009 "Climate change and the adaptation deficit," E. L. F. Schipper and I. Burton (Eds.) *The Earthscan Reader on Adaptation to Climate Change* London: Earthscan, 89–98.

Byravan, S. and S. C. Rajan 2006 "Providing new homes for climate change exiles," *Climate Policy* (6), 247–252.

Caney, S. 2005 "Cosmopolitan justice, responsibility, and global climate change," *Leiden Journal of International Law* (18): 4, 747–775.

Caney, S. 2008 "Human rights, climate change, and discounting," *Environmental Politics* (17): 4, 536–555.

Castles, Stephen 2001 "Environmental change and forced migration: making sense of the debate" Refugees Studies Centre Working Paper No. 70 University of Oxford at www.unhcr.org/3de344fd9.html (13.06.2013).

Christian Aid 2007. *Human Tide: The Real Migration Crisis* London, May. At www.christianaid.org.uk/images/human-tide.pdf (13.06.2013).

Cernea, M. M. 1997 "The risks and reconstruction model for resettling displaced populations," *World Development* (25): 10, 1569–1589.

Cernea, M. and Schmidt-Soltau, K. 2006 "Poverty risks and national parks: Policy issues in conservation and resettlement," *World Development* (34): 10, 1808–30.

Clemens, Michael 2010 "To help Haiti's earthquake victims, change US immigration laws" *Washington Post* 24 January 2010 at www.washingtonpost.com/wp-dyn/content/article/2010/01/22/AR2010012202274.html (13.06.2013).

Corbridge, S. 1983 "Urban bias, rural bias, and industrialization: an appraisal of the work of Michael Lipton and Terry Byres," in J. Harriss (Ed) *Rural development: theories of peasant economy and agrarian change.* London: Hutchinson, 94–118.

Couldrey, M. and Herson, M. 2008 (Eds.) "Climate change and displacement," *Special Issue of Forced Migration Review* Issue 31, October 2008.

de Haan, A. 1999 "Livelihoods and poverty: The role of migration—a critical review of the migration literature," *The Journal of Development Studies* (36) 2, 1–47.

de Sherbinin, A., VanWey, L. K., McSweeney, K., Aggarwal, R., Barbieri, A., Henry, S., Hunter, L. M., Twine, W. and Walker, R. 2008 "Rural household demographics, livelihoods and the environment," *Global Environmental Change* (18): 1, 38–53.

de Sherbinin, A. K. Warner and C. Ehrhart 2011a "Casualties of Climate Change," *Scientific American* January, 2011: 32–39.

de Sherbinin, A. et al. 2011b "Preparing for Resettlement Associated with Climate Change," *Science* Vol. 334 28 October 2011, 456–7.

Deshingkar, P. 2005 Maximizing the benefits of internal migration for development, at www.preventtraffickingchina.org/english/Website_Files/Deshingkar.pdf (29.06.2009).

Finnemore, M. and K. Sikkink 1998 "International norm dynamics and political change," *International Organization,* (52): 4, 887–917.

Foresight: Migration and Global Environmental Change 2011 *Final Project Report: Executive Summary* The (UK) Government Office for Science, London.

GEF (Global Environment Facility), 2007 *Linking Adaptation to Development* at www.thegef.org/gef/sites/thegef.org/files/publication/Adaptation_brochure.pdf (18.12.2012)

Gemenne, F. 2011 "Why the numbers don't add up: A review of estimates and projections of people displaced by environmental changes," *Global Environmental Change* (21S), S41–S49.

Hartmann, B. 2010 "Rethinking climate refugees and climate conflict: Rhetoric, reality and the politics of policy discourse," *Journal of International Development* (22): 2, 233–246.

Hayter, T. 2003 "No borders: the case against immigration controls," *Feminist Review* 73, 6–18.

Hulme, M. 2008 "Commentary—Climate refugees: Cause for a new agreement?" *Environment* 50 (6) at www.environmentmagazine.org/Archives/Back%20Issues/November-December%202008/hulme-full.html (13.06.2013).

Huq, S. and H. Reid, 2004 [2009]. "Mainstreaming adaptation in development," in E. L. F. Schipper and I. Burton (Eds.) *The Earthscan Reader on Adaptation to Climate Change.* London: Earthscan, 313–22.

IOM (International Organization for Migration), 2009. "Migration, climate change and the environment," IOM Policy Brief, May 2009 at www.iom.int/jahia/

webdav/shared/shared/mainsite/policy_and_research/policy_documents/policy_brief.pdf (29.06.2009).

IPCC WG-I (2007) "IPCC, 2007: Summary for Policymakers," in S. Solomon et al. (Eds.). The Physical Science Basis: Contribution of Working Group I to the Fourth Assessment Report of the Intergovernmental Panel on Climate Change. Cambridge: Cambridge University Press at www.ipcc.ch/pdf/assessment-report/ar4/wg1/ar4-wg1-spm.pdf (13.06.2013).

IPCC WG-II 2007 "IPCC, 2007: Summary for Policymakers," in M. L. Parry et al. (Eds.). "Climate Change 2007: Impacts, Adaptation and Vulnerability." Contribution of Working Group II to the Fourth Assessment Report of the Intergovernmental Panel on Climate Change. Cambridge: Cambridge University Press, 7–22.

IPCC 2012 *Managing the Risks of Extreme Events and Disasters to Advance Climate Change Adaptation*. Cambridge: Cambridge University Press.

Johnson, C. and K. Krishnamurthy 2010 "Dealing with displacement: Can 'social protection' facilitate long-term adaptation to climate change?" *Global Environmental Change* (20): 648–55.

Johnson, C., 2012. "Governing climate displacement: The ethics and politics of human resettlement," *Environmental Politics* (21): 2, 308–28.

Joppke, Christian, 1999. *Immigration and the Nation-State: The United States, Germany, and Great Britain*. Oxford: Oxford University Press.

Khagram, S., Clark, W. C. and Raad, D. F., 2003. "From the environment and human security to sustainable security and development," *Journal of Human Development* (4): 2, 289–313.

Kolmannskog, Vikram Odedra, 2008. Future Floods of Refugees: A Comment on Climate Change, Conflict and Forced Migration. Oslo: Norwegian Refugee Council at www.nrc.no/arch/_img/9268480.pdf (13.06.2013).

Levant, E., 2010. "Tamils playing us for fools," *The Toronto Sun*, 22 August 2010 at www.torontosun.com/comment/columnists/2010/08/20/15091486.html (18.12.2012).

Lipton, M., 1983. "Why poor people stay poor," J. Harriss (Ed). *Rural development: theories of peasant economy and agrarian change*. London: Hutchinson, 66–81.

Mali, Republique du, 2007. *Programme d'Action National d'Adaptation aux Changements Climatiques* at www.adaptationlearning.net/sites/default/files/mli01f.pdf (18.12.2012).

McAdam, J. and B. Saul, 2010. "Displacement with Dignity: International Law and Policy Responses to Climate Change Migration and Security in Bangladesh," *University of New South Wales Faculty of Law Research Series 2010*. Working Paper 63 at law.bepress.com/cgi/viewcontent.cgi?article=1258&context=unswwps-flrps10 (13.06.2013).

McGranahan, G., Balk, D., and Anderson, B., 2007. "The rising tide: assessing the risks of climate change and human settlements in low elevation coastal zones," *Environment and Urbanization* (19): 1, 17–37.

McLeman, R. and B. Smit 2006. "Migration as adaptation to climate change," *Climatic Change* (76): 31–53.

Moench, M. 2007 [2009]. Adapting to climate change and the risks associated with other natural hazards: Methods for moving from concepts to action,"

E. L. F. Schipper and I. Burton (Eds.). *The Earthscan Reader on Adaptation to Climate Change* London: Earthscan, 249–82.

Myers, N. 1993. "Environmental refugees in a globally warmed world" *BioScience* 43. 752–61.

Nelson, D. R., W. A. Adger and K. Brown (2007). "Adaptation to environmental change: Contributions of a resilience framework," Annu. Rev. Environ. Resour. (32) 395–419.

Neumayer, E. 2006 "The environment: One more reason to keep immigrants out?" *Ecological Economics* (59): 204–7.

Nicholls, R. J., P. P. Wong, V. R. Burkett, J. O. Codignotto, J. E. Hay, R. F. McLean, S. Ragoonaden and C. D. Woodroffe, 2007. "Coastal systems and low-lying areas," Climate Change 2007: Impacts, Adaptation and Vulnerability. Contribution of Working Group II to the Fourth Assessment Report of the Intergovernmental Panel on Climate Change, M. L. Parry, O. F. Canziani, J. P. Palutikof, P. J. van der Linden and C. E. Hanson, Eds., Cambridge: Cambridge University Press, 315–356.

Nordas, R. and N. P. Gleditsch, 2007. "Climate change and conflict," *Political Geography* (26): 6, 627–638.

Oxfam, 2007. "Adapting to climate change: What's needed in poor countries, and who should pay," Oxfam Briefing Paper 104. Oxfam, UK at www.oxfam.org/en/policy/briefingpapers/bp104_climate_change_0705 (13.06.2013).

Perch-Nielson, S.L., Battig, M.B. and Imboden, D., 2008. "Exploring the link between climate change and migration," *Climatic Change* (91): 375–93.

Peskett, L., Grist, N., Hedger, M., Lennartz-Walker, T. and Scholz, I. 2009 "Climate change challenges for EU development cooperation: emerging issues," *EDC 2020 Working Paper No. 3* Bonn: European Association of Development Research and Training Institutes at www.edc2020.eu/fileadmin/Textdateien/EDC2020_WP03_ClimateChange_online.pdf (13.06.2013).

Raleigh, Clionadh, Lisa Jordan and Idean Salehyan 2008 Assessing the Impact of Climate Change on Migration and Conflict Washington DC: World Bank at siteresources.worldbank.org/EXTSOCIALDEVELOPMENT/Resources/SDCCWorkingPaper_MigrationandConflict.pdf (13.06.2013).

Reuveny, R. 2007 "Climate change-induced migration and violent conflict," *Political Geography* (26): 6, 656–673.

Riddell, R., 2008. Does Foreign Aid Really Work? Oxford: Oxford University Press.

Risse, M., 2009a. "The Right to Relocation: Disappearing Island Nations and Common Ownership of the Earth," *Ethics and International Affairs* (23): 3, 281–300.

Risse, M. 2009b. "Immigration, Ethics and the Capabilities Approach." *Human Development Research Paper (HDRP) Series*, Vol. 34, No. 2009. at mpra.ub.uni-muenchen.de/19218/ (05.07.2011).

Scott J. C. 1998. *Seeing Like a State: How Certain Schemes to Improve the Human Condition Have Failed*. New Haven: Yale University Press.

Solberg, M. 2010 "Tamil 'refugees' get the best of Canada," *The Toronto Sun*, 23 August 2010, at www.torontosun.com/comment/columnists/monte_solberg/2010/08/21/15097166.html (18.12.2012).

Sudan (Republic of the, 2007) *National Adaptation Programme of Action* at unfccc.int/resource/docs/napa/sdn01.pdf (18.12.2012).
UNDP, 2008. "Human Development Report 2007/08: Fighting Climate Change: Human Solidarity in a Divided World" New York: Oxford University Press.
UNDP, 2009. "Summary: Overcoming barriers: Human mobility and development," *Human Development Report 2009*. at: hdr.undp.org/en/media/HDR_2009_EN_Summary.pdf (13.06.2013).
UNFCCC, 2007. *Climate Change: Impacts, Vulnerabilities and Adaptation in Developing Countries* at unfccc.int/adaptation/items/4159.php (10.06.2009).
Uganda (Republic of, 2007). *National Adaptation Programme of Action* at unfccc.int/resource/docs/napa/uga01.pdf (18.12.2012).
Viotti, P. and M. Kaupi 2010 *International Relations Theory Fourth Edition*. Toronto: Longman.
WBGU (German Advisory Council on Global Change) (2008). "Climate Change as a Security Risk." London: Earthscan at www.wbgu.de/fileadmin/templates/dateien/veroeffentlichungen/hauptgutachten/jg2007/wbgu_jg2007_kurz_engl.pdf (13.06.2013).
Warner, K., C. Ehrhart, A. de Sherbinin, S. Adamo and T. Chai-Onn 2009. In Search of Shelter: Mapping the Effects of Climate Change on Human Migration and Displacement CARE International and CIESEN at ciesen.columbia.edu/publications.html (17.03.2010)
Werz, M. and L. Conley 2012 *Climate change, migration and conflict: Addressing complex crisis scenarios in the 21st century*. The Center for American Progress at www.americanprogress.org/wp-content/uploads/issues/2012/01/pdf/climate_migration.pdf (16.11.2012)
Westra, L. 2009. "Environmental Justice and the Rights of Ecological Refugees." London: Earthscan.
World Bank 2010. *World Development Report 2010: Development and Climate Change* Washington DC: The World Bank.
Young, R. and O. Pilkey 2010 "How High Will Sea Levels Rise? Get Ready for Seven Feet," at e360.yale.edu/content/feature.msp?id=2230 (29.10.2010).
Zetter, R. (2007). "More labels, fewer refugees: Remaking the refugee label in the era of globalization." *Journal of Refugee Studies* (20): 2, 172–92.

5 Part I

China and Climate Change

Environmental Impacts, Human Security, and Migration Policies and Actions

Li Jian

Since 1979, with the introduction of socioeconomic reforms and opening up to global trade and investment, China's Gross Domestic Product (GDP) has grown an average of 9.8 percent per year, nominal per capita income has increased approximately fifty-fold, and more than 500 million Chinese people have been lifted out of poverty (UNDP 2010: 1; Morrison 2012: 2). Today, China has risen to become the world's largest manufacturer, largest merchandise exporter, and second-largest economy (IMF2012: 1).

In the past three decades or so, however, China's unprecedented economic growth has been mainly driven by fossil fuel and resource consumption, and the country is now also the world's largest energy consumer and leading emitter of greenhouse gases (GHGs), which vitally contribute to global climate change (BP 2011: 1; International Energy Agency 2012: 11; Yale Center for Environmental Law and Policy 2012: 1). While China's average per capita GHG emissions remain far below those of the United States, the European Union, and Japan, the country's per capita emissions have already surpassed the global average, placing its emissions well above those of most developing countries in the world (Table 5.1). In 2008, two-thirds of the total global increase in GHG emissions came from China (Harris 2010: 10858). In 2010, China contributed

Table 5.1. Top 8 CO_2 Emitting Countries in 2011

Country	The country's total CO_2 emissions / thousands of metric tons	The country's CO_2 emissions / per capita / metric tons
World	33,376,327	4.9
China	9,700,000	7.2
United States	5,420,000	17.3
India	1,970,000	1.6
Russia	1,830,000	12.8
Japan	1,240,000	9.8
Germany	810,000	9.9
South Korea	610,000	12.6
Canada	560,000	16.2

Sources: EDGAR CO_2 Time Series 1990-2011 per Region/Country and EDGAR CO_2 Time Series 1990-2011 per Capita for World Countries (database created by European Commission and Netherlands Environmental Assessment Agency).

approximately 24 percent of the world GHG emissions (International Energy Agency 2012: 12). Economists predict that such a trend will likely continue until China's per capita GDP reaches approximately $25,000 from its current status of some $5,000 in 2030 (Kaiman 2012). In view of the sheer volumes of GHGs emitted by China, it is clear that the country plays an important role in global climate change. Carbon dioxide (CO_2) is a major GHG that contributes to air pollution and climate change.

To understand global climate change and help develop policies and actions to mitigate its negative impacts, this chapter investigates the major environmental impacts of global climate change upon China and how such impacts affect human security in the country. In addition, it examines China's core environmental policies and actions and scrutinizes how such policies and actions contribute to climate change mitigation. The chapter is organized in the following manner: (i) Looks into the statistical facts and observed phenomena that indicate that climate change has been happening in China in recent decades, (ii) concentrates on the major environmental problems associated with climate change and how such problems affect the human security in China; (iii) analyzes China's major environmental policies and actions to mitigate climate change, with an emphasis on the strength and limits of such policies and actions. The chapter ends with an overview of the challenges and future prospects of China's climate change mitigation program.

CLIMATE CHANGE IN CHINA: STATISTICAL FACTS AND OBSERVED PHENOMENA

Regardless of the overwhelming evidence and scientific data, up till now, global climate change still remains a somewhat debatable topic in some circles around the world. In some cases, for example, even substantial public doubt persists about the anthropogenic cause and scientific agreement about the role of GHGs in climate change (Dunlap 2008: 26; Anderegg 2010: 12107). Despite that, by and large, the Chinese environmental scientists and activists have long reached a general consensus that climate change has been happening in China in the past decades (China Center for Climate Change Communication 2012). Such a consensus has mainly been reached on five categories of statistical facts and observed phenomena.

Above all, China's annual average temperature has been rising nationwide since the early 1950s. In the past fifty years or so, China's annual average temperature increased by 1.1 °C, which was higher than the average global rise of 0.4–0.8°C (Dou 2010: 6). In terms of geographic distribution, the colder zones, such as northeast region and northwest region have been becoming warmer. During the period of 1951–1990, the temperature and rainfall showed negative correlation over the most areas. It turned warmer and drier over the northern China area while colder and wetter over the area between the Huai River and the Yangtze River (Chen et al., 2004: 545). Seasonal distribution of the temperature increase frequently occurred in winter and China experienced twenty consecutive warm winters between 1986 and 2005 (Climate Group 2006: 5). Besides, China's leading meteorologists predict that such a trend will continue and China's annual average temperature will increase by 1.7°C in the period of 2030 and 2.2°C by 2050 (Qin 2004: 5).

Moreover, China's annual precipitation trends show noteworthy changes. The country's total annual precipitation has decreased with an average rate of 2.9 mm per decade since the 1950s. Besides, the regional distribution of rainfalls was highly uneven. The long-term decrease in mean precipitation and uneven distribution contributed to inadequate and excessive precipitation in different regions (Nicholls et al., 1996). Over the last fifty years of the twentieth century, specifically, the number of rain days all over the country has decreased but the rainfall intensity has increased. Western China experienced a substantial increase in annual and seasonal total precipitation. In contrast, the precipitation in northeastern China and northern China was decreasing (Liu et al., 2010: 5). Prominent trends have been found in annual, winter, and summer precipitation, in general, with downward trends in northern China and upward trends in southern China, including the Yangtze River region.

Such changes in precipitation have resulted in more recurrent droughts in the north and more frequent and sometimes severe floods in the south, resulting in an increase in the number of extreme events. (Zhai et al., 2005: 1106). Indeed, northern China had eight droughts in the past twenty years, and in 2009, even Yunnan Province and Guizhou Province in southwest China unexpectedly suffered from a severe drought that was not seen once in the past hundred years (Dou 2010: 7). Yet, the Yangtze River Valley became frequently flooded, a pattern that has intensified since the 1990s (Tong, Su, and Hartmann 2007: 144)

Thirdly, the frequency and intensity of extreme climate events in China have increased. The general tendency is that droughts in northern China and floods in the middle and lower reaches of the Yangtze River have become more severe. According to Ren et al. (2011: 107), the most frequently observed extreme events in China in the past fifty years included: (1) high temperature, (2) low temperature, (3) intense precipitation, (4) dust storms, (5) strong wind events, (6) meteorological droughts, and (7) land-falling tropical cyclones. Specifically, for example, in spring 2004, parts of the northeastern China experienced the worst drought conditions since 1951, and southern China received the lowest autumn rainfall since 1951. In 2008, China witnessed the worst severe winter weather in five decades in January, with over seventy-eight million people affected by the freezing temperatures and heavy snow. In many parts of China, high temperatures broke historical extremes. Floods, landslides, and mud-rock flows also caused serious economic damage. In August of 2010, Zhouqu County, Gansu Province, was hit by the most devastating flood and mud-rock flow in the country's history, killing more than 1,500 people (World Meteorological Organization 2011: 2–3). The Yangtze River basin floods of 1998 were considered to be among the worst in the twentieth century (Qian and Glantz 2005: 159). Yet, amazingly, in 2011, the Yangtze River Basin experienced a record drought (Coumou and Rahmstorf 2012: 1).

Fourth, the sea level rose along China's coasts in the past fifty years. Regarding sea areas, a study reported that the rise of sea level was nonhomogeneous, with a highest rate of 2.7 mm/a in the deep basin west of Luzon and generally low rates over the shallow continental shelves between 1993 and 1999. It also suggested that the observed rising of sea levels was mainly caused by warming of the upper layer of the South China Sea, which showed a bulk warming rate of 0.15°C/a in the same period (Li, Xu, and Cai 2002: 5). In a recent study reported in the Bulletin on the Sea Level in China, the sea level rise rate in the East China Sea is the highest among the sea areas of China, reaching 3.1 mm/a. in 2003 (China's National Climate Change Program 2004: 61). Regarding coastal areas, the general trend is that the change along the southern coast was relatively large, while that along the northern coast the change was small

Map 5.1. Map of China: Provinces, Municipalities, Autonomous Regions, and Major Rivers

Source: Flavor & Fortune, http://www.flavorandfortune.com/dataaccess/article.php?ID=582.

(Xu 2012). An earlier publication estimated that relative rates of sea level rise in China were 0.5 mm per year in the north and about 2 mm per year in the south (Zhen and Wu 1993: 99). A more recent study indicated that from 1980 to 2011, China's coastal sea level climbed by about 85 mm or 2.7 mm per year. Among key marine areas, importantly, the coastal sea level rise of the Yangtze River delta and the Pearl River delta is at an average rate of 3.1 and 1.7 mm/a, respectively (China's National Climate Change Program 2004:61). The rising sea levels have threatened China's sinking coastal cities. For example, waters off the industrial port city of Tianjin climbed by 20 centimeters over the past three decades. The Seas off Shanghai rose by 11.5 centimeters over the same period (Bodeen 2008).

Fifth, major mountain and plateau glaciers in China have all retreated in recent past decades. Under the impact of global warming, the glaciers in China have been retreating continuously with negative glacial mass balance in recent decades. However, the retreat became more intensive in the past decade. In addition, most of advancing glaciers gradually shifted to retreat. The smallest magnitude of retreat is in the inland of the Qinghai-Tibetan Plateau, the magnitude increases from the inland to the margin of the Plateau, and the largest magnitude at the margin of the Qinghai-Tibetan Plateau. Several recent studies offered specific data about these retreats. The total glacier area in Pumqu basin has decreased by 9 percent and the ice reserve has reduced by 8.4 percent in western China (Jin Rui et al., 2004). The retreat of the Large and Small Dongkemadi Glaciers in the Tanggula Mountains caused glacial melt-water runoff to increase by more than 5.5 percent in the 1990s in Northwest China. In the Tarim River Basin where glaciers are most concentrated, the net water supply from glacial retreat reached 13 percent in the 1990s (Yao et al., 2004: 1065). Over the past ten years in western China, 82.2 percent of glaciers were in retreat, and the areas of glaciers shrank by 4.5 percent. In Xinjiang, glacier runoff shows a clear trend of increase (Liu et al., 2006: 762). In northwestern Yunnan, a comparative study of historical and current images indicated that all glaciers have been shrinking (Moseley 2006: 217).

IMPACTS OF CLIMATE CHANGE ON HUMAN SECURITY IN CHINA

Human security is a people-centered paradigm for understanding global vulnerabilities. The 1994 Human Development Report defined human security as people's "safety from chronic threats and protection from sudden hurtful disruptions in the patterns of daily life" (UNDP 1994). In its broadest sense, human security includes economic security, food security, health security, environmental security, personal security, community security, and political security (Vajpeyi 2013). While climate change directly or indirectly affects all aspects of human security, research literature suggests that its negative impacts are at present most evident on food security, environmental security, and health security. In the Chinese case, specifically, current data point at three vital aspects: agricultural production, water resources, and human health. In this section, I will concentrate on these three aspects.

As a Chinese proverb puts it, "the masses regard food as their heaven." China has 1.32 billion people and food security is one of the most important issues in China. So far, climate change has already made tangible impacts on China's agricultural production and posed a noteworthy threat to the country's food security by dragging down farmers' crop yields. Due to the complexity of agricultural production, estimates of the impacts of climate change on crop yields vary widely depending on different assumptions and approaches. Despite that, agricultural scientists in general confirm that the impacts of climate change on crop yields are significant (Wang, Huang, and Rozelle 2010: 5).

Agricultural production is highly sensitive to climate change. On the whole, the impacts of climate change on crop yields were associated with climate extremes such as droughts, floods, and insect plagues rather than merely with the rise of the average annual temperatures (Zhao et al., 2010: 201). While the evidence and data are not yet conclusive, the frequency and intensity of extreme climate events throughout China have increased meaningfully during the past fifty years. Areas of northern China, on the whole, saw severe decreases in rainfall. Beginning in the 1950s, rainfall decreased between 20 to 40 mm per decade on average. At the same time, precipitation significantly increased in southern and southwestern China. Since the 1950s, rainfall levels have risen 20 to 60 mm per decade on average in these areas (Ren 2007). Under such climatic impacts, agricultural production suffered severe losses, especially in 1997, 2000, 2001, and 2003, in which years the disaster-affected areas were over 50 million hectares and the inundated areas were 30 million hectares, and drought and flood, the two climate-related disasters, counted for between 70 percent and 85 percent of the country's total agricultural losses (Lu et al., 2009:1578). In addition, in 2007, the freezing disaster alone resulted in an economic loss of 186.5 billion Yuan (Liu, Liu, and Guo 2010:908).

Climate change also affected the crop yields by causing crop plagues as well. The armyworm (*Mythimna separata*) offered a case in point. The armyworm infection can damage young plants of rice, corn, and wheat and lower the yields of these crops drastically. In China, when the annual average temperature rises by 2.69°C, the wintering boundary for the armyworm can shift northward for approximately 3° northern altitude, which helps spread the plagues to previously-unaffected regions (Li 1993: 41). In 2012, for example, an armyworm plague broke out in major grain producing regions in northern China and northeastern China. The infestation had affected 1.8 million people and 481,600 hectares of farmland, causing direct economic losses of 1.44 billion Yuan (US $226 million) in Heilongjiang Province and Jilin Province (Food and Agriculture Organization 2012: 6).

Besides, climate change put adverse constraints on livestock sector by droughts and dramatic fluctuations of cold and warm temperatures, which contributed to grassland degradation and desertification. Most of the plants on grasslands in west China and northwest China are C_3 plants and are often negatively affected by the rise of the regional temperatures. In recent decades, China's grasslands experienced a warming trend, particularly in Inner Mongolia during the winter months. As a result, spring droughts in these grassland areas became more severe. In the past decades, under the combined impacts of warming, population pressure, and pastoral overgrazing, the area of grassland degradation reached 23 percent, 41 percent, and 68 percent in the Hulun Buir Grassland, Xi Lingole Grassland, and the wilderness grasslands in Inner Mongolia respectively (Ying et al. 2011: 1133). The productivity of these grasslands (in terms of its biomass production) has gone downward since 1993 (Wang, Huang, and Rozelle 2010: 13). In the grasslands on the Qinghai-Tibetan Plateau, even the pasture height declined by 30 percent to 50 percent between 1980 and 1990 (Zhang et al., 2007: 66). In the past ten years, in Xilingole League County, the productivity of grassland declined by nearly 50 percent (Ying et al., 2011: 1134). Such degradation and decline resulted in meaningful drops in China's meat supply.

Water is a vital component of environmental security and has been deeply affected by the global climate change. An adequate supply of clean water is essential to the health, economy, and environment of any country and its people. In the past decades, climate change had visible effects on China's water resources by several ways, one of which was the alteration of the country's water availability and water demand. In different regions, the climate change effects either escalated or reduced the total number of precipitation days and flood frequencies. The frequency and severity of droughts worsened as a result of a regional decrease in total precipitation days and more frequent dry spells, and greater evapotranspiration (Parry et al., 2007; Cheng, Hu, Zhao 2009: 241). Largely compatible to the precipitation fluctuations in the past decades, China's water distribution patterns have also changed, with increased river runoff in the south and reduced water resources in the north. Specifically, the Pearl River and Yangtze River basins became wetter, and the Yellow River and Songhua River basins became drier (Wang and Zhang 2011: 78).

Historically, China has long been prone to floods. Since the 1990s, many rivers in China have flooded. The Yangtze River and the Huai River flooded in 1991. In 1996, the Hai River flooded. In 1998, in addition to the Songhua River Flood and the Min River Flood, the Yangtze River flooded again at an unprecedented scale (Zhang et al., 2009). In fact, the 1998 Yangtze River Flood was among the worst floods in China in the twentieth century and one of the most devastating floods in human history. Regardless of the nationwide efforts to control the flood and rescue people, the official death toll was 3,600, with 13.2 million people made homeless. In addition, about five million hectares of cropland were inundated. The flood destroyed or damaged much infrastructure and social services facilities. Economic losses were estimated at over US $36 billion (Ye and Glantz 2005: 161). In 2003, the Huai River, the Wei River and the Han River flooded; and the Huai River and the Han River flooded again in 2005. Furthermore, the largest whole basin flood since 1954 occurred in the Huai River basin in 2007. Besides, the increasing number of high intensity rainstorms have also struck Chongqing, Nanjing, and a number of large cities on the banks of the Yangtze River since the 1990s. During 1990 and 2005, China's annual flood losses exceeded 110 billion Yuan, accounting for between 1 percent and 2 percent of China's GDP in the same period (Zhang et al., 2009).

In contrast to floods in south china, climate change has posed challenges for northern China by water shortage. One of the most noticeable indicators is the dry-outs of the Yellow River. The Yellow River is China's second longest river, with a total length of 3,395 miles. Some 110 million people live in the Yellow River Basin and depend on the Yellow River for their livelihood. During the last half of the twentieth century, the discharges of the Yellow River declined continuously. Since 1972, a drying up of the Yellow River along the lower reach has been observed, and the situation has become more and more serious during the 1990s. In 1997, the river bench even kept dry for 226 days and no water flowed into the Bohai Sea for more than 300 days (Cui, Huang, Chen, and Morse 2009: 476). Some of the contributing factors for the river's drying-up included the dramatic decreases in precipitation in the middle and lower reaches of the Yellow River and the increase in evaporation due to warming conditions (Yang et al., 2004).

In addition to floods and droughts, China suffered from water shortage as well. China's annual per capita renewable freshwater availability is only 2,300 cubic meters, or just approximately 25 percent of the world's average. Besides, its limited water resources are unevenly distributed to different regions in the country and such imbalances became worsened in the north and the south due to climate change in recent decades. The water shortage is now at its worst in the north of the Huai River, where some 64 percent of China's cultivated land has access to only about 19 percent of the country's water. The water supply problem has even grown around big cities in northern China since the 1950s. It is estimated that about two-thirds of China's 660 cities are water short and 110 of those suffer severe shortage (Edmonds 2008: 274). Severe water distress is also found in rural areas, some 20 million people have a shortage of drinking water supplies (Cheng Hu, Zhao 2009: 240).

Water shortage is probably one of the most upfront challenges for the environmental security in China. Especially in the areas with high population density, water shortages can even lead to social unrest. In eastern Shandong Province, for example, a large scale of village conflict occurred over the access to water for irrigation in 2000 (Keidel 2005: 3). Besides, agricultural use presently claims more than 60 percent of China's water and most of China's agricultural provinces are in northern China nowadays. Water shortages in these important grain producing provinces may significantly compromise China's agricultural production potential. Despite some significant progress in water conservation in the past decades, there is still a big gap in water supply and demand in agricultural production in China (China Water Resources Bulletin, 2006; Cheng Hu, Zhao 2009: 240).

Climate change can fatally affect human health through the increased frequency and intensity of climate extreme events, including storms, floods, droughts, Typhoon, Tsunamis, hurricanes, and so on. Such events can degrade the residential environment, pollute water, and help enlarge breeding grounds for various disease agents. The impacts include those caused by changes in direct exposure to different climate extremes, such as heat waves or icy rain, and increase in certain pollutants in the air, or presence of certain harmful elements such as molds inside and outside residence (WHO 2003: 52).

The city of Nanjing offered a case in point. Nanjing, a city of about four million people at the time, encountered an unusual heat wave between July 4 and July 20, 1988. The maximum daily temperature exceeded 36°C continuously. The health impact was astounding: according to hospital records, the number of people suffered from heat stroke totaled 4,500 during the heat wave, 9.2 percent of the patients developed severe heat stroke, and 124 people died, rendering a very high mortality rate of 30.2 percent.

GHGs not only contribute to climate change but adversely affect human health in the form of air pollution (e.g., CO_2, N_2O, etc.). Air pollution is now one of the top environmental concerns in China and its drastic impacts on health security has been mounting up in recent years (Chan and Yao 2008: 1; Chen et al. 2012: 33).

Air pollution in the form of particulate matter (PM) offers a case in point. PM can come from any activity which involves burning of materials or dust generating activities. Except for natural sources, PM are predominantly from combustion sources like vehicles, diesel engines, and industrial facilities (Nevada Division of Environmental Protection 2013: 2). Epidemiological studies have shown a consistent association between particulate matter pollution and premature deaths in China (Wong et al. 2001; Kan et al., 2007; Qian et al., 2007). There was abundant evidence associating PM pollution with increase in hospital admissions or outpatients for respiratory and cardiovascular diseases (Peng et al., 2008).

A recent study in the Pearl River Delta in China, a region with a population of about 45.5 million, has found that in terms of avoidable deaths, for short-term exposure, 2,700 premature deaths would be prevented annually if the PM_{10} daily concentrations were reduced to below the value of WHO 24-Hour Air Quality Guideline. Much larger benefits would be gained if the annual concentration did not exceed the value of WHO Air Quality Annual Guideline. For PM_{10}, 42,000 premature deaths would be prevented annually and for $PM_{2.5}$, the number of deaths avoidable would be 40,000. In terms of gain in life expectancy, the avoidable person-years lost would be 68,000,000 and the average lifespan would prolong 2.57 years for PM_{10}. For $PM_{2.5}$ the avoidable person-years lost would be 63,000,000 and the average lifespan would prolong 2.38 years, if the annual concentrations of $PM_{2.5}$ reduced to below the values of WHO Air Quality Guideline (Xie et al., 2011: 360).

Based on what has already been learned, climate change can also render a host of impacts on human health indirectly. Acting via less direct mechanisms insistently, climate change can affect the transmission of many infectious diseases, especially water, food, and vector-borne diseases. In the long run, such indirect impacts are likely to pose even greater extent of risks to human health than the more direct. Many of the major killers such as diarrheal diseases, malnutrition, malaria, and dengue are highly climate-sensitive and become worsened as the climate changes. An epidemic study in Hubei province offered an example to illustrate the point. In Hubei Province in central China, a number of infectious diseases, digestive diseases, and natural endemic diseases showed higher incidences than normal years in the regions suffered from 1996 and 1998 floods showed (Zeng, Chen, and Zhang 2005: 249).

Climate change helps spread vector-borne diseases by expanding breeding grounds for the intermediate hosts and / or vectors. For instance, an increase in ambient temperature would likely cause net increases in the geographical distribution of certain vector organisms (e.g., malarial mosquitoes). Besides, temperature related changes in the life-cycle dynamics of both the vector species and the pathogenic organisms (flukes, bacteria, and viruses) would likely further increase the potential transmission of such vector-borne diseases, including malaria (mosquito), dengue fever (mosquito), and Japanese encephalitis (mosquito) (WHO 2003).

In China, schistosomiasis offered an example. Schistosomiasis, also known as the "big belly disease," is a parasitic disease transmitted by a waterborne fluke (*Schistosoma japonicum*), with a snail (*Oncomelania hupensis*) as its intermediate host. Schistosomiasis had been endemic in central, south, and southwest China for many centuries and has just been brought under control in most of previously endemic regions in recent years. Climate change may help this dreadful disease make a comeback. In a recent endemic disease study, Chinese medical scientists reported that as winter temperatures continued to rise due to global warming, *Oncomelania hupensis*, the intermediate host of *Schistosoma japonicum* in China may have extended their breeding range, thereby spreading schistosomiasis to the northern part of China (Zhou et al., 2010).

In addition to schistosomiasis, climate change can affect the transmission and enlarge the endemic region for a number of vector-borne diseases in Chin. These include malaria, dengue fever, epidemic encephalitis B, angiostrongyliasis cantonensis, and leptospirosis (Yang et al., 2006: 182).

CHINA'S MAJOR POLICIES AND ACTIONS FOR CLIMATE CHANGE MITIGATION

In 2007, China's National Development and Reform Commission issued the country's first national climate policy initiative. In it, the government outlined measures ranging from laws, economy, administration, and technology which aim to reduce GHG emissions and prepare the country for both mitigation and adaptation. In 2008, the government released a white paper which highlighted China's ongoing effort to combat climate change and clarified China's position in international climate negotiations (Dale 2009: 12).

China's general strategy has been to "address climate change in the process of pursuing sustainable development," so that it can gradually build up a resource-conserving and environmentally-friendly society (National Development and Reform Commission 2009: 6). In recent years, China attempted to reduce its GHG emissions and increase its capacity to adapt to climate change through implementing a number of policies, regulations, and institutional and technological instruments. The key actions taken have included encouraging energy conservation, consummating ecological environment, enhancing adaptation capacity, developing research capacity, raising public awareness, and improving mechanisms for climate change administration.

In terms of governance, in the beginning, the Chinese Meteorological Administration, along with the Chinese Academy of Sciences and the Ministry of Foreign Affairs, were responsible for issues related to climate change. Since 1998, the National Development and Reform Commission became the hub for climate change policies (it was named "the State Planning Commission" prior to 2003) (Wang 2012: 7). In 2006, the National Coordination Committee on Climate Change was established. Also, within the National Development and Reform Commission, the Department of Climate Change was created to analyze climate change's impact on China's socioeconomic development and organize the formulation of climate change mitigation strategies, planning, and policies (China Climate Change Info-net 2013). Importantly, in the institutional restructuring in 2008, China's State Environmental Protection Administration was upgraded to the Ministry of Environment, which gave it the status of a ministry and allowed it to participate in national policy making (Rommeney 2008: 15). Such an event has been helpful for China's efforts to tackle its worsening environmental problems and mitigate climate change.

Retrospectively, climate change mitigation was an issue brought to China from the international arena, when the United Nations founded the Intergovernmental Panel on Climate Change in 1988. The Chinese policies and actions have evolved since then. For the convenience of discussion, such an evolution may be divided into three phases.

From the late 1980s to the mid-1990s, generally, climate change and GHG emissions were merely seen as a developed-countries' issue. Only a few Chinese scientists studied climate change and policies were essentially limited to scientific investigations. From the late 1990s to the mid-2000s, China's economy developed rapidly. In the meantime, environmental problems and climate-related disasters also began to mount up. It became obvious that the government must respond to the environmental problems, while trying to maintain China's economic growth. Under such circumstances, the Chinese view shifted from viewing climate change mainly as an international issue to a national concern of China. In 2001, for the first time, China mentioned climate change in its tenth Five-year Plan (2001–2005). In the international arena, China's ratification of the Kyoto Protocol in 2002 confirmed its new position that all countries in the world, developed and developing countries, had a role to play in mitigating climate change.

From the mid-2000s to the present, climate change has been elevated to a national priority of China. In 2007, the National Development and Reform Commission prepared China's national climate change mitigation program and specified its main corresponding objectives, policies, and actions. Also, the domestic climate change mitigation policies were developed and a low carbon-intensity target was set up as the goal. The government began to see "sustainable development" as a desirable direction for economic growth, and the term "low-carbon" began to appear in official statements, reports, and policy texts (Zhao and Hou 2013: 30). In 2009, China further adopted its first carbon specific goal, according to which China would have to lower its carbon intensity by 40–45 percent by 2020 compared to 2005 levels. Besides, in 2011, preparations were undertaken for formulating a climate change mitigation law in China (Legal Daily 2011).

In essence, China's current climate change mitigation policies and actions have five important aspects. First, climate change mitigation has now been defined and included in China's sustainable development program (Qi and Ma 2007: 9) and China proposes to address "the issue of climate change within the framework of sustainable development" (The National Development and Reform Commission 2012: 24). As a result, in 2006, climate change mitigation was included in China's eleventh Five-Year Plan and the country's GDP energy consumption was set to decrease by 20 percent (China's National Climate Change Program 2007). Specifically, China attempted to reach its goals by upgrading or eliminating traditional energy industries and by supporting or developing more environment-friendly new energy industries.

Furthermore, in China's twelfth Five-year Plan, energy conservation and energy efficiency were highlighted as key areas for climate change mitigation and the fiscal and tax incentives were offered to encourage corresponding actions. The government also pledged to promote clean utilization of fossil fuel and develop non-fossil fuel energies such as hydropower, wind power, solar power, and biomass energy (National Development and Reform Commission 2012: 3–8). According to the government's plan, when implemented, China's consumption of renewable energy will increase to 15 percent of the total energy consumption by 2020 (Olsson 2012: 3).

Second, the government has tried to develop policies to strengthen the capability of key sectors to adapt to climate change and reduce the negative impacts of climate change on socioeconomic development and the people's welfare. The highlighted areas included agriculture, forestry, water resources, marine resources, public health, meteorology, and disaster prevention and mitigation (NDRC 2012: 11–14).

Third, the government has proposed to strengthen capacity building to cope with environmental problems related to climate change. The emphasis is given to strengthen top level planning of low-carbon development. The National Development and Reform Commission drafted a legal framework for addressing climate change. The government pledged to establish statistical and accounting systems for measuring GHG emissions (National Development and Reform Commission 2012: 16–17; Yao, Lib, and Steemersa 2005: 1973).

Fourth, China's climate change mitigation program has attempted to utilize the participation of the Chinese public in mitigating climate change. The emphasis is given to the creation of a social atmosphere favoring green and low-carbon development. It is hoped that proactive participation by the public will help popularize low-carbon lifestyle choices in eating habits, housing, transport, and tourism, and encourage people to adopt a generally moderating and less-waste consumption style. The public are encouraged to avoid extravagance and waste. Besides, energy-saving and carbon reduction activities are publicized and the low-carbon lifestyle and environmental protection through various activities.

Fifth, China's climate change mitigation program has tried to promote a low-carbon life style by launching pilot projects in selected communities, industrial zones, cities, and provinces. Methods to calculate product carbon emissions were developed and standards to identify low-carbon products were also established. In 2011, the Ministry of Transport initiated pilot projects for the construction of low-carbon transport systems, with an emphasis on road and river transport and urban passenger transport. Ten cities were selected to carry out the initial trials. A further sixteen cities were named the second batch of pilot cities in February, 2012 (National Development and Reform Commission 2012: 16–17).

Sixth, the Chinese government has initiated a carbon tax of ten Yuan per ton of carbon to start in 2016 and pilot carbon trading program are currently underway in a number of major cities and provinces such as Beijing, Shanghai, and Guangdong (Yale Center for Environmental Law and Policy 2012: 1).

The carbon tax has long been proposed by economists as a functional policy for GHGs mitigation (OECD 1996). A number of European Union countries have implemented such a tax policy since the 1990s. Despite its limits, the results from the European Union countries suggested that the carbon tax can help reduce GHGs emissions (Baranzini, Goldemberg, and Speck 2000: 395).

However, carbon tax faces several obstacles in China. It will cause GDP losses for China's economy, because of the increased energy prices and disappeared job opportunities. The closure of one old power plan, for example, often means that thousands of workers would lose their jobs. It will likely meet with resistance from local officials and enterprises, especially the large GHGs emitters. Besides, in one way or another, the tax burden will eventually fall on the consumers and the prices of consumer goods will go up.

Despite that, carbon tax not only helps reduce GHGs but helps improve energy efficiency. It also promotes technical innovations and industrial updates and increases tax revenues. For example, when the carbon tax rate is set at twenty Yuan/tC, the CO_2 emission reduction could reach 90 million tons which is equal to 4.5 percent of China's CO_2 emission per year. In the meantime, the effect of energy saving rate can reach approximately 3 percent and the carbon tax revenue will reach 40 billion Yuan (Wang 2009: 102–104). So far, the carbon taxation policy has met with support from the public. As indicated in a 2012 national survey, 98.1 percent of the participants agreed that the government should take the efficient measures to mitigate climate change and that 87 percent stated that they would be willing to pay 10 to 30 percent more for purchasing environment-friendly low-carbon products (Outlook Weekly 2012: 22). Carbon taxation is a policy option for China to mitigate climate change.

As a whole, to some degree, these policies and actions have helped reduce GHG emissions. In 2006, China's GDP energy intensity declined by 1.8 percent, 3.7 percent in 2007 and 4.6 percent in 2008. Also, according to the government, China's climate mitigation program resulted in 950 million tons of avoided CO_2 emissions by 2010 (Olsson 2012: 2).

Despite the achievements, there is much room in Chinese climate change mitigation policies and actions for improvement. Above all, the

Chinese climate change mitigation policies should include more specific and efficient means and measures to involve local governments and large energy-consuming enterprises in GHGs reduction actions. Large enterprises consume approximately one-third of China's energy and emit the bulk of China's pollution, including CO_2 and other GHGs (Leggett, Logan, and Mackey 2008: 22). Yet, current policies lack clear-cut regulations on the specific responsibilities of those enterprises in GHGs reduction (Qi and Ma 2007: 9–11). Also, the role of local governments needs to be more clearly specified so that they can actively participate in mitigation actions, especially in implementing regulations in state-owned enterprises.

That is crucial, because the environmental governance has now become considerably decentralized in China. Most pollution control and enforcement is now in the hands of local officials who often prefer economic growth to pollution control. While China does have a comprehensive set of environmental laws now, implementation is frequently impeded by decentralization of power and weak local enforcement. The problem is often the wide discretion local agencies have in addressing environmental issues.

In fact, the local governments in China have now gained considerable administrative autonomy and fiscal autonomy. However, while enjoying more autonomy, the local officials are also responsible for generating most of their own revenue and balancing their own budget. Such an economic situation generates extensive pressure at the local level to compete in attracting and promoting economy-building companies. Besides, local governments often sponsor or even own industries and consider environmental regulations to be incompatible with economic growth. In addition, since local environmental protection agencies often obtain their funding from the local governments of which they are part, the enforcement of environmental policies understandably faces weighty financial constraints and is frequently undermined by economic pressure. Although the Ministry of Environmental Protection has nominal authority over lower-level agencies, it does not really have much administrative or financial leverage in ensuring that national policies and standards are strictly enforced at the local level (Beyer 2006: 209–210).

Evidently, climate change mitigation is a difficult endeavor. For conventional environmental actions, both costs and impacts are often local. A local government will probably meet less resistance utilizing its resources to alleviate a local problem for local benefits. Yet, in climate change mitigation, for example, for GHG emissions reduction, the costs are often local, while the impacts are national or even global. Thus, even when national officials genuinely want to implement mitigation actions, local officials may seek to minimize enforcement in order to maximize local economic growth and employment (Leggett 2011: 5). To diminish such "a tragedy of the commons," it is necessary to include mandatory requirements and coordinated actions in current policies and oblige local government agencies and large enterprises to take their responsibilities.

Moreover, there needs to be more specific and adequate policies and actions to reduce the role of coal combustion in China's energy consumption. As of 2008, all coal-fired power plants built in China must use commercially available technology, or better. As a result, most of the world's cleanest and most efficient coal-fired power plants are now located in China and the average efficiency of its entire coal-fired power plant fleet is now better than that of U.S. plants (International Energy Agency 2009: 1). Coal is the most carbon intensive of the major fossil fuels. When combusted, it emits almost twice as much CO_2 per unit of energy released as natural gas and around 33 percent more than oil (Salovaara 2011: 17). Besides, China continues to be the world's largest consumer of coal and much of China's greenhouse gas emissions, as well as its emissions of conventional pollutants, continue to be associated with its high reliance on coal (Yale Center for Environmental Law and Policy 2012: 1). China's consumption of coal declined between 2001 and 2005 and increased again since 2007(BP 2012: 33). In 2011, coal continued to supply as high as 69.7 percent of China's primary energy. In 2012, in absolute figures, China consumed 3.7 billion tons of coal, the amount almost half of the world's total coal consumption in the year. According to the China Electricity Council, China's demand for coal will continue to increase to 4.3 billion tons by 2015 (China Daily 2012). To mitigate greenhouse gas emission efficiently, China needs to continue to consummate its coal combustion technologies and cut down its coal consumption.

Thirdly, China's climate change mitigation policies should involve more participation and cooperation from nongovernmental organizations (NGOs). It is commendable that China's National Climate Change Program has proposed to involve the whole society in mitigating climate change. Also, a number of NGOs such as the China Environmental Protection Foundation and the China Green Carbon Foundation launched various social activities to promote participation of the public in energy conservation and GHGs reduction and promote low-carbon lifestyles (National Development and Reform Commission 2012: 23). Also, in recent years, several Chinese NGOs dedicated to climate change mitigation, such as the China Civil Climate Action Network, China Green Carbon Foundation, and Friends of Nature, have been established. These NGOs devote solely to their goal to form a wider coalition of stakeholders to address climate change issues (China Civil Climate Action Network 2011).

Despite that, up to now, few Chinese NGOs are able to make substantial influences on mitigation policies and rely mainly on media and information campaigns to influence policy decisions on climate change

(Lo 2010: 5690). In fact, most Chinese NGOs' current efforts are largely limited to promoting and facilitating information sharing and organizing stakeholders to participate in some sociocultural activities related to climate change mitigation, such as energy saving, providing technical assistance, and elucidating the causes and effects of global climate change. The government should encourage NGOs and relevant civil societies to move a step further and actively provide their proposals which can complement and supplement the government's own efforts to respond to the climate change challenge. However, to allow NGOs and civil societies to participate more meaningfully, it is vital for the government to recognize the important roles of these players and actively involve them in the process of policy making.

Fourth, China's climate change mitigation program should continue to encourage and promote more research on climate change and related issues, especially in areas of social sciences. Chinese scientists and researchers have long played a key role in China's climate policy making and have made remarkable achievements in research on climate change in recent decades. For example, they have warned of China's vulnerability to the threat of climate change and have presented policy recommendations that go far beyond media reports and governmental statements. To varying degrees, China's top scientific institutions, such as the Chinese Academy of Sciences, Beijing University, People's University, and Tsinghua University, have all participated in research on climate change (Wübbeke 2010: 27–36). In 2011, the Ministry of Environmental Protection and the Nanjing Institute of Environmental Sciences jointly founded the Environment and Climate Change Center and the Ecological Protection and Climate Change Response Research Center. Also in 2011, the Energy Conservation and Emission Reduction Research Center was established in China Civil Aviation University (National Development and Reform Commission 2012: 21). The number of research articles on climate change in Chinese publications has grown from about 1,000 articles published in 2006 to more than 8,000 in 2010 (Stensdal 2012: 10).

Despite the achievements, more contributions from China's scientists, especially social scientists are clearly needed. For example, however, the government (and to a large degree, the Chinese public) so far continues to view climate change mainly as a natural phenomenon and a natural scientific issue and concentrates on natural scientific research, and devotes only inadequate attention to the sociocultural implications of climate change.

Yet, in addition to its natural components, climate change possesses vital sociocultural implication and its mitigation policies require guidance and inputs from social sciences. To a large degree, climate change is a type of "structural violence" against human security in the world (Soron

2007). Like other types of violence, climate change is an "avoidable impairment of fundamental human needs" and is a violence that constrains on "human potential due to economic and political structures," using Johan Galtung's terms (Galtung 1969: 169/182). Clearly, climate change is far more than a subject solely reserved for natural sciences and requires more attention from social scientists.

Building on what has been accomplished, the social scientists have a great deal to contribute to China's climate change mitigation policies and actions. Far beyond merely surveying public opinions on climate change, social scientists can conduct both quantitative and qualitative research on essential issues related to climate change (Crate and Nuttal 2009). Focusing on human-environment relationships, social scientists can address the climate change issues vigorously. They can investigate sociocultural implications of climate change and facilitate collaborative, community-based programs focused on mitigation and adaptation. They can also develop culturally sensitive strategies for communicating climate change to diverse audience (Hirsch et al., 2011: 267). At present, however, such a gap is still waiting to be filled.

Fifth, China's mitigation policies should devote even more efforts to expanding and deepening the public awareness of climate change issues and encourage its citizens to participate in mitigation actions. This is important, because public understanding and participation directly affect GHG emissions of individuals and households and their support of climate change mitigation policies. For example, between October 28 and October 29, 2008, despite the unfavorable weather conditions, the farmers in the central and northern parts of Jiangsu Province tried to burn crop residues in the fields and unknowingly created a serious air pollution event in the city of Nanjing (the city's total land area: 2,548 square miles) and surrounding regions, accompanying with sharply increasing of pollutants such as PM_{10}, CO_2, and SO_2 in the air (Zhu et al., 2010: 585).

For a positive example, during the 2008 Beijing Olympic Games, utilizing the Chinese national pride to encourage its citizens "to give the world a wonderful Beijing Olympics" with a "blue sky," Beijing government was able to successfully mobilize its 17 million citizens to follow an alternate-day driving restriction. The restriction took almost half of Beijing's 3.3 million cars off the road by limiting car owners to driving on alternate days depending on the last number of their license plates during the games (Eimer 2008). Such an action not only cut down emissions of carbon dioxide by 24,000 to 96,000 metric tons during the event but clearly demonstrated what the public participation can do to reduce GHG emissions (Worden et al., 2012: 1).

Despite the efforts from the scientific circle, government, media, and NGOs, the overall levels of the Chinese public awareness of climate change remain to be improved. In 2007, for example, the Chinese Academy of Social Sciences conducted a survey of the Chinese public opinion on China's major environmental problems. In the survey, while ranking serious environmental problems in China, climate change was ranked only the fourth most serious environmental problem, behind consumer waste pollution, declining arable land, and air pollution. Besides, the respondents only perceived climate change as an environmental problem between "quite serious" and "not so serious" (Chinese Academy of Social Sciences 2007: 5). More recently, at a micro-level, a study of public awareness of climate change in Yinchuan City of Ningxia Hui Autonomous Region showed that the cognition on climate change has basically taken shape at both the public and policy making levels. Nonetheless, the awareness of the public is weaker than that at policy making level, and so is the cognition accuracy (Zhou and Feng 2011: 61).

The 2012 Survey on Public Climate Change Awareness and Climate Change Communication in China has provided a notable point. The survey interviewed 4,169 Chinese citizens to investigate the state of public climate change awareness, beliefs, attitudes, policy support, and environmental behaviors. This nationwide survey on climate change was the first of its kind in China and was conducted jointly by the Center for China Climate Change and the Statistics Institute of People's University. (Outlook Weekly 2012: 22).

Survey results showed that the Chinese people pay close attention to climate change and public awareness of climate change has been improving, the outcomes were quite mixed as well. Encouragingly, 93.4 percent of the respondents reported that they knew at least a little about climate change. Yet, 28.4 percent reported that they knew just a little about it, 53.7 percent knew something, and 6.6 percent had never heard of it. Only 11.4 percent stated that they knew a lot about climate change. Regarding the causes of climate change, 55.3 percent believed that climate change was caused mostly by human activities, yet 38.1 percent believed that climate change was caused mostly by natural changes in the environment. For the impact of climate change, 23 percent of the respondents said they were very concerned, 54.7 percent were only somewhat concerned, 14.1 percent were not very concerned, and 8.2 percent were not at all concerned. When being asked about who should be mainly responsible for climate change mitigation, 88.9 percent chose the government among four other entities: the public, media, enterprises, and nongovernmental organizations. The respondents considered scientific institutions to be the most reliable source for information regarding climate change, and the government and media, were ranked as the second and third respectively. Paradoxically, NGOs and enterprises were ranked the least reliable sources for information (China Center for Climate Change Communication 2012).

CONCLUSION: CHINA'S CLIMATE CHANGE MITIGATION PROGRAM: CHALLENGES AND FUTURE PROSPECTS

The United Nations Development Program (UNDP 2010: 99) has argued forcefully that overall impacts of climate change are and will be negative, with adverse consequences for human security. Given its vulnerability to the negative impacts of climate change, China needs to accelerate and further strengthen its climate change mitigation actions.

Strategically, China should continue to address climate change issues in the context of sustainable development, with an emphasis on "a convergence between the three pillars of economic development, social equity, and environmental protection" (UN 2012: 2). While China has made remarkable socioeconomic achievements in the past decades, it is crucial to recognize that its present model of growth is not sustainable. Additionally, if climate change impacts are not effectively tackled, there is a growing danger that many of China's socioeconomic achievements in the past decades may be jeopardized or even reversed (UNDP 2010: 101). In China and elsewhere, the impacts of climate change are already visible and substantial. To diminish the problems, China must shift away from the traditional growth models that depend heavily on fossil fuel and resource consumption. Furthermore, China needs to move toward the low carbon growth and sustainable development that will preserve and increase its socioeconomic achievements in the long run.

Climate change mitigation is consistent with China's sustainable development and can make significant contribution to the country's efforts to enhance human security. To be viable, China's climate change mitigation program needs to entail win-win-win opportunities for all and work on GHGs reduction and environmental protection as a multiple-benefit program. For instance, GHGs mitigation should also help improve public health by monitoring and reducing air pollution consistently. Agriculture has always been a vital component in Chinese economy. For millennium, famine was a persistent socioeconomic problem in China despite the rise and fall of imperial dynasties. Much progress has been made in food production in the past decades, and China's grain production has increased from about 200 kg per capita in 1949 to about 400 kg in the early 1990s (Zhang 2011: 3). However, food security remains a formidable task for China, given the potential crop failures due to climate related droughts, floods, and other natural disasters. Along with other actions, China's climate change mitigation program can contribute to China's food security by GHGs reduction which helps diminish climate extreme events and by strengthening the country's agricultural adaptive capacity to climate change.

To ensure human security in China or any country, the low carbon and sustainable development is a necessary requirement. In December, 2012, encouragingly, China has made its decisive commitment to reduce GHG emissions per unit of GDP in 2020 by 40 to 45 percent compared with 2005 levels. Meanwhile, Chinese companies are seizing the global opportunities for the low carbon growth models and have so far developed a renewable energy sector worth US $17 billion and employing one million workers (UNDP 2010: i).

Despite all efforts and achievements, China's climate change mitigation program still faces a zigzag and rough road. Today, China's market economy is both a strength and a weakness when it comes to the GHGs reduction and sustainable development. SunTech, the biggest solar energy producer in China, has just declared bankruptcy (Business Week 2013). Yet, PetroChina, China's largest oil company, has already surpassed Exxon in oil production and has become the world's biggest publicly-traded producer of oil (Kahn 2012: 10). Ultimately, China's GHGs reduction and climate change mitigation program are no easy tasks and will require the country to take hard steps to conserve a clean environment and ensure the full range of human security in China.

REFERENCES

Anderegg, William R., L. James W. Prall, Jacob Harold, and Stephen H. Schneider, "Expert Credibility in Climate Change, 2010, " Proceedings of the National Academy of Sciences of the United States of America, 107 (27): 12107–12109.

Baranzini, Andrea, and José Goldemberg, Stefan Speck, 2000, "A future for Carbon Taxes, Ecological Economics, 32: 395–412.

Beyer, Stefanie, 2006, Environmental Law and Policy in the People's Republic of China. Chinese Journal of International Law (2006), Vol. 5, No. 1, 185–211.

Bodeen, Christopher, 2008, "Rising Seas Threaten China's Sinking Coastal Cities," National Geographic. news.nationalgeographic.com/news/2008/01/080117-AP-china-risin.html.

BP, 2011, Statistical Review of World Energy, London, UK.

BP, 2012, BP Statistical Review of Word Energy, London UK.

Business Week, 2013, "Biggest Solar Collapse in China Imperils $1.28 Billion," March 21, 2013, www.businessweek.com/news/2013-03-20/biggest-solar-collapse-in-china-imperils-1-dot-28-billion-energy.

Chak K. Chan and Xiaohong Yao, 2008, "Air pollution in mega cities in China," Atmospheric Environment, 42:1–42.

Chen, Longxun, Zhou Xiuji, Li Weiliang, Luo Yunfeng, and Zhu, Wenqin, 2004, "Characteristics of the Climate Change and Its Formation Mechanism in China in Last 80 Years," Acta Meteorologica Sinica, 52 (5): 534–545.

Chen, Renjie, Evangelia Samoli, Chit-Ming Wong, Wei Huang, Zongshuang and Wang, Bingheng, 2004, Associations between short-term exposure to nitrogen dioxide and mortality in 17 Chinese cities: The Air Pollution and Health Effects Study (CAPES). Environment International 45 (2012) 32–38.

Chen, and Haidong Kan, 2012, "Associations between Short-Term Exposure to Nitrogen Dioxide and Mortality in 17 Chinese Cities: The China Air Pollution and Health Effects Study," Environment International, 45: 32–38.

Cheng, Hefa, Yunan Hu, and Jianfu Zhao, 2009, "Challenges Meeting China's Water Shortage Crisis: Current Practices and Challenges," Environmental Science and Technology, 43: 240–244.

China Center for Climate Change Communication, 2012, China Public Awareness Survey Report on Climate Change and Climate Communication, Beijing, China.

China Civil Climate Action Network (CCAN) Brochure, 2011, Beijing, China.

China Climate Change. Brief Introduction of National Coordination Committee on Climate Change. Source en.ccchina.gov.cn/.

China Daily, 2012, "China's Coal Consumption Continues To Rise," February 27.

China's National Climate Change Program, 2004, Initial National Communication on Climate Change. Beijing, China.

China Water Resources Bulletin, 2006, 2007 China Water Resources and Hydropower Press: Beijing.

Climate Group, 2006, China Briefing, Issue 1: 1–6.

Coumou, Dim, and Stefan Rahmstorf, 2012, "A Decade of Weather Extremes," NATURE OLOGY, March 25.

Crate, Susan, and Mark Nuttall, 2009. Anthropology and Climate Change: From encounters to actions. Walnut Creek, CA: Left Coast Press.

Cui, X., G. Huang, W. Chen, and A. Morse, 2009, "Threatening of Climate Change on Water Resources and Supply: Case Study of North China," Desalination 248: 476–478.

Dale, Jiajun Wen, 2009, "Climate Change and China: Technology, Market and Beyond: A Report for Focus on the Global South." Focus On The Global South, Chulalongkorn University, Bangkok, Thailand.

Dou, Guangyu, 2010, Climate Change Affects China. Science in 24 Hours, V5: 6–9.

Dunlap R., McCright A. M., 2008, "A Widening Gap: Republican and Democratic Views on Climate Change," Environment 50: 26–35.

Edmonds, Richard L., 2008, "China's Environmental Problems." In: Understanding Contemporary China (4th Edition), Robert E. Gamer, ed. Pp. 271–298 Lynne Rienner Publishers, Boulder, CO.

Eimer, David 2008, "Beijing Olympics: China to Celebrate Week of 'Fresh Air,'" Telegraph, London, August 2, 2008.

Food and Agriculture Organization (UN), 2012, "Army Worm Outbreak Brought under Control," Monthly News Report on Grains, Issue 85–August 2012.

Galtung, Johan, 1969, "Violence, Peace, and Peace Research," Journal of Peace Research, V6(3): 167–191.

Harris, Paul G., 2010, "China and Climate Change: From Copenhagen to Cancun," Environmental Law Reporter, 9: 10858–10863.

Hirsch, Jennifer, Sarah Van Deusen Phillips, Edward Labenski, Christine Dunford, and Troy Peters, 2011, "Linking Climate Action to local Knowledge and Practice: A Case Study of Diverse Chicago Neighborhoods." In Environmental Anthropology Today, Helen Kopnina and Eleanor Shoreman-Ouimet, ed. New York: Routledge.

International Energy Agency, 2009, "Cleaner Coal in China," OPEC Report.

International Energy Agency, 2012, Co_2 Emissions from Fuel Combustion *Highlights,* 9 rue de la Fédération 75739 Paris Cedex 15, France.

International Monetary Fund, 2012, China. April, 2012.

Jin Rui, Che Tao, Li Xin, Wu Lizong, 2004, "Glacier Change in Pumqu Basin Based on Remote Sensing Data and GIS Technique," Journal of Glaciology and Geocryology, 26 (3).

Kahn, Chris, 2012, "PetroChina surpasses Exxon in oil production," the Ledger Independent, March 30, 2012, A10, the Ledger Independent.

Kaiman, Jonathan, 2012, "China's Emissions Expected to Rise until 2030, Despite Ambitious Green Policies," November 26, 2012. The Guardian, London, UK.

Kan, H., London, S., Chen, G., Zhang, Y., Song, G., Zhao, N., et al., 2007, Differentiating the Effects of Fine and Coarse Particles on Daily Mortality in Shanghai, China. Environment International, 33 (3): 376–384.

Kan, H., London, S. J., Chen, G., Zhang, Y., Song, G., Zhao, N., 2008, "Season, sex, age, and education as modifiers of the effects of outdoor air pollution on daily mortality in Shanghai, China: the Public Health and Air Pollution in Asia (PAPA) study." Environ Health Perspective, 116: 1183–1188.

Kan, Haidong, 2011, "Climate Change and Human Health in China," Environmental Health Perspectives, 119 (2): 60.

Keidel, Albert, 2005, "The Economic Basis for Social Unrest in China, "Carnegie Endowment for International Peace.

Leggett, Jane A., 2011, "China's Greenhouse Gas Emissions and Mitigation Policies," CRS Report for Congress, Congressional Research Service, USA.

Leggett, Jane A., Jeffrey Logan, and Anna Mackey, 2008, "China's Greenhouse Gas Emissions and Mitigation Policies," CRS Report for Congress, Congressional Research Service, USA.

Li, Xu Jindian, and Cai Rongshuo, 2002, "Trends of sea level rise in the South China Sea during the 1990s: An altimetry result," Chinese Science Bulletin, 47 (7): 5.

Li, Shihua, 1993, "Impacts of Climate Warming on Crop Diseases and Pests in China," Chinese Journal of Agrometeorology, 14 (01): 41–43.

Liu, Yingjie, and Lin Erda, 2007, "Effects of Climate Change on Agriculture in Different Regions of China," *Advances in Climate Change,* vol. 3 (4): 229–233.

Lo, Alex, 2010, "China's Response to Climate Change," Environmental Science and Technology, VOL. 44 (15): 5689–5690.

Liu, S.Y. Liu, Y.J. Ding, J. Li, D.H. Shang Guan, and Y. Zhang, 2006, "Glaciers in Response to Recent Climate Warming in Western China," *Quaternary Sciences,* 26 (5): 762–771.

Liu, Yansui, Liu Yu, and Guo Liying, 2010, "Impact of Climatic Change on Agricultural Production and Response Strategies in China," *Chinese Journal of Eco-agriculture,* vol. 18 (4): 905–910.

Liu, Yanxiang, Xiang Li, Qiang Zhang, Yufu Guo, Ge Gao, and Jianping Wang, 2010, "Simulation of Regional Temperature and Precipitation in the Past 50 Years and the Next 30 Years over China, Quaternary International, vol. 212: 57–63.

Lu, Liping, Cheng Conglan, Liu Weidong, and Qin Zhihao, 2009, "Effect of the Agricultural Meteorological Disasters on Agricultural Production and Its Spa-

tial Distribution Characteristics during the Last 30 Years in China," 2009, Ecology and Environmental Sciences, 18 (4): 1573–1578.
Morrison, Wayne M., 2012, China's Economic Conditions. US Congressional Research Service, 7–5700, www.crs.gov.
Moseley, Robert K., 2006, "Historical Landscape Change in Northwestern Yunnan, China," Mountain Research and Development, 26 (3): 214–219.
National Development and Reform Commission, 2009, China's Policies and Actions for Addressing Climate Change—The Progress Report, 2009, Beijing, China.
National Development and Reform Commission, 2012, China's Policies and Actions for Addressing Climate Change, Beijing, China.
Nevada Division of Environmental Protection, 2013, Particulate Matter Pollution Fact Sheet.
Nicholls, N. N., and Coauthors, 1996, Observed Climate Variability and Change, In: Climate Change 1995: The Science of Climate Change, J. T. Houghton et al., Eds., Cambridge University Press, 135–192.
OECD, 1996. Implementation Strategies for Environmental Taxes. OECD, Paris.
Olsson, Marie, 2012 "China acting on climate change," Policy Briefing, Stockholm Environment Institute, Stockholm, Sweden.
Outlook Weekly, 2012, "The First Mass Survey on Climate Change," Outlook Week, 2012 (48): 22.
Parry, M. L., Canziani, O. F., Palutikof, J. P., van der Linden, P. J., Hanson, C. E., 2007, *Climate Change 2007: Impacts, Adaptation and Vulnerability*, Cambridge University Press.
Peng, R. D., Chang, H. H., Bell, M. L., McDermott, A., Zeger, S., L., Samet, J. M., 2008, "Coarse Particulate Matter Air Pollution and Hospital Admissions for Cardiovascular and Respiratory Diseases among Medicare Patients," Journal of the American Medical Association, 299 (18): 2172–2179.
QI, Ye and MA Li, 2007, "Towards Proactive Climate Change Policy," China Population Resources and Environment, Vo1 (17/2):8–12.
Qian, Z., He, Q., and Lin, H., 2007, "Association of Daily Cause-Specific Mortality with Ambient Particle Air Pollution in Wuhan, China," Environmental Research, 105 (3): 380–389.
Qin, Dahe, 2004, "Climate Changes into the 21 Century: Facts, Impacts, and Strategies Addressing Climate Change," Climate Change and Its Strategies, Vol. 7: 4–7.
Ren, G., 2007. *Climate Change and Water Resources in China,* China Meteorological Press, Beijing.
Ren, G, Z. Guan, X. Shao, D. Y. Gong, 2011, "Changes in Climatic Extremes over Mainland China," Climate Research, Vol. 50: 105–111.
Richerzhagen, Carmen, and Imme Scholz, 2007, China's Capacities for Mitigating Climate Change, Discussion Paper, Deutsches Institut für Entwicklungspolitik.
Rommeney, Dirk, 2008, Climate and Energy Policy. Heinrich Böll Foundation China.
Salovaara, Jackson, 2011, Coal to Natural Gas Fuel Switching and CO_2 Emissions Reduction, Thesis, Harvard College, Cambridge, Massachusetts, USA.
Soron, D., 2007, Cruel Weather, Natural Disasters, and Structural Violence. Transformations. Issue 14.

Stensdal, Iselin, 2012, China's Climate-Change Policy 1988–2011: From Zero to Hero? Fridtjof Nansen Institute Report, Issue 9/2012.

Tanakam T., R. Jayakumar, B. Erdenechimeng, 2008, "UNESCO Chair Workshop on Sustainable Groundwater Management in Arid and Semi-arid Regions." UNESCO Publications.

Tong, Jiang, Buda Su, and Heike Hartmann, 2007, "Temporal and Spatial Trends of Precipitation and River Flow in the Yangtze River Basin, 1961–2000," Geomorphology 85 (2007) 143–154.

UN, 2008, "China's Progress Towards the Millennium Development Goals 2008 Report." UN, New York.

UN, 2012, Sustainable Development: From Brundtland to Rio 2012, UN, New York.

Vajpeyi, Dhirendra K., 2013, Climate Change, Sustainable Development, and International Security.

Wang, Jinxia, Jikun Huang and Scott Rozelle, 2010, Climate Change and China's Agricultural Sector: *An Overview of Impacts, Adaptation and Mitigation,* International Centre for Trade and Sustainable Development.

Wang, Jinnan, Yan Gang, Jiang Kejun, Liu Lancui, Yang Jintian, Ge Chazhong, 2009, "The Study on China's Carbon Tax Policy To Mitigate Climate Change," China Environmental Science 29 (1): 101–105.

Wang, Jinnan, Ge Chazhong, and Gao Shuting, 2006, Environmental Taxation Policies and Implementing Strategy. Beijing: China's Environmental Science Press.

Wang, Shourong, and Zuqiang Zhang, 2011, "Effects of Climate Change on Water Resources in China," Climate Research, V47: 77–82.

Wang, Wen, 2012, Overview of Climate Change Policies and Prospects for Carbon Markets in China, Les Cahiers de la Chaire Economie du Climat, 2012 (July Issue).

WHO, 2003, Climate Change and Human Health: Risks and Responses. WHO Fact Sheet No. 266, October 2003.

WHO, 2012, "Climate Change and Health." WHO Fact Sheet No. 266, October 2012.

Worden, H. M., Y. Cheng, G. Pfister, G. R. Carmichael, Q. Zhang, D. G. Streets, M. Deeter, D. P. Edwards, J. C. Gille, and J. R. Worden, 2012, "Satellite-Based Estimates of Reduced CO and CO_2emissions Due to Traffic Restrictions during the 2008 Beijing Olympics," Geophysics Research Letters, 39, L14802.

World Bank, 2009, "From Poor Areas to Poor People: China's Evolving Poverty Reduction Agenda—An Assessment of Poverty and Inequality in China," Poverty Reduction and Economic Management Department, East Asia and Pacific Region, World Bank.

World Meteorological Organization, 2011, Weather Extremes in a Changing Climate, New York: United Nations.

Wong, C. M., Ma, S., Hedley, A. J., and Lam, T. H., 2001, Effect of Air Pollution on Daily Mortality in Hong Kong, Environmental Health Perspectives, 109 (4): 335–340.

Wübbeke, Jost, 2010, The Power of Advice. Experts in Chinese Climate Change Politics, FNI report 15/2010, Lysaker: Fridtjof Nansen Institute.

Xie, Peng, and Xiaoyun Liu, Zhaorong Liu, Tiantian Li, Liujv Zhong, and Yunrong Xiang, 2011, "Human Health Impact of Exposure to Airborne Particulate Matter in Pearl River Delta, China," Water Air Soil Pollution, 215: 349–363.

Xu, Rui, "China's Sea Level Rising, Coast Sinking," 2012, CNS, Beijing, China, www.ecns.cn/cns-wire/2012/07-05/18829.shtml.

Yale Center for Environmental Law and Policy, 2012, Webinar Series Climate Policy and Emissions Data Sheet: China. Yale University Press.

Yang, Dawen, Chong Li, Heping Hu, Zhidong Lei, Shixiu Yang, Tetsuya Kusuda, Toshio Koike, and Katumi Musiake, 2004, "Analysis of Water Resources Variability in the Yellow River of China During the Last Half Century Using Historical Data," Water Resources Research, 40 (6).

Yang, A. Kun, W. Xianhong, Lv Shan, Zhang Ling, Jia Tieu, Li Lianhua, Dong Yao, Zhou Xiaonong, 2006, "Impact of Global Warming On Transmission of Vector-Borne Diseases in China, International Journal of Medicine and Parasitic Diseases," Vol. 33 (4): 182–187.

Yao, R., Lib, B., Steemersa, K., 2005, "Energy Policy and Standard for Built Environment in China," Renewable Energy, Vol. 30: 1973–1988.

Yao, Tandong, Wang Youqing, Liu Shiying, Pu Jianchen, Shen Yongping, and Lu Anxin, 2004, "Recent Glacial Retreat in High Asia in China And Its Impact on Water Resource in Northwest China," Science in China / Ser. D Earth Sciences, Vol. 47 (12): 1065–1075.

Ye, Qian, and Michael H. Glantz, 2005, "The 1998 Yangtze Floods: The Use of Short-Term Forecastsin the Context Of Seasonal to Interannual Water Resource Management," Mitigation and Adaptation Strategies for Global Change, 10: 159–182.

Ying, Yanting, et al., 2011, "Advances in the Climate Change Influencing Grassland Ecosystems in Inner Mongolia," Pratacultural Science, 28 (6): 1132–1139.

Zeng Guang, Chen Wei, and Zhang Xian-Feng, 2005, The Influence of 1998 Flooding on Infectious Diseases Epidemiology in Hubei Province. Practical Preventive Medicine (2005): 2.

Zhai, Panmao, Xuebin Zhang, Hui Wan, Xiaohua Pan, 2005, "Trends in Total Precipitation and Frequency of Daily Precipitation Extremes over China," Journal of Climate, V18: 1096–1107.

Zhang, Jianhua, 2011, "China's Success in Increasing Per Capita Food Production," Journal of Experimental Botany, 5: 1–5.

Zhang, Jianyun, Wang Guoqing, Yang Yang, He Ruimin, and Liu Jiufu, 2009, "Impact of Climate Change on Water Security in China," Advances in Climate Change Research, 34 (7): 1673–1719.

Zhang, Qiang, Kebin He, and Hong Huo, 2012, "Cleaning China's Air," Nature, V484 (4): 161–162.

Zhang, Xiuyun, Yao Yubi, Deng Zhenyong, and Yin Dong, Wang Runyuan, and Chen Changping, "Climatic Change Over the Grassland in the Northeast Border of Qinghai Tibetan Plateau and Its Influence on Animal Husbandry," Pratacultural Science, V124 (16): 66–73.

Zhao, Jun-Fang, Guo Jian-Ping, Zhang Yan-Hong, and Xu Jing-Wen, 2010, "Advances in Research of Impacts of Climate Change on Agriculture," Chinese Journal of Agrometeorology, 31 (2): 200–205.

Zhao, Zhenyu, and Liying Hou, 2013, "How China Responds to the Carbon Emissions: Challenges, Commitment and Action," Journal of Low Carbon Economy, 2: 30–37.

Zhen, W. Z., and R. H. Wu, 1993: Sea Level Changes of the World and China. Mar. Sci. Bull., 12, 95–99.

Zhou, Jingbo, and Feng Xiangzhao, 2011, "Cognition of Adaptation to Climate Change and Its Policy Evaluation," China Population, Resources and Environment, V21 (7): 57–61.

Zhou, Y. B., Zhuang, J. L., Yang, M. X., Zhang, Z. J., Wei, J. G., and Peng, W. X., 2010, "Effects of Low Temperature on the Schistosome-Transmitting Snail *Oncomelania Hupensis* and the Implications of Global Climate Change. Molluscan Research, 30 (2):102–108.

Zhu, Bin, Su Jifeng, Han Zhiwei, Yin Cong, Wang Tijian, 2010, "Analysis of a Serious Air Pollution Event Resulting from Crop Residue Burning over Nanjing and Surrounding Regions," China Environmental Science, 30 (5): 585–592.

5 Part II

Climate Change and Sustainable Development in Western China's Minqin Oasis

Joining Forces with Society[1]

Maria Bondes and Ding Li

In recent years, China's environmental problems have been rapidly growing. While society at large has profited from China's economic opening and the introduction of private market reforms, the downturns of the rapid economic development are starting to pose a serious challenge to the legitimacy of the ruling Chinese Communist Party (Holbig 2009). China's society is not only facing growing income disparities between urban and rural areas and between the winners of reform and the less privileged social strata,[2] but decades of economic policies focusing on fast rather than sustainable development have also led to serious ecological problems that pose a stern threat to human security.[3]

In its recent Five-Year plan, the Chinese Communist Party has thus declared sustainable development and the solution of China's growing environmental problems a top priority of the country's domestic policy. The twelfth Five-Year-Plan was welcomed by many as the "greenest" Five-Year-Plan to date (Lam 2011, Xinhua 18.10.2010). However the effective implementation of such national policies at the local level has to draw on local expertise and the engagement of social forces at the grassroots. Lately the Chinese party-state is thus reducing some of its functions under the slogan "small state, large society" (*xiao zhengfu, da shehui*), turning over responsibility to the newly evolving social and political forces, and explicitly encouraging them to get involved in realms such as social service provision, poverty alleviation, education, health and environmental protection as "'partners' of the government."[4]

Induced by climate change and human factors, some of China's most pressing environmental problems are rapidly growing water shortages and wide-spread desertification that threaten the country's vast arid and semi-arid regions in the Western part of Northeast China, the Northern part of Northern China and the country's extended Northwestern regions. These aggravating problems are severely impeding the populace's livelihood and human security and have led to poverty and large-scale environmental migration. In recent years, China has been losing an annual amount of 3,400 square kilometers to deserts, causing an annual economic loss of approximately 54 billion dollars and affecting more than 170 million people (Zhang 2006; Chen 2009; Zhao 2000).[5] While the factors causing environmental degradation in northwestern China's arid and semi-arid regions are largely man-made, climate change is likely to intensify these problems in the future. It is thus of utmost importance that the Chinese party-state finds a sustainable solution to meet these problems.

Also in modern times, studies on the effects of global climate change on desertification have found a significant increase in both temperature and precipitation over the past fifty years that are likely to accelerate desertification and the problems associated with the fragility of the ecosystem of Northwestern China's arid and semi-arid regions (Wang et al. 2012, Chen et al. 2007, Zhao 2000). Wang et al. (2006) have further argued that two climatic indices, drift potential and the frequency of sand-driving winds, have had a significant impact on desertification in semi-arid China. Desertification might further be aggravated by the melting of mountain glaciers—by as much as 21 percent over the past fifty years—in Northwestern China (Zeng et al. 2008), which have been a major source of water. Zeng et al. (2008: 730) have thus termed climate change one of China's most significant future challenges, arguing that the country will be one of the world regions most severely impacted should the climate change as predicted (Smit and Cai 1996).

While natural and climatic factors are seen as central in the origination of desertification analyses of the relative role of climate change and human activities on Chinese desertification in the past decades have found that human activities are the major cause of desertification. Rapid population growth and consequent over-cultivation, over-grazing, over-reclamation of land, as well as the excessive cutting of trees and brush for fuel and utilization of water resources have led to sinking ground water levels, soil erosion and degradation, a significant reduction of the vegetation cover and thus the desertification of large tracts of land, the formation and activation of sand dunes and the intensification of sand and dust storms (Wang and Zhu 2001, Chen et al. 2007, Kang et al. 2004, Zhao 2000, Wang et al. 2012).

Since the 1970s, the Chinese government has undertaken comprehensive measures to combat these pressing ecological problems, including massive afforestation projects such as China's "Great Green Wall."[6] Centralized state measures have, however, been frequently criticized by both environmental experts and the affected populations to be lacking sustainability and being badly suited to the local situation. Environmental experts moreover criticize that Chinese environmental policy and poverty alleviation are thus far two largely separate policy domains that need to be combined to a "green poverty alleviation" strategy in order to generate sustainable results (Tan and Guo 2007; Yu 2007; Zheng and Qian 2004; Liu et al. 2002b). The engagement of social forces and the local populace in formulating and implementing state policies is thus of utmost importance in order to guarantee their effectiveness and contribute to finding long-term sustainable solutions.

This chapter takes a closer look at the case of Minqin County in northwestern China's Gansu Province and how many economic problems in the region are being impacted by changing environment—human and natural. Trapped between two fast-approaching deserts, the remote oasis region is one of China's most arid regions and severely affected by desertification, water shortages and poverty that have led to a large-scale migration from the area. Due to the severity of these problems and the region's specific geographic characteristics, the environmental problems of Minqin County have in recent years attracted major national attention. The region has not only been declared a priority national pilot region in China's combat against desertification and poverty, but has also attracted the engagement of a variety of social forces.

The following sections will first outline the origins of Minqin County's environmental degradation, their extent and impacts on human security, before turning to the state measures undertaken to combat desertification and poverty in the region and their shortcomings, which are representative of China's approach to solving its environmental problems. The fourth section looks at how social actors at the grassroots level are not only joining the party-state in its fight against environmental degradation and poverty, but also attempt to impact both regional and national decision making, and offering a more sustainable and greener poverty alleviation approach more attuned to the local situation and the populace's needs. The chapter will conclude with some policy prescriptions.

THE CASE OF MINQIN

Minqin County in Northwestern China's Gansu Province is an arid oasis region in the hinterlands of China's vast Northwestern deserts. The county is located in the Northeast of Western China's Hexi Corridor, part

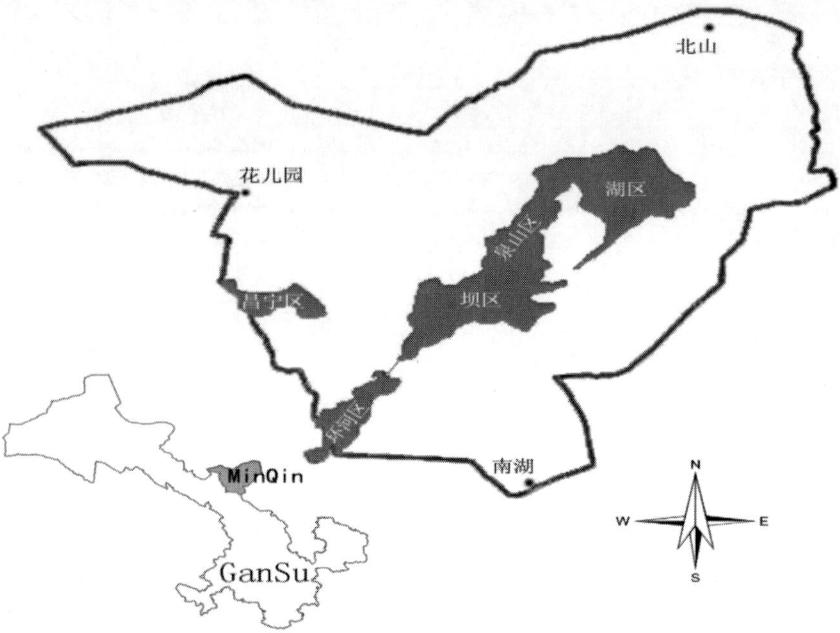

Map 5.2. Geographic Distributions of Minquin County and Minquin Oasis

of China's old Silk Road, and encompasses the Minqin Basin at the very lowest reaches of the Shiyang River drainage area. In the East, North, and West, the county is surrounded by the fast-approaching Tengger and Badain Jaran Deserts. The heartland of the county is Minqin Oasis, which makes up about 9 percent of the county's total area of 16,000 square kilometers. The county comprises eighteen townships and 249 villages with a total population of 3.15 million people. Minqin County is an agricultural region with a relatively small proportion of the second and third sectors. 85 percent of the county's inhabitants work in agriculture (Map 5.2).

Minqin County is one of the most arid regions, not only in China but in the world. Rainfall in the region is little, and evaporation very high. For many years, the average annual precipitation has been 110 millimeters, with an annual evaporation rate of 2644 millimeters, or twenty-four times the amount of precipitation. Sandy winds are frequent in Minqin, blowing across the county at an average of 139 days a year. The eco-system is very fragile, with 94.51 percent of the total area already turned into desert land (Ma et al. 2005a, b; Ma and Edmunds 2006; Ma et al. 2008).

Climatic changes and the excessive exploitation of the region's natural resources have led to the drying-up of most of the region's surface water, rapidly sinking ground water levels, the deterioration of the natural vegetation, desertification, and rapid salinization. Since the foundation of the People's Republic in 1949 and especially since the 1990s, the deterioration of the ecosystem of Minqin County has accelerated and the tension between the fragility of the ecosystem on the one side and population growth and economic development on the other have become more and more acute. This severely threatens the security, livelihood, and development of the local population and has led to poverty and large-scale migration from the region. If no timely solution for a sustainable development of Minqin County is found, the region will likely fall prey to the deserts.

The costs of letting Minqin be engulfed by the sands are high. For thousands of years, the oasis has prevented the merging of Northwestern China's three large deserts, the Tengger, Badain Jaran, and Wulanbuhe Deserts, by forming a natural green protective shield. The oasis thus plays a crucial role for Northwestern China's environmental stability. According to scientists from the Cold and Arid Regions Environmental and Engineering Institute in Lanzhou, the complete abandonment and dry-up of Minqin Oasis would have detrimental climatic impacts for the whole of China (Xue et al. 2005).

In 2001, the Hexi region with Minqin at its heart was identified as one of the four major bases of origin for the sand storms that frequently lash across Northern China towards Beijing (Chinanews 30.03.2001). These sand storms have gained in frequency and fierceness in recent years, and received major media attention at the time of the Olympic Games in 2008, when they threatened to impede the Games' smooth implementation.[7] The drying-up of Minqin Oasis would not only have a major detrimental impact on the environmental stability of Northwestern China, but also lead to a significant intensification of the storms.

The socioeconomic and ecological problems of Minqin County have thus attracted high-level of national attention. In 2001, China's Premier Minister Wen Jiabao visited the region and pledged, under no circumstances, to "let Minqin turn into a second Lop Nor" (*Yue bu neng rang Minqin chengwei di'er ge Luobubo*)—a once-blossoming ancient caravan town, which disappeared in western China's Taklamakan Desert after the nearby lake had dried up around 330 AD. The "rescue" of Minqin (*zhengjiu Minqin*) has since received major Chinese media attention and been declared a national priority by the party-state (Nanfang Daily 18.10.2009; People's Daily 31.10.2007; Guangming Daily 19.11.2003). The national government and the State Council have given explicit directions to research and solve the region's environmental problems, declaring Minqin a national pilot region to fight against desertification and poverty.

As part of China's national environmental protection and poverty alleviation programs, the county and local administrations in Minqin have in recent years been undertaking comprehensive measures to improve the situation, explicitly calling for support from local residents, the media and social actors. As in other regions not all of these state measures have been welcomed by the local population. Some of the measures have been criticized by environmental experts and the populace as short-sighted, not attuned to the local situation and prejudicial to the people's livelihood. Recently a variety of local and regional social forces have thus stepped in not only to support the government in its fight against desertification and poverty, but also to work on a more sustainable approach to solving the ecological and socioeconomic problems of China's arid and semi-arid regions.

Origins of the Environmental Changes in the Minqin Region

Environmental changes in the Minqin region primarily take the shape of water shortages and desertification having a detrimental effect on the region's fragile ecosystem. Factors responsible for these changes are mostly man-made, such as population growth, economic development and the excessive use of the region's natural resources. These have led to the drying-up of the area's surface water, and sinking ground water levels, accelerating the withering of the vegetation cover on large plots of land, desertification and the salinization of soil. They are also responsible for the deterioration of the ecosystem in the Minqin region and can be roughly divided into factors external to the region and factors originating within the region itself.

External Factors: Declining Water Supply From the Upper and Middle Reaches of the Shiyang River

Because Minqin is an exceptionally arid region with high evaporation levels, the region's primary source of water is the supply from the middle and upper reaches of the Shiyang River. Changes in the water volume of the river's upper reaches immediately impact upon the stability of the oasis. The Shiyang River originates in the Qilian Mountains in the Southwest of the Minqin Basin, bordering on Qinghai Province. The mountainous upper river area has eight major river flows. After emerging from the mountains, they enter the Yongchang-Wuwei Basin with the two major municipalities of Wuwei and Jinchang, where most of the water is used for irrigation, thus draining into the ground. In the Hongjishan border region, the water resurfaces as numerous spring water rivers that converge to the middle reaches of the Shiyang River, which cuts northwards through the Honya Mountains and enters the Minqin Basin at the river's very lowest reaches. Here, the river stretches and evaporates, gradually disappearing. Throughout history the water volume entering the Minqin Basin from the upper and middle reaches of the Shiyang River has rapidly declined due to two major reasons: decrease in the amount of water flowing down from the Qilian Mountains, and the expanding use of water resources in the upper and middle reaches of the Shiyang River.

According to the Hexi Annals (*Hexi Zhi*), more than 2,000 years ago the Qilian Mountains were covered by six million hectares of natural forests. By the time of the foundation of the People's Republic of China in 1949 the forest area was reduced to about 150,000 hectares. Since the early 1950s, logging, overgrazing, mining, and cultivation led to the clearing of nearly 1,500 square kilometers of forests and grasslands. Less than 550 square kilometers of the forests in water recharge areas have remained. The vegetation cover in the mountain areas has been reduced to 40 percent of its original and the leftover vegetation cover is heavily damaged. The water conservation capacity has thus significantly declined and the snow line is permanently getting higher (Ma et al. 2004).

The rising tree line and desertification of woodlands have resulted in a reduced water retention capacity, degraded buffer function, growing erosion areas and large amounts of sand and boulders being flushed down by the floods that silt up river beds, water reservoirs and channels. More than ten reservoirs in the mountainous upper reaches of the drainage area are, to different degrees, affected by siltation, which reduces their effective volume by one fifth to one eighth. The annual runoff fluctuation increased by 30 percent from the 1950s to the 1980s, with the result that some of the reservoirs do not have sufficient buffer to balance the runoff fluctuation of the rivers. Between 1972 and 2010 the Lenglong Mountain Range Glacier, where the Shiyang River has its source, has been permanently shrinking. In the last decade, the recession has accelerated. The shrinking glacier significantly reduces the supply of water to the Shiyang River, exacerbating the tense water situation in the Minqin region (Tian et al. 2012).

Further downstream, in the oases at the upper and middle reaches of the Shiyang River, the continuous expansion of the water volume diverted by irrigation ditches and the growing amount of water used by the permanently developing industry and agriculture have drastically reduced the amount of surface water reaching the Minqin region at the lower reaches of the Shiyang drainage area. While the annual runoff of the rivers at the Siyang River's upper reaches remained relatively stable at around 1200 billion cubic meters between the 1960s and the 1990s, the annual inflow at the Hongyashan Reservoir at the border of Minqin County declined dramatically by 63.11 percent from 2129.1 billion cubic meters in the 1960s to 1150.8 billion cubic meters in the 1990s (2012 Compilation of Economic and Social Development Statistics of Minqin County). The distribution of water use between the oases at the middle reaches and those at the lower reaches is thus heavily unbalanced.

In addition, the increasing efficiency of the irrigation system, the transformation of river beds into a channel system and the establishment of hydraulic storage and conduction facilities right where the rivers flow out of the mountains, have reduced the groundwater recharge level, thus leading to a great reduction of groundwater resources in the river's middle reaches. Fountain water resources in the Liangzhou area have dropped from 816.3 billion cubic meters in 1956 to 210.3 billion cubic meters in 1978 to 16.4 billion cubic meters in 1982 (Li et al. 2010). Since emerging fountain water used to make up a large part of the overall water that the river transported to the lower reaches, this development has directly aggravated the water scarcity at the lower reaches, leaving Minqin Oasis in a situation of severe water shortage.

Population Growth and the Excessive Use of Natural Resources at the Rivers' Lower Reaches

In addition to the drastic decline of the amount of water reaching the Minqin Basin at the lower reaches of the Shiyang River, also the growth of the population and the excessive use of the natural resources in Minqin itself have severely contributed to the growing pressure on the oasis' fragile ecosystem. Part of the problem lies in the political system itself. Since the evaluation and promotion of local officials in China depends, to a large degree, on increasing economic output rather than on improving the environmental situation, the local government has long desperately tried to preserve and expand Minqin's farming and agricultural production rather than opting for a sustainable management of the region's natural resources. Government-led cultivation and reclamation of land, deforestation and irrigation have thus significantly contributed to the degradation of the region's ecosystem.

The total population in Minqin has constantly grown between 1990 and 2008 by a total of 19.7 percent but decreased during the last years due to environmental migration from the region, which has slightly eased environmental situation. While the population working in the agricultural sector has been decreasing since 1998, the major ecological impact associated with the population development in Minqin has been the increasing need for cultivated land during the long period in which the agricultural population was growing. This has pushed the region's fragile ecosystem over the limits of its population bearing capacity. The short period during which the agricultural population has now been declining has not led to a sufficient reduction of cultivated land. Moreover, the nonagricultural population has also been constantly growing.

Diachronic analyses of the change of land use and vegetation cover in the Minqin region, drawing on remote sensing technology combined with GIS and correlation data, show that both the living space and the area of cultivated land have significantly expanded between the 1960s and 2005, and decreased again since 95 percent of the inhabitants of Minqin County live within the oasis, which makes up about 9 percent of the county's total area (Maps 5.3–5). Using remote sensing and Arcgis software, these figures

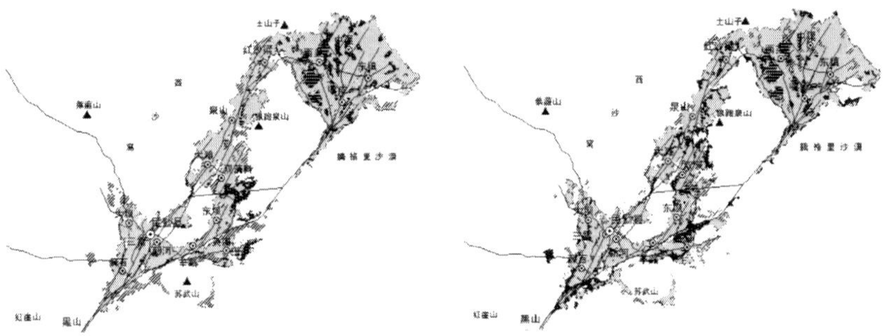

Map 5.3. Extension of Minquin Oasis 1960–1987

Source: Li 2008: 35–36.

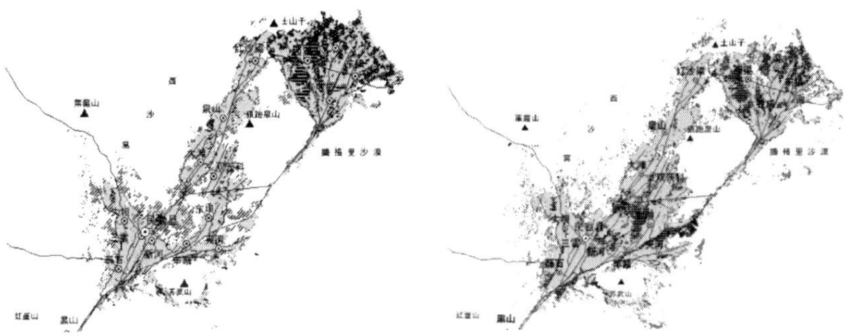

Map 5.4. Extension of Minquin Oasis 1987–1998.

Source: Li 2008: 35–36.

Map 5.5. Extension of Minquin Oasis 1998–2001[8]

Source: Li 2008: 35–36.

show that the oasis area doubled from 865 square kilometers in the 1960s to 16335 square kilometers in 2001, and remained stable in the last years.

The 1970s were a baby boom period for the region caused by an improvement of the living situation. The local population reclaimed large amounts of wasteland that led to the expansion of the outskirts of the oasis. The expansion stopped in the 1980s with the implementation of China's "Birth Control Policy," which led to a relatively stable population growth. In the 1990s, the oasis area continued to expand due to economic growth and a further improvement of the people's living standard. More desert land was reclaimed both along the brink of the Southern Ba District area and outside the oasis. Remote sensing shows that the area of cultivated land increased at an average rate of 2.21 percent per year between 1990 and 2005 (Li 2008: 35–36). At the same time, the environmental situation in the Northern Hu District area, located at the lowest reaches of the Shiyang River, deteriorated with intensifying water shortages. Much of the cultivated land in this area could no longer be irrigated and the vegetation on large tracts of land withered and died, leading to the shrinking of the Northern oasis area. Since 2000, the size of the oasis has remained relatively stable due to various state policies, and growing local awareness of the environmental perils associated with the excessive reclamation and cultivation of land and the associated increase in water use. Since 2008, the amount of cultivated land has thus decreased.

Population growth and the expansion of cultivated land have led to an increase in the amount of water needed for agriculture, industrial production and residential use. Since the water volume supplied by the upper and middle reaches of the Shiyang River has been dramatically declining since the end of the 1960s, the growing water need can only rely on ground water, leading to a further decrease of ground water levels and rising mineralization, thus aggravating desertification and the deterioration of the ecosystem. Figure 5.1 shows the declining water supply from the upper reaches of the Shiyang River and the growing area of cultivated land in Minqin between 1956 and 2010.

While this development was stopped several years ago due to various state activities, including a water transfer project, the exploitation of ground water continued. The construction of large-scale irrigation systems that use up ample water resources has put severe pressure on the ground water level. In the 1970s, due to expanded utilization of ground water motor-pumped wells were introduced in large numbers. By the 1980s, the number of wells in Minqin County had reached 8250. Since then, the tempo of development has somewhat slowed down (Yang 2003: 44–45). Nevertheless, the volume of exploited ground water has

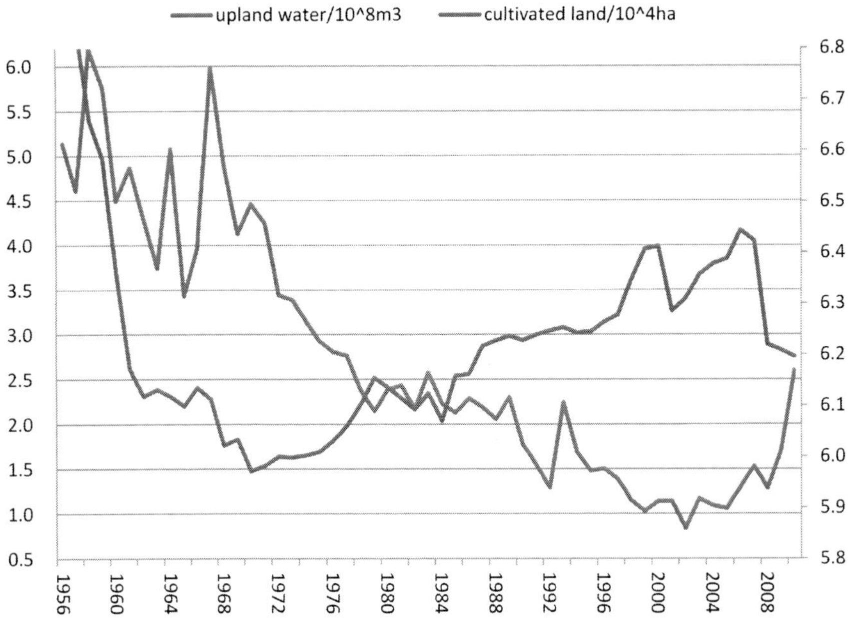

Figure 5.1. Water supply from the upper reaches of the Shiyang River and area of Cultivated Land in Minquin 1956–2010

Source: Authors' calculations.

increased tremendously from 118.5 billion cubic meters in 1955 to 617.6 billion cubic meters in 1977. By 2000 it had reached 1214.2 billion (Zhong et al. 2002: 10–11).

In addition, the improvement of the irrigation system and the construction of the high-standard Chenqi Channel in the late 1990s that transfers surface water directly from the Hongyashan Reservoir to the oasis have reduced the extent of exchange between surface and ground water. This has not only led to a further decline of the natural ground water level. The consequent shrinking of the river's natural drainage area has also led to a further exploitation of groundwater in the areas along the channel.

Moreover, the growth of the population has led to excessive grazing and wood-cutting thus further accelerating the depletion of the region's scarce vegetation cover and consequently its desertification. Minqin severely lacks energy resources. The seasonal temperature difference in the region is very high. The amount of energy resources needed for the population to stay warm during the biting winters is accordingly large. The local population traditionally relies on firewood for cooking and heating. With the growth of the population the amount of logged wood and brushes increased, enlarging the deforested and desertified land.

Environmental Change and the Deterioration of the Ecosystem in Minqin Oasis

Above factors have had a detrimental impact on the fragile ecosystem of Minqin Oasis, causing environmental changes in the region that severely impede the population's security and livelihood. The ground water level in the oasis is declining by one to three meters annually and has already sunk by 35 meters in some areas (Ma et al. 2005a). The sinking ground-water level, salinization of soil, over-grazing, and excessive wood-logging, have led to the withering and dying-off of large parts of the ground vegetation and a substantial reduction of cultivated land, aggravating desertification and the deterioration of the ecosystem. Since the 1950s, over 50 percent of the brush forests in the oasis and 44 percent of the grasslands have died off (Ma et al. 2005a, b, 2006, 2008; Xinhua 29.06.2007).

The decline of the vegetation cover at the fringe of the oasis has made the natural protective screen against the desert sands more and more porous, exacerbating wind erosion and the invasion of the desert. Since 1949 the construction of an oasis shelter system managed to keep the desert at the fringe of and within the oasis largely in place, and forest shelter belts have been built around about two-thirds of the farmlands in the county. However, the farmlands that still do not have shelter belts and those at the brim of the oasis suffer from severe harm caused by wind-blown sand. Moreover, due to the degeneration of the forest shelter belts and the dying-off of the vegetation cover that functioned as a green shield against the desert, the sand stabilization capacity has severely weakened and the fierceness of the desert expansion has increased significantly.

The area of shifting sand dunes crisscrossing the inner part of Minqin Oasis is 560 square kilometers. Sand dune activation and the dune's continuous convergence have led to the evolution of crescent shape dunes both along the brim of the oasis and within its inner parts. The sand is rapidly encroaching the farms at an average annual speed of 3 to 4 meters, peaking at 10 meters along the major wind gateways. In some areas, the sand dune movement has accelerated to more than 20 meters per year. Today, the total area of drifting sand that needs to be urgently administered in the county is more than 40,000 hectares about one-third of the cultivated land within the oasis is threatened by wind-blown sand. Since the 1960s, the total area of land that can no longer be cultivated due to desertification is more than 867 hectares. The cultivated area that has turned into desert amounts to 631 hectares and cultivated land covered by sand totals 520 hectares (Zhang 2001: 27–29). This increase in wasteland and desertified area further accelerates the process of desertification and permits the wind-blown sands to become even more devastating.

The desertification of land is also one of the major causes for the emergence of sand storms. The fierceness and frequency of sandy winds have significantly increased. The winds blow sand across the oasis on 139 days per year, amounting to full-grown sandstorms on thirty-seven days. Since the 1950s, the speed of these storms has nearly doubled to 17.5 meters per second (Ma et al. 2005a).

IMPACTS ON HUMAN SECURITY

The above environmental changes severely impede the security and livelihood of the local population. Most importantly, they have led to a typical vicious circle between poverty and environmental degradation.

Poverty

Minqin is primarily an agricultural county contributing more than 40 percent to the county's economy. The deterioration and downsizing of cultivated land in Minqin and the salinization of soil have significantly reduced the output of crops. In addition, also the sand storms have had a detrimental impact on the farmlands both by covering them with sand and through wind erosion. The fertile soil on the top layer of the farmlands is easily scraped off, and wind erosion can reach a depth of 1–10 centimeters, revealing the crops' root system or carrying away the young plants altogether. Also the sand particles carried by the storms severely damage the crops. Total crop failures that severely threaten the farmers' livelihood have thus been frequent in the region. The poverty level in Minqin County, particularly in the Hu District area of Shiyang River is already low and further declining. Since 1997, 1262 households with a total of 5853 people in the Hu District area have fallen into poverty, bringing the poverty rate to above 80 percent (Yang 2009: 19–26). The increasing rate of poverty has led to a further deterioration of the ecosystem. With declining income, more land had to be cultivated in order to guarantee the people's livelihood, further impeding the sustainable development of the region. Moreover, the development of the industry in the region has stagnated due to scarcity of water resources, and decreasing economic income. Traditional agricultural production with high cost and low water utilization efficiency rates due to the use of inferior technology is the dominant source of income. This poses a tremendous environmental burden, causing a further deterioration of the ecosystem that again intensifies the poverty in the region.

Migration

The deterioration of the environment and the lack of natural resources have led to a large-scale environmental migration from the region. A common saying in Minqin is that "in the whole world there are people from Minqin, but in Minqin there are no more people" (*tianxia you Minqin ren, Minqin wu tianxia ren).* While reliable data is almost unavailable it is estimated that migrants from Minqin are more than one million people (Li and Wei 2005: 293–294). Specially the young who leave the county to study and work outside the region.

The most important settlement regions are some districts in neighboring Inner Mongolia and Xinjiang Provinces and some places in Gansu Province. Full-grown "Minqin villages" have developed in some of these areas. In Qitai County in Xinjiang Province for instance, more than 80 percent of the 280,000 inhabitants come from Minqin. The numbers are most striking in the Hu District area: With the rapid deterioration of the living situation, 7,972 households with a total of 31,982 people migrated in 1990 (Li and Wei 2005: 289–292). Many schools in the county's Northern villages had to be closed down due to low student enrollment and decreasing population in the region. However the remaining population still exceeds the region's population bearing capacity. Moreover, the migration causes some severe environmental problems both within and outside the region. In Minqin, the homelands of those families who have migrated are engulfed by the sand and many villages have turned into permanent ruins, accelerating the region's desertification. Moreover, since large parts of arid Northwestern China have a similarly fragile ecosystem, the resettlement of environmental migrants in these regions causes a deterioration of their environment. Environmental migrants can thus easily turn into environmental fugitives. This has also been a major impediment for large-scale state migration projects, including the Minqin region.

Drinking Water Shortages and Health Impediments

The environmental degradation in Minqin has also led to severe drinking water shortages. The lack of clean drinking water is one of China's most pressing environmental problems. According to the World Bank, China has only one third of the average global amount of water per person. About 34 percent of the rural population has no access to clean drinking water (Chen 2009). In Minqin County, the depth of wells in some of the border regions of the oasis has reached up to 300 meters. The quality of ground water is rapidly declining due to mineralization and salinization. Also the water from the Hongyashan Reservoir, which completely dried up in 2005 for the first time, is heavily polluted due to industrialization, urbanization, and over-fertilization in the upper and middle reaches of the Shiyang River, particularly around the larger municipalities of Jinchang and Wuwei. The water is classified with the highest level of Chinese pollution standards and cannot be used for drinking (Ma et al. 2005a).

Together with the sandy winds, this is an impediment to the health of the local population. Epidemological research shows that sand storms are a serious threat to human health. The high density of inhalable particulate matter (IPM) carried by sand storms significantly increases the rate of respiratory diseases. Own research in the area has shown high levels of cerebral embolism, rheumatism, hemiplegia, high blood pressure, asthma and other respiratory diseases among the local population in 2012. Apart from a high average age of the remaining population, this can be contributed to the region's environmental deterioration.

WHAT IS BEING DONE: STATE MEASURES AND THEIR SHORTCOMINGS

Due to the severity of the ecological and socioeconomic problems in Minqin, which are representative for the wider problems of Northwestern China's vast arid regions, and because of the specific role of Minqin Oasis for the country's larger environmental stability, the case of Minqin County has in recent years attracted high-level national attention.

In 2001 China's former Prime Minister Wen Jiabao (2003–2012) visited the region and pledged that under no circumstances the government will "let Minqin turn into a second Lop Nor" (*Yuebu neng rang Minqin chengwei di er ge Luobubo*)—a once-blossoming ancient caravan town, which disappeared in Western China's Taklamakan Desert after the nearby lake had dried up around 330 AD. The 'rescue' of Minqin (*zhengjiu Minqin*) has since received major Chinese media attention, and been declared a national priority by the party-state (Nanfang Daily 18.10.2009; People's Daily 31.10.2007; Guangming Daily 19.11.2003). The national government and the State Council have given explicit directions to research and solve the region's environmental problems, declaring Minqin a national model county (*shifan xian*) to combat desertification.

As part of China's national environmental protection and agricultural development programs, the county and local administrations in Minqin have been undertaking comprehensive measures to improve the situation by soliciting the support of local residents, the media and social actors. According to government reports the ecological pressure on Minqin Oasis has somewhat relaxed and the situation is slowly improving due to comprehensive measures. However some of the state measures have been met with fierce criticism from environmental experts, the local population, and social actors, however.

China's Centralized Approach to the Combat against Desertification and Environmental Degradation

Three measures undertaken to combat desertification in Minqin represent China's centralized approach to environmental protection and the combat against environmental degradation.[9] First, in 1996, China adopted a "China National Action Program to Combat Desertification" to implement the United Nations Convention to Combat Desertification. The strategies at the core of the program include afforestation, dune stabilization, measures for the conservation and treatment of water and soil, and further measures for the regeneration of the vegetation cover. Since large parts of the desertified areas are located in China's poor regions the program also encompasses a plan for poverty alleviation with three central approaches: Educational measures about desertification combat among the local population, introduction of alternative methods of agricultural production, and a resettlement of the local population to regions with less fragile ecosystems (China National Committee for the Implementation of the UN Convention to Combat Desertification 1996).

Second, the "Three-North Shelterbelt Project" (*Sanbei fanghulin gongcheng*), better known as China's "Great Green Wall," a massive afforestation project that started as early as 1978 with the goal of planting a total of 1.7 million square kilometres of protective forests along China's Northern border regions by 2050. The project encompasses 324 counties, including Minqin, and details concrete afforestation targets for each county (Zheng and Qian 2004). The implementation of the project strongly relies on the engagement of the local population across Northern China, which is assigned with the plantation of the forest's trees and brushes.

Third, the fourth phase of the "Great Green Wall" project is a central part of China's "Great Western Development Strategy" (*Xibu da kaifa*), implemented since 1999 and targeted at the economic development of China's less-developed Western regions. The strategy is a prominent aspect of China's recent Five-Year-Plans (2001–2005, 2006–2010, and 2011–2015) and focuses on environmental protection, the fight against ecological degradation, and poverty alleviation as central priorities. Similar to the "Program to Combat Desertification," the "Western Development Strategy" encompasses various concrete strategies and project goals that are implemented by the county and local administrations. Core strategies include the conversion of farmland to forest and grassland, afforestation, the stabilization of sand dunes and measures for the sustainable use of water and energy resources. With regards to poverty alleviation, the program primarily slates resettlement and infrastructure projects (Western China Development Report 2008; Zheng and Qian 2004).

Implementation at the County and Local Levels

The measures implemented at the county and local levels largely follow the strategies outlined in national programs. Since 2003 central environmental measures include the large-scale afforestation of protective forest belts, sand dune stabilization, the conversion of farmland to forestry and grasslands, closure of wells, recovery of wetlands, introduction of water and energy saving technologies, and implementation of a water self-administration system through the foundation of agricultural water user associations on the village level, a restructuring of the local agricultural production, and a water transfer project and environmental migration.

Under the Green Wall afforestation project, the county plans to plant 250,000 hectares of brush forests and a green protective shelter belt of 342 kilometres along the oasis. According to the state news agency Xinhua, 10 percent of the trees have already been planted by 2007 (Xinhua 29.06.2007). According to county records 2,000 hectares of sand dunes have been annually stabilized by the population with wheat stalk and more than 500 wells were closed throughout the county preventing the further cultivation of 4,000 hectares of land.

A central feature of the state measures has been the restructuring of the local economy and agricultural production. With an agricultural output worth about 50 percent of the total GDP, of the country, and agricultural water consumption amounts to 95 percent of the total water consumption in the county. The restructuring of the agriculture is primarily based on changes in the proportion of land used for the production of grain crops such as wheat, corn, hops, and potatoes, and cash crops such as cotton and melons. The amount of water needed for the cultivation of cash crops is much smaller than the amount needed for the cultivation of grain crops. According to the 2010 Statistical Data on Minqin County's Economy and Development, the percentage of grain crops produced in Minqin County was reduced from 80.83 percent in 1980 to 29.9 percent in 2010 (1990: 71–4; 2000: 59–28), however the amount of cash crops increased from 10.44 percent in 1980 to 45.25 percent in 2010 (1990: 19.48, 2000: 30.28).

The production of water-intense crops was reduced and the cultivation of drought-resistant and more water economic crops such as cotton, fennel and sunflowers expanded. Moreover, large tracts of land have been converted from farmland to grasslands and the breeding of herbivorous livestock was vigorously promoted. Measures also included the improvement of the irrigation technology, the construction of new irrigation ditches and greenhouses to prevent evaporation and the introduction of energy saving technologies such as solar water heaters.

Also the Minqin Water Diversion Project implemented by the Gansu Provincial Committee and the Gansu Provincial Government is aimed at relaxing the tense water situation in the oasis. The project is designed to transfer an annual amount of sixty-one million cubic meters of water to the Hongyashan Reservoir along a 260 kilometers long transfer route. Construction of the project began in 1995 and the diversion channel formally started transporting water in 2001. In 2010 the amount of transferred water reached 653 million cubic meters. In 2007, the State Council further approved a "Shiyang River Basin focal administration plan" project.

As discussed above the population of Minqin County significantly exceeds the bearing capacity of the oasis. According to official statistics, about 34,000 people have left Minqin County between 2001 and 2007, both individually and as part of state migration projects. By 2020, 10,000 people are supposed to be resettled annually in regions with less fragile ecosystems. The number of people to be resettled from the total Shiyang River drainage area were 240,000 between 2008 and 2010. While the numbers are hard to verify, the county government has reported some success in recent years. Most importantly, the annual exploitation of ground water has markedly declined since 2004. In 2009 and 2010 the reduction in the exploited water volume was more than 100 million cubic meters. Also the mineralization of ground water has declined and the water quality has improved. The water volume reaching Minqin via the water diversion channel is annually growing and has contributed to an expansion of the area covered by ground vegetation, afforested plants have a slow recovery of the ecosystem. Despite the incessant fluctuation of the market price of cash crops, the income of the agricultural population is reportedly increasing. This is partly due to compensation money paid by the state to the inhabitants in the border regions of the oasis.

SHORTCOMINGS OF THE STATE MEASURES

These measures have been criticized by environmental experts and the populace as fragmentary, short-sighted, not attuned to the local situation, and in prejudicial to people's livelihood. A central criticism is that the combat against desertification and poverty are still two separate policy approaches and an integrative long-term strategy for Minqin is lacking. While various projects and state measures undertaken in recent years have resulted in large amounts of funding and technology, they each tackle specific problems. Without a comprehensive and more integrative approach, the vicious circle between poverty and environmental degradation cannot be broken, however, and measures against one problem continue to aggravate the other. The environmental projects severely lack in sustainability, and economic intervention can only produce temporary results.

Second, the state measures mainly focus on large-scale projects that do not sufficiently tackle the actual problems and needs of the local population. After the projects' conclusion, the results thus rapidly vanish since they cannot evoke popular support, but rather produce conflicts between the government and the local populace. Policies such as the conversion of farmland, afforestation, measures for the sustainable use of water resources such as closures of wells, and environmental resettlement projects often ignore the well-being of the local population and unintentionally reinforce their poverty, which in turn further aggravates environmental degradation. A central problem is that thus far no sufficient compensation system for those inhabitants negatively affected by these measures has been developed "[I]n reality, it can be perceived that, in the eco-environment construction, the right of development and well-being of the local people are often prejudiced" (Tan and Guo 2007: 22; Zheng and Qian 2004: 35–39; Liu et al. 2002b: 146). Zheng and Qian (2004: 35). Particularly the resettlement projects, which are at the core of the state measures undertaken in Minqin, have met with fierce criticism by local environmental experts and the population alike. Despite clear-cut guidelines for compensation payments these payments often only exist on paper or are too low to guarantee the livelihood of the resettled population. Also China's residency permit (*hukou*) system poses a serious impediment to the resettled inhabitants unless they obtain a residency permit at their new site of settlement. For instance, of 10,000 people slated to leave Minqin County to work outside between 2006 and 2020 under a work migration program, only 1,000 people are ensured a permanent residency permit outside the county. The number is much lower for those migrating on their own terms. The most severe problem for the large-scale environmental resettlement projects in Northern China goes beyond Minqin County. Usually they lack clear guidelines and suitable target cites. Since most of arid and semiarid Northwestern China has similar fragile ecosystems, the new settlement cites often do not provide a significantly better living situation, and after a few years their ecosystem are severely damaged due to the growing population. Environmental migrants thus easily become environmental nomads.

A third major criticism is that many of the measures in Minqin County are short-sighted and not attuned to the local situation. Both environmental experts from the region and the local population criticize the state measures as opting for fast results rather than a sustainable and long-term solution of problems. While the government has spent vast amounts of money on afforestation and the reclamation of land, environmental experts suggest that the only viable option for the regeneration of the local ecosystem in Minqin is to abandon at least parts of the oasis. As in other regions along China's "Great Green Wall," a large number of trees planted during the course of the large-scale afforestation projects have died in recent years, and lack any capacity to protect the soil, because these plants were ill suited to the local climate (cf. Cao 2008). Moreover, they increased the water demand in the region rather than relaxing the tense water situation (Jiang 2009; Bennett 2008). In order to solve the environmental and socioeconomic problems of Minqin County, a comprehensive and long-term strategy for the region's sustainable development is needed that is not driven by short-term success and economic output. Such a strategy requires integration of needs and grievances of the local population rather than focusing on large-scale top-down projects.

JOINING FORCES WITH SOCIETY: THE ENGAGEMENT OF SOCIAL ACTORS IN MINQIN COUNTY

In recent years, a variety of local and regional social forces have stepped in not only to support the government in its fight against desertification and poverty, but also to work on a more sustainable approach to solving the ecological and socioeconomic problems of China's arid and semiarid regions. Their engagement reflects the broader development in China towards the emergence of a nascent civil society, and the increasing engagement of social forces in the environmental realm.[10]

Faced with rapidly growing social and environmental problems, the Chinese party-state is in recent years turning over responsibility to the newly evolving social and political forces by actively encouraging them to get involved in social service projects, poverty alleviation, education, health and environmental protection. Social and nonprofit organizations are recognized as "'partners' of the government" (People's Daily 28.10.2001; People's Daily 31.10.2007; Chen 2001). However their number and scope of activities are still restricted by political environment and legal framework. Each organization negotiates its own niches for action (Saich 2000). Since 2003 in Minqin County several types of social organizations have become involved in these activities (China National Committee for the Implementation of the UN Convention to Combat Desertification 1996).

In 2003 the College of Earth and Environmental Sciences (CEES) at Lanzhou University became active in Minqin. It was dissatisfied with the "inadequate" measures undertaken by the government. The Institute's major objectives are the promotion of a sustainable environmental and poverty alleviation policy for China's arid and semiarid regions as well as local community development and promotion of local public participation. In cooperation with the international NGO Hong Kong Oxfam, the Institute runs various projects in different villages in Minqin County that are targeted to combat both poverty and further environmental degradation.[11] Now it is a major advisor to the government for the formulation and implementation of a sustainable development strategy for China's arid and semiarid regions.

In 2004 five Minqin residents directly affected by the region's socioeconomic and environmental problems founded the Save Minqin Volunteer Association (*Zhengjiu Minqin zhiyuanzhe xiehui*), a local grassroots association with the aim of organizing collective social action to improve the local living conditions, to protect the rights of the local population adversely impacted government measures, and to raise nationwide public awareness for the county's precarious situation. According to the Association's records it has 190 registered—and about fifty active—members, most of them Minqin natives. The Association is not officially registered due to problems in finding a supervisory institution and collecting the mandatory start-up capital. Its major activities involve afforestation projects in two Minqin villages, environmental education work at local elementary schools, and information platform effort targeted at raising the income of local farmers by providing market information and coordinating the sale of agricultural products from Minqin. The Association severely suffers from a lack of funding and local participation.

In 2006 two Lanzhou-based environmental organizations further took up project work in Minqin County. The province-level environmental organization Green Camel Bell (*Lü tuoling*), which was established in 2004 as the first province-level environmental organization in Gansu province, is active in Minqin with an environmental education project at local schools. The organization has more than 300 registered—and about thirty active—members and a nationwide reputation among Chinese environmental organizations. The organization is primarily financed by international funds.

A student environmental association the Lanzhou University Green Team (*Lanzhou daxue lüdui*), was founded in 1999, to impart a school environmental education program and an afforestation project in the region. The student association closely works with the above-mentioned local association for both projects. The Green Team has about 500 registered—and around one-hundred active—members, mostly Lanzhou University students. It also enjoys a nationwide reputation due to its active engagement in various national level environmental campaigns.[12] It's funding is derived from Lanzhou University and other national and international funds.

In 2006, the state-run newspaper *Lanzhou Morning Post* (*Lanzhou chengbao*) set up the "Worldwide Initiative for the Rescue of Minqin" (*Zhengjiu Minqin quanqiu xingdong*) in close cooperation with various local, regional, and national state run media institutions—as well as the Gansu Province and local administrations—in order to support the government in its fight against desertification in Minqin.[13] To launch the initiative, the participating media called on the public to participate in the "large-scale charitable initiative" (*daxing gongyi yundong*) focused on afforestation, and the collection of charitable public funds for Minqin. While the editorial of the *Lanzhou Morning Post* claimed that the initiative was socially-initiated, the composition of its Action Committee reflected close ties with various Gansu Province government departments and the Gansu branch of the China Charity Federation (*Gansu sheng cishan zonghui*), one of the so-called "governmentally-organized nongovernmental organizations," or GONGOs. The initiative is an example of the blurred lines between government and social action in China.

While some of the actors criticize lacking government support and a general distrust towards social organizations, both the organizations and the local and county governments nonetheless report overall good mutual relations. This is not surprising since a large part of the projects directly compliments the state measures and adheres to the official call for a stronger engagement of social actors. For example the organizations' afforestation and public service provision projects are subsidized by state funds or official institutions—including financial subsidies for the environmental education materials developed by Green Camel Bell and the Green Team. Water and seedlings for the afforestation projects of the local association and newspaper initiative, and poverty alleviation projects by CEES receive official financial support.

The organization's activities go beyond mere service provision. Their ultimate goal is to impact policy change both at the local and national level. As representatives of the people they aim to protect the interests and rights of the local population and try to engage the population in the local political process.[14] These organizations compile residents' needs and grievances through social surveys and pass them on to the responsible government departments to invoke policy change through deliberative means. Another local grassroots Association, Save Minqin Volunteer Association was founded in response to the grievances of Minqin residents negatively affected by policies in the region. According to this Association, many of the state measures have caused more harm than good. In an interview with the members of the association made the following observation:

> This is a typical case of only treating the symptoms. [. . .] The work of the government is targeted at fast success; they want to see immediate results. But the populace has a diverging opinion. [. . .] Thus, like in many other places in China, a conflict between the government and the people emerges.

With the help of students, the Association conducted a large-scale survey among the local population in 2008 to collect information on the residents' level of income and living standards and their understanding of, and attitude towards government policies and the restructuring of the local economy. The study was provided to the county administration hoping that its results would raise the official awareness on shortcomings of policies and be included in the future policy formulation process. In a similar fashion, the CEES conducted a comprehensive study on the state and origins of the water pollution in the Hongyashan Revervoir between 2003 and 2005, which was presented to the county government during various official meetings to illustrate the urgency of better control and prevention measures both for the Hongyashan Reservoir and for other water reservoirs along the Hexi Corridor. The CEES further compiled various expert reports and recommendations for a more sustainable resettlement policy based on original research.

At the local level organizations with close ties to the local administrations and state representatives, and those in the official position of government consultants, appear to be more effective in bringing about policy change, therein relying on a combination of formal and informal channels. All organizations reported receiving regular invitations to official meetings and forums on specific subjects in their area of expertise as well as their participation in national conferences. However, according to CEES, government bodies usually pay more attention to social needs referred in "scientific packages" with concrete recommendations than to the voices of the populace or volunteer associations. This is confirmed by the Save Minqin Volunteer Association, which reported that it did not receive any official reply from the Minqin County administration to its comprehensive survey on the local residents' grievances, nor had the critical measures been stopped or reduced at the time of research.

Opportunities are increasing for the active engagement of social forces in the environmental policy formulation and decision-making process. Some of the social actors in Minqin County achieved concrete results in advocating policy change, also on a supra-regional level. CEES advocates a more comprehensive sustainable development model for Minqin and other arid regions that is better suited to the local situation and based on the people's needs and grievances instead of the uncoordinated closure of wells, conversion of farmland and resettlement. The Institute (CEES) advocates for a fundamental restructuring of the local economy, a transformation of the technology of production, a comprehensive model for the sustainable use of Minqin's natural resources, and the resettlement of those parts of the population that exceeds the region's bearing capacity.

In recent years, the CEES have been increasingly engaged in the region's environmental policy formulation process and have become a crucial advisor to the government. Various projects based on the Institute's recommendations have been implemented together with the international NGO Hong Kong Oxfam and in cooperation with the County and local governments. These include the development of a comprehensive environmental poverty alleviation and sustainable development model in close collaboration with the local population and a resettlement strategy that is focused on the people's well-being by enforcing better residency permit (*hukou*), compensation and land right conditions. In March 2009, a contract was signed between CEES and the Minqin County administration, cementing the Institute's active engagement in the development of a sustainable regional migration and poverty alleviation policy with the prospect of turning Minqin County into a national pilot project for other regions with fragile ecosystems. Beyond Minqin, also the two environmental organizations, the student green team, and Green Camel Bell, have pointed to the political impact of national environmental campaigns, in which they played an active role.[15]

Moreover, understanding themselves as part of a budding Chinese civil society, the social organizations active in Minqin County started to form comprehensive regional networks, to join national alliances, and to participate in national environmental campaigns. Also in their Minqin projects they frequently cooperate and organize regular discussion forums on such subjects as Minqin desertification drawing local and regional organizations and media representatives. One of the organizations' central objectives is the promotion of regional "civil society structures." Green Camel Bell, and the Save Minqin Volunteer Association are engaged in "local NGO capacity-building" efforts, such as holding various training units for student associations, and regional social organizations, and organize frequent "NGO symposia" and social organization exchange meetings that attract the participation of over thirty regional organizations. In 2008 various organizations tackling the problems in northwestern China's arid regions joined forces formed a "Gansu Association for a Green Home and Desert Combating" (*Gansu lüse jiayuan shamo zhili cujinhui*), as an "alliance of welfare organizations." Its members—including the Save Minqin Volunteer Association and Green Camel Bell, as well as various national-level environmental organizations—implement joint research, develop collective strategies, and mutually support each other in the case of a shortage of personnel or project participants.

POLICY PRESCRIPTIONS

In recent years, China's growing environmental problems have become a serious threat to the legitimacy of the ruling Communist Party. For the party-state, it is thus of utmost importance to find a timely solution to meet these problems. Desertification and water shortages are two of China's most pressing environmental challenges that affect large parts of the country and severely impede the livelihood of the affected population.

For a sustainable solution of these problems, a shift away from the centralized approach to environmental policy making towards the effective integration of social forces and local expertise is needed. Rather than large-scale centralized projects and uncoordinated measures that tackle single problems and are targeted at short-term success and economic output, and which run the risk of further impeding the local people's livelihood, comprehensive strategies for the affected regions' sustainable development are needed that are attuned to the local situation, integrate the needs and grievances of the populace, guarantee their rights, engage them in the policy process and gain their support and cooperation. Most importantly, the vicious circle between environmental degradation and poverty has to be broken through an integrative "green poverty alleviation" strategy.

While Chinese social organizations have to adapt to the restrictive political framework, the engagement of the variety of social actors in Minqin County, their growing impact on local and supra-regional environmental policy formulation and their efforts to engage the local populace in the policy process mark first steps towards such a shift from the traditional centralized approach to environmental policy making towards a stronger integration of social actors and local expertise. A further development in this direction and the expansion of the responsibilities attributed to China's social forces are crucial for a sustainable approach to combating China's environmental problems.

NOTES

1. The authors thank Shenglan Wang, Xiaomei Ye, Qing Ma and Qiang Chen for invaluable research assistance.

2. In 2007, the Gini Index for China as a metric for social polarization, ranging from 0 (absolute equity) to 1 (all wealth in one hand), was set at just below 0.5. This is above the critical margin for social stability at 0.4 (Ru et al. 2008).

3. On China's ecological problems see among others Economy (2004), Chen (2009), Shapiro (2012).

4. Among others explicated in People's Daily (28.10.2001).

5. The Food and Agriculture Organization of the United Nations (FAO) (1997) sets the number of people adversely affected by desertification as high as 400 million. Numbers of the economic loss caused diverge.

6. The massive afforestation project slates the afforestation of a gigantic green shelterbelt planted along 4,480 kilometres of Northern China's border. Until 2050 the "green wall" is supposed to cover a total area of 1.7 square-kilometres.

7. According to media reports, a major sand storm in March 2002 lasted for fifty-one hours and dumped 30,000 tons of sand on the capital (CNN 21.03.2002).

8. The legend, in order listed, translates to: county government area, public road, rivers and ditches, non-oasis region, deforested region, oasis conservation region, oasis growth region.

9. An overview over China's policies for environmental protection and desertification combat is given in Chen (2009), Lu and Neilson (2004) and Jodha et al. (2002).

10. On the emergence of a Chinese civil society and the development of environmental organizations see among others Lu (2009), Schwartz and Shieh (2009), Xie (2009), Mertha (2008), Sun and Zhao (2008), Ma (2006), Ho and Vermeer (2006), and Ho (2001).

11. Local measures include providing the farmers with sheep from Shandong Province that have a higher fertility rate and can be held within the courts to prevent overgrazing, the construction of rain water reservoirs, afforestation and dune stabilization, environmental awareness and financial management trainings for the local population and the foundation of various local groups and associations including an environmental protection, a sales and management and a women's awareness group.

12. In 2005 the student association was listed as one of China's hundred most important social organizations (*quanguo baijia shetuan*) and one of its ten most important environmental organizations (*quanguo shijia huanbao shetuan*) by the Chinese government. In 2006 the Green Team received a prize from the Jane Goodall Intitute in the United States.

13. Media institutions involved include: the Beijing Morning Post (*Beijing chengbao*), Shanghai Youth Daily (*Shanghai qingnian bao*), the Chongqing Daily (*Chongqing shibao*), Shanghai Television, and the national broadcast institution China Central Television (CCTV).

14. This includes the promotion of social empowerment and democratic self-governance, as well as measures to increase government accountability and transparency at the local level. Concrete measures include capacity-building activities such as the foundation of social groups (including a democratic awareness and women's advocacy team), regular trainings for the residents, the active involvement of the local populace in project implementation and decision-making processes, and the introduction of regular meetings between popular and local administration representatives during which local political decisions are discussed with the residents.

15. These include the campaigns to save the snub-nosed monkey and the Tibetan antelope. Comprehensive studies of national environmental campaigns and their impact on national environmental policy were conducted, for example, by Mertha (2008) and Sun and Zhao (2008).

REFERENCES

Bennett, Karen 2008: "Swallowed by Sand: China's Billion-Dollar Battle against Desertification," World Resources Institute, at www.wri.org/stories/2008/08/swallowed-sand-chinas-billion-dollar-battle-against-desertification (17.12.2012).

Cao, Shixiong 2008: "Why Large-Scale Afforestation Efforts in China Have Failed To Solve the Desertification Problem," *Environmental Science & Technology*, 42, 6 1826–1831.

Chen, Yaning et al. 2007, "Effects of climate change on water resources in Tarim River Basin, Northwest China," *Journal of Environmental Sciences*, 19 (4), 488–493.

Chen, Gang 2009: *Politics of China's Environmental Protection: Problems and Progress* (Singapore et al.: World Scientific).

Chen, Guangyao 2001: "China's Nongovernmental Organizations: Status, Government Policies, and Prospects for Further Development," *The International Journal of Not-for-Profit Law*, 3, March 2001.

China National Committee for the Implementation of the UN Convention to Combat Desertification 1996: "China National Action Program to Combat Desertification (Abstract)," Beijing, at www.unccd.int/actionprogrammes/asia/national/2000/china-eng.pdf (01.10.2009).

Chinanews 30.03.2001: "Experts identify four major bases for sandstorms in Western China," at www.chinanews.com.cn/2001-03-30/26/82182.html (17.12.2012) [in Chinese].

CNN 21.03.2002: "Sandstorm blankets East Asia," at edition.cnn.com/2002/WEATHER/03/21/china.sandstorm/index.html (17.12.2012).

Economy, Elizabeth 2004: *The River Runs Black: The Environmental Challenge to China's Future* (New York: Cornell University Press).

Food and Agriculture Organization of the United Nations (FAO) 1997: "Drylands development and combating desertification. Bibliographic study of experiences in China," in *FAO Environment and Energy Paper (FAO)*, no. 15, Rome (Italy), Forest Resources Division, at www.fao.org/docrep/W7539E/w7539e03.htm (23.11.2010).

Guangming Daily 19.11.2003: "Do Not Let Minqin turn into a second Lop Nor" at www.gmw.cn/01gmrb/2003-11/14/07-D37D00B9556D186148256DDD0082FB46.htm (28.09.2009).

Ho, Peter 2001: "Greening Without Conflict? Environmentalism, NGOs and Civil Society in China," *Development and Change*, 32, 893–921.

Ho, Peter and Eduard B. Vermeer (eds.) 2006: *China's Limits to Growth: Greening State and Society* (Malden et al.: Blackwell Publishing).

Holbig, Heike 2009: "Ideological reform and political legitimacy in China: Challenges in the post-Jiang era," in: Gunter Schubert (ed.): *Regime Legitimacy in Contemporary China: Institutional Change and Stability* (Oxon, New York: Routledge), 13–34.

Jiang, Hong 2009: "China's Great Green Wall Proves Hollow: Tree planting damages environment in northern China," *The Epoch Times*, 29.07.2009, at www.theepochtimes.com/n2/china-news/desertification-in-china-20291.html (17.12.2012).

Jodha, Narpat S. et al. (eds.) 2002: *Poverty Alleviation in Mountain Areas of China.* Proceedings of the International Conference held from 11–15 November, 2002, in Chengdu, China (Feldafing: InWEnt gGmbH).

Kang, Shaozhong et al. 2004: "The impacts of human activities on the water–land environment of the Shiyang River basin, an arid region in northwest China," *Hydrological Sciences Journal*, 4 (3), 413–427.

Kassiola, Joel and Sujian Guo (eds.) 2010: *China's Environmental Crisis: Domestic and Global Political Impacts and Responses* (New York: Palgrave Macmillan).

Lam, Cory 2011: "12th Five Year Plan Hailed as 'Greenest FYP in China's History,'" *China Briefing*, at www.china-briefing.com/news/2011/04/05/12th-five-year-plan-hailed-as-greenest-fyp-in-chinas-history.html (27.11.2011).

Li, Ding 2008: *An Empirical Research on the System of Comprehensive Intervention of Ecological Poverty Alleviation in Arid Regions* (Lanzhou: Lanzhou University Press) [in Chinese].

Li, Hanlin and Fenzi Wei 2005: "Research Report on Minqin County Ecological Migration," *Northwest China and Development Forum Collection* [in Chinese].

Li, Mei et al. 2010: "Evolution of Water Supply and Demand in the Shiyang River Basin over nearly 40 Years," *Arid Land Geography*, 33 (4), 651–658 [in Chinese].

Liao, Jie 2007: *Research on the LUCC in Minqin based on Remote Sensing and GIS* (Lanzhou: Lanzhou University Press) [in Chinese].

Liu, Hongyan et al. 2002a: "Holocene History of Desertification along the Woodland-Steppe Border in Northern China," *Quarternary Research*, 57, 2, 259–270.

Liu, Jiyuan et al. 2002b: "Management and Use of Natural Resources for Poverty Alleviation in Mountainous Areas of Western China," in Jodha, Narpat S. et al. (eds.): *Poverty Alleviation in Mountain Areas of China.* Proceedings of the International Conference held from 11–15 November, 2002, Chengdu, China (Feldafing: InWEnt GmbH).

Lu, Ding and William A. W. Neilson (eds.) 2004: *China's West Region Development: Domestic Strategies and Global Implications* (Singapore et al.: World Scientific).

Lu, Yiyi 2009: *Non-governmental Organizations in China* (London, New York: Routledge).

Ma, Jinzhu et al. 2005a: *The Shiyang River Drainage Area: The Evolution of the Water Environment and Sustainable Use of the Water Resources* (Lanzhou: Lanzhou University Press) [in Chinese].

Ma, Jinzhu et al. 2005b: "The characteristics of ground-water resources and their changes under the impacts of human activity in the arid Northwest China: A case study of the Shiyang River Basin," *Journal of Arid Environments*, 61, 277–295.

Ma, Jinzhu and Michael W. Edmunds 2006: "Groundwater and lake evolution in the Badain Jaran Desert ecosystem, Inner Mongolia," *Hydrogeology Journal*, 14, 1231–1243.

Ma, Jinzhu et al. 2008: "Sources of water pollution and evolution of water quality in the Wuwei basin of Shiyang River, Northwest China," *Journal of Environmental Management*, 1–10.

Ma, Qiusha 2006, *Non-Governmental Organizations in Contemporary China: Paving the way to civil society?* (London, New York: Routledge).

Ma, Quanlin et al. 2004: "Environmental Problems of the Shiyang River Basin, its Root Causes and Rehabilitation Measures," *Journal of Safety and Environment*, 4 (5), 64–68 [in Chinese].

Mertha, Andrew 2008: *China's Water Warriors: Citizen Action and Policy Change* (New York: Cornell University Press).

Nanfang Daily 18.10.2009: "Wen Jiabao: Do not let Minqin and Dunhuang turn into a second Lop Nor," at china.nfdaily.cn/content/2009-10/18/content_6023475.htm (13.10.2009) [in Chinese].

People's Daily 28.10.2001: "Chinese NGOs to Work with Government for Poverty Reduction," at english.people.com.cn/200110/28/eng20011028_83359.html (14.12.2010).

People's Daily 31.10.2007: "Chinese government praises NGOs' role in making environment-friendly policies," at english.people.com.cn/90001/90776/90882/6293643.html (14.10. 2010).

People's Daily 28.10.2001: "Chinese NGOs to Work With Government for Poverty Reduction," at english.people.com.cn/200110/28/eng20011028_83359.html (17.12.2012).

People's Daily 02.10.2007: "Wen Jiabao makes on-the-spot investigation: Do not let Minqin turn into a second Lop Nor," at politics.people.com.cn/GB/1024/6340112.html (13.10.2009) [in Chinese].

Ru, Xin et al. 2008: *Bluebook of China's Society: Society of China Analysis and Forecast* (Beijing: Social Science Academic Press).

Saich, Tony 2000: "Negotiating the State," *The China Quarterly*, 161, 124–141.

Schwartz, Jonathan and Shawn Shieh (eds.) 2009, *State and Society Responses to Social Welfare Needs in China: Serving the People* (London, New York: Routledge).

Shapiro, Judith 2012: *China's Environmental Challenges* (Cambridge, Malden: Polity Press).

Smit, Barry and Yunlong Cai 1996, "Climate change and agriculture in China," *Global Environmental Change*, 6 (3), 205–214.

Sun, Yanfei and Dingxin Zhao 2008, Environmental Campaigns, in Kevin J. O'Brien (ed.): *Popular Protest in China* (Cambridge/London: Harvard University Press), 144–162.

Tan, Yan and Fei Guo 2007: "Environmental Concerns and Population Displacement in West China," paper presented at the 8th APMRN Conference, 26–29 May, 2007, Fuzhou, China, at apmrn.anu.edu.au/conferences/8thAPMRNconference/26.Tan%20Guo.pdf (01.10.2009).

Tian, Hongzhen et al. 2012: "The Relationship between Climate Change and Glacier Retreat over the last 40 Years in the Lenglongling Range of Eastern Qilian Mountains," *Research of Soil and Water Conservation*, 19 (5): 34–38 [in Chinese].

Wang, Tao and Zhenda Zhu 2001: "Some Problems of Desertification in Northern China," *Quarternary Sciences*, 21 (1), 56–65.

Wang, Tao et al. 2012: "The relative role of climate change and human activities in the desertification process in Yulin region of northwest China," *Environmental Monitoring and Assessment*, 184 (12), 7165–7173.

Wang, Xunming et al. 2006: "The relative role of climatic and human factors in desertification in semiarid China," *Global Environmental Change*, 16 (1), 48–57.

Western China Development Report 2008 (Beijing: China Statistics Press).

Xie, Lei 2009: *Environmental Activism in China* (London, New York: Routledge).

Xinhua 29.06.2007: "Minqin and Duanhuang implement the earmarked measures to combat desertification," at www.gs.xinhuanet.com/news/2007-06/29/content_10435701.htm (19.09.2009) [in Chinese].

Xinhua 18.10.2010: "The 12th Five-Year-Plan and Green Development," at news.xinhuanet.com/observation/2010-10/18/c_12670507_2.htm (13.02.2013) [in Chinese].

Xue, Xian et al. 2005: "The Desertification Status and Regional Development in the Shiyang River Basin, *China Desert,* 25 (5), 682–688 [in Chinese].

Yang, Guobin 2005, Environmental NGOs and Institutional Dynamics in China, in: *The China Quarterly*, 181, 46–66.

Yang, Yifei 2009: *A Study of Ecological Poverty in Minqin County* (Lanzhou: Gansu Agricultural University Press) [in Chinese].

Yang, Yongchun 2003: "Analysis of the Environmental Changes in Oases at Lower River Reaches in Arid Regions and their Reasons; A Case Study of Minqin County at the Lower Reaches of the Shiyang River on the Hexi Corridor, Gansu Privince," *Human Geography,* 18 (4), 42–47 [in Chinese].

Yu, Gengqiu 10.01.2007: "Green Poverty Alleviation," *Sina Finance,* at finance.sina.com.cn/g/20070110/15273234204.shtml (13.12.2012) [in Chinese].

Zeng, Ning et al. 2008: "Climate Change—the Chinese Challenge," *Science Policy Forum Sustainable Development,* 319, 730–731.

Zhang, Hucai et al. 2000: "A Holocene Climatic Record from Arid Northwestern China," *Palaeogeography, Palaeoclimatology, Palaeoecology,* 162 (3), 389–401.

Zhang, Chunxia 2006: *Challenging the Deserts: Stories of China* (Beijing: Foreign Languages Press).

Zhang, Haiyuan 2001: "Desertification and Countermeasures in the Oases along the Hexi Corridor in Gansu Province," *Gansu Agriculture,* 2001 (2), 27–29 [in Chinese].

Zhao, Xueyong 2000: "Desertification Research in China: History, status and challenge," paper presented at the Korea Long-Term Ecological Reserch (KLTER) Network Korea-China Seminar, Institute of Forest Science, Kookmin University, Seoul, 20 November, 2000, at www.klter.org/EVENTS/Conference00/html/Xueyong%20Zhao.htm (04.12.2012).

Zheng, Yuxin and Yihong Qian 2004: "Eco-Environmental Protection and Poverty-Alleviation in West China Development," in Lu, Ding and William A.W. Neilson, William (eds.): *China's West Region Development. Domestic Strategies and Global Implications* (Singapore et al.: World Scientific), 27–40.

Zhong, Huaping et al. 2002: "On Water Resources and Environmental Protection of the Downstream Shiyang River in Minqin," *Northwest Water Resources & Water Engineering,* 13 (1), 10–13 [in Chinese].

6

Climate Change

Challenges for Brazil

Eduardo D. Assad, António Márcio Buainain,
Hilton Pinto, Miguel Rocha de Sousa,
and Vanessa Duarte

The rapid economic development in Brazil has lifted millions of its people from poverty, and has propelled Brazil into global limelight along with other emerging economies of the world such as Russia, India, and China (BRIC). It has also started a lively debate as how to balance industrialization and urbanization with traditional sectors of its economy such as agriculture, desertification, and issues related to deforestation. Following discussion analyzes major areas of Brazilian economy and their impact on quality of human life in Brazil with particular reference to climate change.

AGRICULTURE

A recent (2012) OECD / FAO joint report on agricultural outlook estimates that global agricultural production needs to increase by 60 percent over the next forty years to meet the rising demand for food, as raw material to bio-fuel (ethanol) and other non-food products such as cotton, timber, wood, and cellulose paste. Most of this additional supply is expected to come from developing countries and the emerging economies including Brazil. Brazil still has a large amount of agricultureal land to be brouht into sustainable cultivation by entreprenurial local farmers. In the last three decades local Research and Development (R and D) institutions have devised technology to improve and increase tropical agricultural produce. However, the success of this optimistic scenario depends on several factors such as infastructure, institutional risks and uncertainties associated with climate change. Brazilian agriculture is shifting from subtropical regions of the south to the tropical areas of the Brazilian savannahs in the Middle West and North West where production is mainly rain fed. In spite of technological progress responsible for nearly 75 percent of output increase in the last twenty years (Gasques, et al. 2012), total productivity is still largely dependent upon the rainfall regime, notably quantity, distribution and frequency of rains throughout the seasons. These rainfall patterns are rapidly changing and will be affected by climate change in the near future. This could have tremendous impact on both local welfare and world food production.

In considering the impacts on local welfare one should pay attention that farming in Brazil is neither an enclave activity as it has been in the past nor itinerant as depicted by the wrong and misleading image of tropical farming as a slash and burn agriculture that leaves only ashes behind after a few years of high yields provided by unsustainable exploitation of natural resources. The new type of agriculture introduced in the frontier regions of the Midwest, North, and Northwest has occupied spaces hitherto almost empty, created many cities where millions of Brazilians now live and are engaged in activities which are mostly associated with agricultural production and agribusiness. This agro-industrial system has an immense destructive potential, and could lead to massive unemployment, migration, and loss of welfare. Of course this is a catastrophic scenario, but it is one that could be related to climate change. What is the expected impact of climate change on Brazilian agriculture, particularly on the tropical agriculture currently responsible for the booming Brazilian agribusiness? What are the possible consequences on the expected ability to actually feed the world and the Brazilian population? This is a dimension we explore using EMBRAPA's data.

DESERTIFICATION

The Brazilian Northeast semiarid has the highest population density in the world. It is also one of the poorest regions of the country, and has the largest poor population in Latin America. The process of desertification is also a fact, and should accelerate with climate change. In 2012 the region experienced the third successive year of drought, with severe social implications. Of course the drought is an old problem, but it seems to be aggravating in recent times, probably due to local climate changes and other anthropic factors. The problem of desertification is thus not new, but is getting worse. The scenario of having a new desert area in the tropics is not completely a science fiction. Following is an analytical presentation of data, related to strategies and policies dealing with desertification and climate change.

DEFORESTATION

So far most of the discourses and debates have focused on the rain forest deforestation. But there are other important dimensions of deforestation, mostly associated with urban life and activities rather than the outskirts of metropolitan areas in different regions of the country.

URBANIZATION

Climate change is also interacting with urban problems such as changes in rainfall patterns, unauthorized occupation of land, illegal and predatory urban soil use by the wealthy middle class and the poor alike. However, nowadays the occupation of hills, traditionally the location of the *favelas*, are led by middle class and rich families rather than by the poor. Migrations of people fleeing from climate change induced droughts in arid areas and caused health issues.

BRAZILIAN AGRICULTURE AND FUTURE CLIMATE SCENARIOS

Sustainability concerns over Brazilian agriculture are not recent and may be traced back to pre-climate change concerns. Although agriculture is only accountable for 7 percent of Brazilian GDP, agribusiness accounts for approximately one third of the wealth generated each year in the country. Agriculture and agribusiness have and will continue to have strategic roles in the development of the country.

Agriculture and agribusiness occupy 28 percent Brazil's total land area. In rural areas, agriculture employs 15 million workers and provides livelihood to people at least three times higher. Many of these areas, as in the case of the Centro-Oeste (Center-West) region are nowadays inhabited by millions in small and medium sized cities, with a standard of living higher than the national average. Without agricultural activities this standard of living would not be viable. It is to be noted that agricultural expansion came as a result of predatory, irrational, and indiscriminate deforestation, often disregarding the indigenous population and small pioneer farmers. However there is a growing awareness in society and in the public sector, of the need for a sustainable occupation, with rational use of natural resources, balancing preservation and development. It is being realized that policies detrimental to a large section of the population should not be supported.

Due to enormous social and economic importance of agriculture in Brazil, the linkage between agricultural practices and environment has attracted much debate and concern. In the beginning these debates emphasized only negative role on the climate. For example burning of tropical forests in the country's Northern region appeared as a major villain in global warming in the international press. Emissions from large metropolitan areas of the world, greenhouse gas emissions from cars and industries, high levels of consumerism, and waste that characterized the developed societies were viewed as coconspirators in the destruction of the Amazon. Extreme suggestions were made to internationalize Amazon territory. The Brazilian livestock industry, which has emerged as a new force in the global meat market, was also criticized for excessive greenhouse gas emissions.

More recently, the debate has become more balanced, based on scientific evidence, recognition of efforts many countries are making to change their development trajectories, and also the difficulties to reconcile climate change concerns with immediate pressures and needs, such as food production and wealth creation to alleviate millions from extreme poverty. In this context, Brazil is emerging as an active participant in government and social endeavors for the promotion of sustainable development and the reduction of primary causes of global warming. It has been promoting measures to avoid and mitigate the local effects of climate change.

AGRICULTURAL RISKS ZONING AND ASSESSMENT OF IMPACTS OF CLIMATE CHANGE IN THE BRAZILIAN AGRICULTURE

One of the main measures adopted to promote sustainable development was the introduction of agroecological zoning. In fact, zoning of agricultural risks was introduced as a public policy in Brazil in 1996, a year before the climate change issue appeared on the forefront of the debate. All 5,560 municipalities in the country have been zoned to cultivate suitable economically viable crops which had 80 percent probability to make profit. The zoning is based on the growth phases of each crop (phenology), drought stress, flood risk, and extreme temperatures at critical phases of crop growth. For instance, drought stress at flowering or grain filling (phenological phase where crop yields are defined) can significantly impact yields. Excessive rain at harvest time can ruin a crop. The incidence of extreme temperatures can cause the loss of production due to flower loss in the case of high temperatures or frost by low temperatures.

In 2001, EMBRAPA (Brazilian Agricultural Research Corporation) and UNICAMP (University of Campinas) developed a simulator to project the agricultural risks as a function of climate and soil (Affholder et al., 1997). This simulator reflects different soil, plants, and climate characteristics of different municipalities in Brazil. It provided advanced knowledge base of the agricultural geography of the country. In addition to information on crop needs, terrain characteristics, soil quality, and weather data, the zoning was further fine-tuned to include specific indices of sensitivity of crops to extreme temperature and moisture events during their critical growth phases based on seasonal agricultural calendars. For example, the crop risk indices are based on agro-meteorological water balance, calculated from crop evapotranspiration. Each crop has optimal soil moisture characteristics for optimal levels of photosynthesis, growth, and yield. Critical climate factors for this process are temperature and soil moisture that can be used to delineate the area in which any crop could be produced in Brazil with associated climate related risks.

By incorporating IPCC global warming scenarios (IPCC 2001, 2007), the projected temperature and any rainfall/soil moisture impacts can be introduced in the simulations on the basis of temperature and moisture risk indices for any given crop. The areas of lowest risk are those where there is no water stress, which guarantees seed germination and especially flowering which are critical to final crop yield. This risk must not exceed 20 percent.

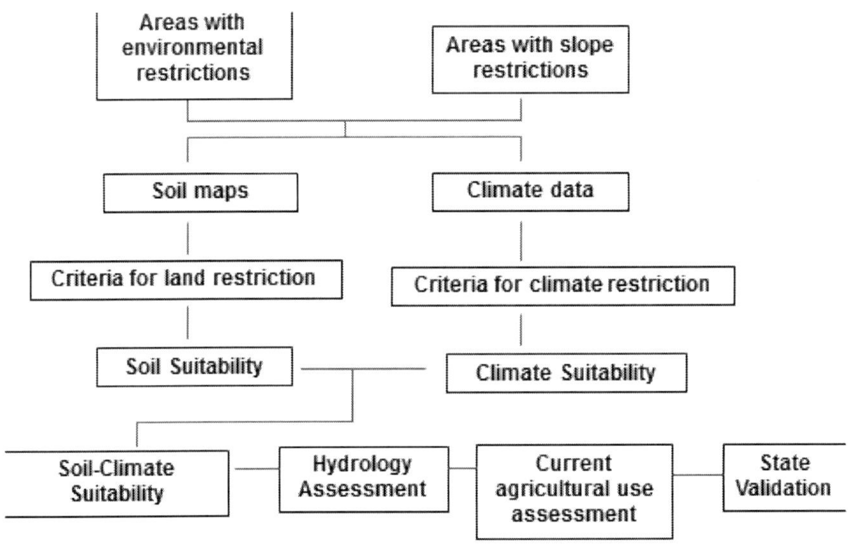

Figure 6.1. Agroecological Zoning Flowchart
Source: EMBRAPA SOLOS 2009.

The principles for determining climate risk are as follows:

a. Areas with the least risk are those that do not have a soil water deficiency that results in good germination as well as flowering and grain filling. This risk should not exceed 20 percent. The risk is based on an evapotranspiration index of the crops.
b. Using the above criteria, it is possible to assess the risk of planting any crop within Brazil. In the case of evaluation of impacts of climate change, in addition to soil moisture, the projected temperatures for 2020 and 2030 are also used to refine risk assessments.
c. The major advantage of the above approach relative to the previous studies is that each low risk agroecological zones are also screened for soil types, steep slopes, legal reserve area, riparian zones (APPs), indigenous areas, and protected areas thereby greatly increasing the precision of the estimates of crop productivity and likely climate impacts.
d. For current modeling efforts, the baseline for the crops planted, area planted, and value of production was 2009/2010.

The Agro Climatic Risk and Vulnerability Zoning Model (Assad and Pinto 2008) developed by EMBRAPA and UNICAMP currently underpins all financial lending to the agricultural sector in Brazil. The Central Bank of Brazil requires mandatory agricultural zoning throughout the country for access to rural credit. The EMBRAPA/UNICAMP model (Figure 6.1) indicates what, where, and when to plant a crop variety according to a zoning system. Three types of zoning are defined:

a. Agroecological—uses the database of soil, topography, climate, and the current land and environmental legal framework. This study used high resolution soil information on the scale of 1:250,000, vegetation, and terrain characteristics data sets and included all restrictions on the types of land use that can be mandated under Brazil's environmental legal framework to produce this high resolution map as a baseline for analyzing future climate change impacts and land use restrictions. This methodology was applied in sugar Cane agroecological zone in Brazil (Mamzatto et. al 2009) and was the basis for the Marin et al. (2012) work.
b. Agroclimate—based simply on climate information without evaluating the potential crop risk.
c. Climate risk—uses climate, soil, and crop culture by assessing the risk analysis taking into account mainly the information about rainfall, temperature and water balance of derivatives that indicate the deficiencies and surpluses of water for agricultural crops. Agro Climatic Zoning integrates crop growth models with refined climate simulations described above and uses a crop risk matrix based on a *state of the art* soil and land quality typology, weather data, crop water needs, and crop phenology.

The basis for the zoning is a crop water supply (Vulnerability) index (ISNA Index of satisfaction of the need for water), (Assad *et al.* 2001; Pinto et al. 2001; Affholder *et al.* 1997) based on the ratio of actual to maximum evapotranspiration per crop and is used to derive a crop risk and suitability zoning. The risk zones set for each municipality in the country indicate which of the nine major food and export crops that are at least 80 percent likely to provide an economically acceptable harvest.

Each variety of crop has a predefined set of climate conditions based on long-term research and field observations. The complete length of a crop cycle is divided into four phenological (growth) phases (Initial Development, Vegetative Growth, Reproduction and Maturity). The third phase is normally considered critical mainly due to the high sensitivity of flowering to dry spells and/or high temperatures. The length of each phenological phase is defined by degree-days or heat units. The incidence of extreme temperatures can cause the loss of production due to flower loss in the case of high temperatures or frost by low temperatures.

Soils are classified into three types—sandy, medium and clayey—or with low, medium, or high capacity for water retention. The crop coefficient (Kc) is defined according to the typical soil and is a measure of water consumption for each phase of the crop development. The ISNA values are based on the rainfall stations and estimated by a specific sowing period produced by the water balance for a fixed combination of soil type and phenological cycle (Figure 6.2).

IDENTIFYING CROP AREAS THAT ARE LESS VULNERABLE TO CLIMATE CHANGE IMPACTS

Since 2001, several measures/techniques were developed to evaluate the climate change impact on Brazilian agriculture. Assad et al. (2004) were first to evaluate the impact of climate change on the Brazilian coffee production, a crop which played an important role in Brazilian economic development. The plantation of coffee in Sao Paulo ceased but later suffered due to industrialization and problems associated with changes in climate. Zullo Jr. et al. (2008) suggested to move the production of coffee to the south of the country, a region with a climate traditionally deemed as too cold for coffee cultivation. While Pinto et al. (2008), using the Hadley Centre model proposed a different area for its production due the increase in temperature. The impact of climate change on agriculture was also assessed by Margulis et al. (2010), Santana et al. (2010), and Fernandez et al. (2012).

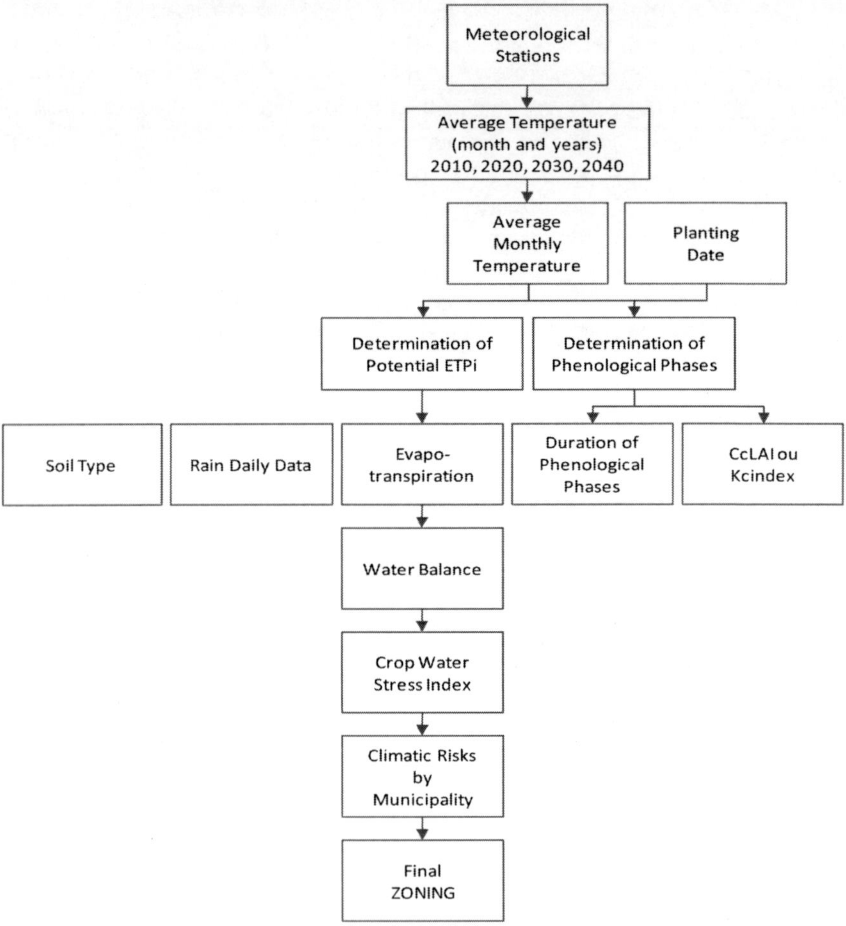

Figure 6.2. Flowchart of Components and Biophysical, Climatic and Plant Growth Processes Used

Source: EMBRAPA SOLOS (2009).

All these studies were to identify the effects of temperature changes in 2010, 2020, and 2030 vulnerable areas. Most of these studies used several regional and global models to compare the results to reduce uncertainties (Santana et al. 2010 and Margulis et al. 2010).

GLOBAL AND REGIONAL CLIMATE MODELS

Atmospheric general circulation models (AGCMs) are useful tools to represent the evolution of atmospheric processes at different time scales ranging from weather to climate, however their scope is quite limited. The typical spatial resolutions of AGCMs are a few hundred kilometers. Therefore AGCMs are not able to handle the large number of feedback processes occurring on sub grid scales controlled by local features such as topography, shorelines, vegetation, and lakes. These small-scale processes as well as sub grid turbulent heat and momentum fluxes cannot be analyzed in detail by AGCMs. The use of regional climate models makes it possible to deal with those scales. Such models can be used for climate simulations on decadal time scales and are able to take into account sub grid scale climate feedback mechanisms. Outside the domain of the regional model surface conditions, the global model generally provides the sea surface temperature (SST), ocean ice and three-dimensional atmospheric fields. During the last decade, the regional climate models with horizontal resolutions in the order of 10–20 km became available to interpret atmospheric conditions. The use of atmospheric or coupled atmospheric-ocean global models to investigate the actual and future climate has increased in the last few years. However, the results obtained from those models lack regional details. For example, at regional scale, precipitation and air temperature can be strongly influenced by topography, different land use type or proximity to the sea. This problem can be addressed through the dynamic downscaling technique by using regional climate models.

Based on previous works by UNICAMP and EMBRAPA following global models were selected for simulating the agriculture scenarios for 2020 and 2030:

CSMK3 (CSIRO-Mk 3.0)—Commonwealth Scientific and Industrial Research Organization considers four principal components of the climate system as basis for the future scenarios: atmosphere, land surface, ocean and marine ice. GIER (GISS-ER)—NASA Goddard Institute for Space Studies—USA.

INCM3 (INM-CM3.0)—Institute for Numerical Mathematics—Russia.

NCCCSM (CCSM3)—National Center for atmospheric Research—USA.

Based on the availability of data, three regional models were chosen (Figure 6.3).

PRECIS (Providing Regional Climates for Impact Studies). The system was developed by the Hadley Center in UK. The model was meant for South America and adjacent oceans. Previous work developed by EMBRAPA and UNICAMP (Pinto and Assad 2008) showed an excellent suitability for temperatures until 2050 but with problems in simulating rainfall.

ETA. The Brazilian Center for Weather Forecasts and Climate Studies (CPTEC) has been using the ETA model since 1996 to provide weather forecasts in South America. Due to its vertical coordinate system, the ETA Model has been able to produce satisfactory results in regions with steep orography such as the Andes range. The CPTEC GCM forecasts in comparisons with ETA showed that the model provided considerable improvement over the initial model. The assessment of the ETA Model's seasonal forecasts climatology showed that in general the model produced additional useful information over climatology. The ETA Model exhibited better results in simulations of upper- and lower-level circulation and precipitation. The ETA Model was developed at Belgrade University and operationally implemented by the National Centers for Environmental Prediction.

BRAMS—Brazilian Developments on the Regional Atmospheric Modeling System. The BRAMS model is based on the Regional Atmospheric Modeling System—RAMS—with specific parameterization for the tropics and subtropics. The model has a complex set of modules to simulate processes of radioactive transfer, water and heat exchange between surface and atmosphere, microphysics of clouds and turbulent transfer in the boundary layer. The main characteristic of the model for Brazil is its ability to estimate regional rainfall. Several tests are being conducted by EMBRAPA/UNICAMP researchers to compare the results of rainfall regional simulation for the actual conditions as well as for the years 2020 and 2030.

It was observed that the mean monthly temperature values of each model were very close. The same behavior registered in the Brasília Station has also been found at other stations, thus validating the use of this information for the creation of scenarios "pessimistic" and "optimistic," respectively, for the years 2010, 2020, and 2030 (Figure 6.4) show one of the results of the station—Brasilia—Federal Capital—to the temperature calculated by all models. Similar trends are observed with a small deviation between the estimated temperatures of all models (DFUNBFAL) is a Federal district, Brasilia University, (Fazenda Agua Limpa Meterological Station).

These scenarios were used to assess the impacts of climate change on nine most important crops of the country (Table 6.1).

These trends are similar to studies by Margulis et al. (2010) and Santana (2010). In terms of biomass energy potential production, global warming has a positive effect on sugar cane production in Brazil in terms of food security. These results are worrisome both for Brazilians and world food security. In Brazil the poorest Northeast region will be very severely affected by expected global warming because the Northeast region demonstrates several unfavorable conditions such as serious desertification of large areas, high rate of poverty, and rural population which depends on subsistence agricultural production to feed itself and to generate income to sustain a miserable livelihood, and high rates of illiteracy, thus render-

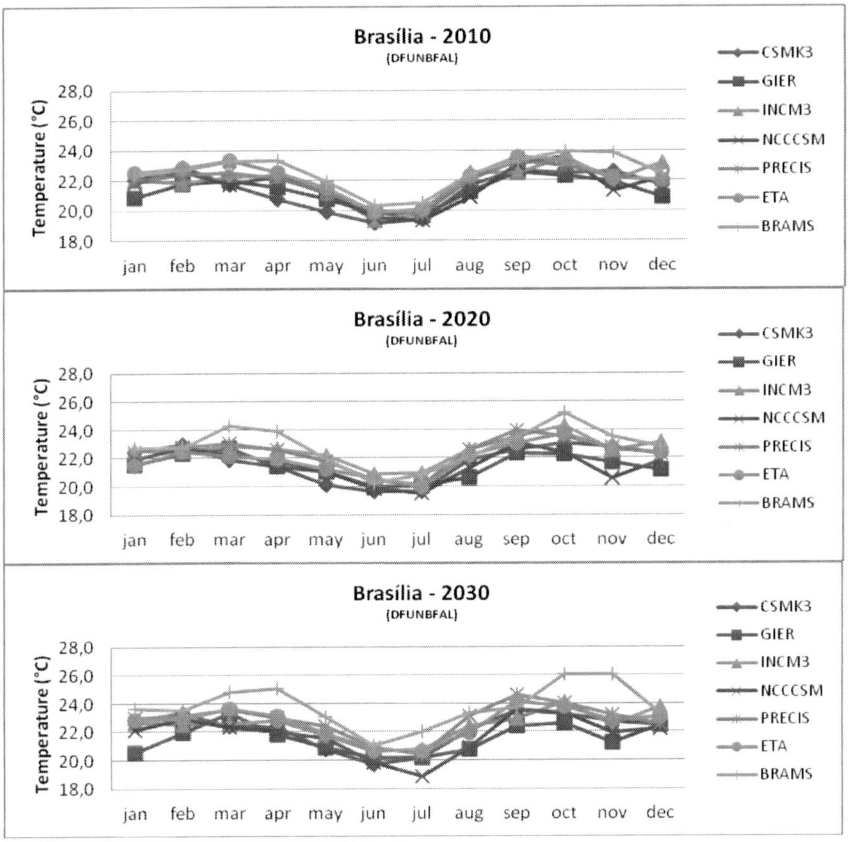

Figure 6.3. Results—Federal Capital, Brasilia—to the temperature calculated to all models

Source: Fernandez et al. (2012).

ing more difficult the alternative of sustainable migration. These implications go beyond Brazil as it will not be able to provide additional food to meet the global food deficit (OCDE/FAO 2012).

Following are the main impacts of climate change on Brazilian food production in the next thirty years:

Global warming could jeopardize food production, leading to 28 percent loss of soybean production, and 22 percent of maize production.

Soybean and bean are expected to be the most affected crops.

Maize, rice, beans, and cotton will suffer extreme reductions in their low risk areas in the Northeast, with a significant fall in production.

Sugarcane production could double in the forthcoming decades.

The global warming could reduce the pasture productivity by 38 percent.

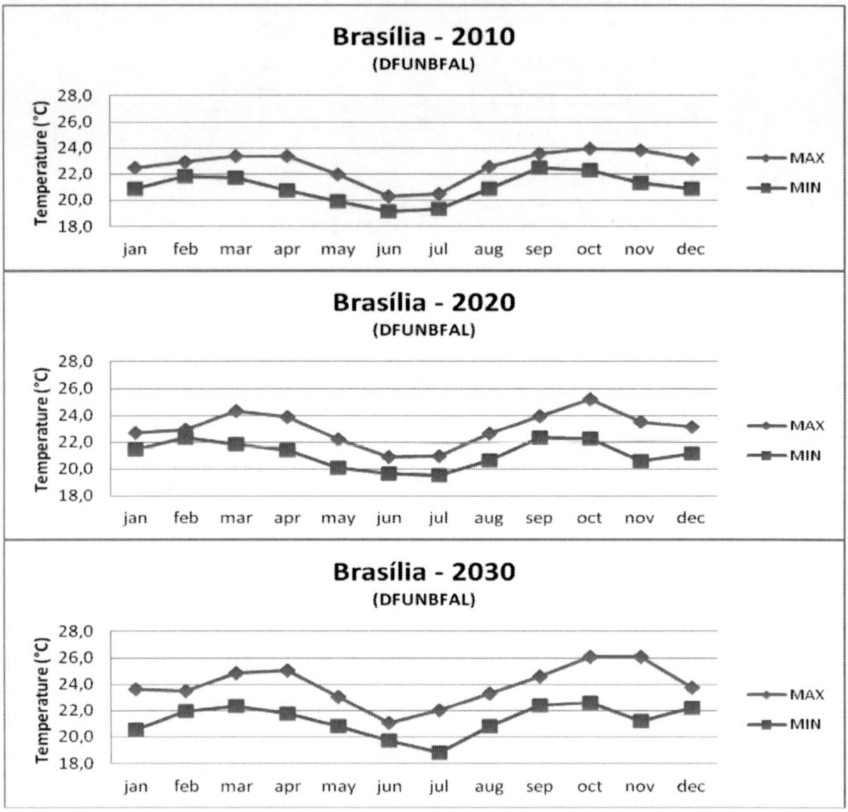

Figure 6.4. Monthly maximum and minimum temperatures estimated from seven models for the weather station "DFUNBFAL" located in Brasilia, DF, Brazil
Source: Fernandez et al. (2012).

This pessimistic scenario presents the worst results of climate change without any technological mitigation reaction. Fortunately this is not the case as a series of medium and long-term solutions to mitigate the problem have been already adopted in 2010 and are currently being fully implemented. One of the pillars of the national strategy in Brazil is the adoption of practices to prevent the advance of deforestation to allow the incorporation of new arable land into production. In addition a series of regulatory, vigilance, and control and punitive measures to prevent further unlawful deforestation are being undertaken. The strategy also includes financial support to alternative production systems in forest zones as well as R and D investments to sustain the increase in Total Factor Productivity (TFP) as the basis for agricultural growth.

It is particularly noteworthy that the Low Carbon Agriculture initiative's main aim is to provide incentives to adopt current technological

Table 6.1. Impacts of Climate Change on Brazilian Crops—Optimistic and Pessimistic Scenarios for 2020 and 2030

Crops	2020		2030	
	Optimistic %	Pessimistic %	Optimistic %	Pessimistic %
Cotton	-4.6	-4.8	-4.6	-4.9
Rice	-10	-7.4	-9.1	-9.9
Sugarcane[1]	107	101	108	91
Soybean	-13	-24	-15	-28
Rainfed wheat	-41	-15.3	-31.2	-20
Bean (summer season)	-54.2	-55.5	-54.5	-57.1
Bean (autumn season)	-63.7	-68.4	-65.8	-69.7
Maize (summer season)	-12	-19	-13	-22
Maize (autumn season)	-6.1	-13	-7.2	-15.3
Pasture[2]	-34.4	-37.1	-34.9	-38.3

[1]Sugarcane includes potential (new) areas not just current areas of production
[2]Pasture value = productivity.

Source: Fernandez et al. (2012, p. 41).

practices to more environmental friendly techniques and also to improve and upgrade the use of degraded pastures both for more intensive cattle raising and crop production. There are approximately 60 million hectares of degraded pasture areas in Brazil today. Some of them are quite suitable for agriculture and if properly explored, can sustain the need for additional land for crop production without further deforestation. There are other measures such as investments and adoption of production systems to capture and store carbon dioxide from the atmosphere. Some technical guidelines and alternatives, such as the integration of livestock and crop production, which are highly feasible on both financial and operational grounds are part of these strategies. Attempts are being made to use agriculture-forest or agriculture-forest-livestock systems whose initial results allow optimistic projections regarding massive adoption in many areas of the country. Increased adoption of the no-tillage system, and reduction of nitrogen-based fertilizers are also being considered.

Other measures include the development and introduction of drought resistant seeds and species—which may help the Brazilian Northeast—to increase the resilience of established production systems to foreseeable climate changes.

Along with studies of mitigation actions, scientists are also developing genetic variants of soybean, maize, beans, coffee, cassava and some fruits that are more tolerant to high temperatures and droughts. It should be emphasized that even if these measures provide more resistant plants there is a limit about how far genetic improvements can go. These measures are capable of handling the problem only to a 2°C temperature rise, hence the actual scientific knowledge is not enough to establish a long-term adaptation strategy. One of the alternatives to this limitation can be the so called "second generation of genetically modified crops." Instead of just being herbicide—tolerant or pest resistant, like the GM crops that exist today, these will be more suited to severe environmental conditions.

CLIMATE CHANGE, DESERTIFICATION, AND POVERTY IN THE BRAZILIAN SEMIARID REGION

Northeast Brazil's semiarid region has been, throughout the country's history, "the problem region." It is characterized by economic, social and political underdevelopment, and massive endemic poverty. Despite Brazil's strong economic growth and improvement in social indicators, no other region depicts the image of "two Brazils" (more than a century ago Euclides da Cunha described in *Os Sertões* literally "The wildernesses," translated as *Rebellion in the Backlands*, in this masterpiece of Portuguese language) as does the northeastern region of the country. It is also a "land of contrasts" as described by the French geographer Roger Bastide in 1938 (Box 6.1).

In fact, the Brazilian northeast is a region rich in natural resources. It was in its fertile and moist soils that the sugar production emerged in 1530, as part of the first cycle of Brazil's colonization. For centuries this region has absorbed large population in search of land and has been the breadbasket of the sugar economy supporting new cities. It was in the northeast that great fortunes were made by small "sugar aristocracy" and cronies however majority of its people suffered from extreme poverty and socioeconomic deprivations not due to its periodic droughts but because of extreme exploitation of its land and people.

The Brazilian semiarid region extends along almost one million square kilometers (969,588 square km), encompassing nine states and 1,133 municipalities, and 11.3 percent of the total Brazilian territory. Its edaphoclimatic characteristics are similar to other semiarid regions of the world. It receives an average rainfall between 250 and 800 mm per year, concentrated in three to five months with high evaporation (between 2,000 and

2,800 mm/year) resulting in drought conditions for the most part of the year. It has 22 million people (11.8 percent of the Brazilian population in 2010), which makes it the most populated semiarid region in the world: 22.7 people per square kilometer. Most of its territory is covered by *Caatinga*, a biome in a state of advance degradation, caused by slow economic growth and general poverty, both in rural and urban areas. During the last three decades, the desertification has been advancing, with very little attention by policy makers.

Historically, its economy was based on livestock, and an extensive farming system combining animal husbandry, subsistence crops, cotton, sisal, and other cash crops. More recently, this situation has changed, both in rural and urban areas. Several systems of irrigated agriculture have been developed: beef cattle, and dairy industries have become more intense, due to easy availability of cactus pear as feed; fruit production has increased and the *Caatinga* cultivation has been intensified. Several cities have attracted labor-intensive manufacturing industries, commerce, and general services contributing to population growth, putting more pressure on the fragile semiarid ecosystem (Buainain and Garcia, 2013).

During recent decades, successive governments, and mostly the local people have been attempting to promote sustainable development in the region keeping in view the unique local conditions of the region. Success of these efforts has been slow and limited. On one hand, public policy actions are not always consistent or robust enough to face these problems. On the other hand, both social needs and short term economic pressures have been forcing the adoption of policies that are increasing population density and anthropogenic pressure on natural resources, which are opposite to the macro-strategy of coping with drought and semiarid conditions. For example, the controversial project to divert water from the São Francisco River and the deployment of hundreds of agrarian reform settlements in the semiarid region have not met with much success. Nevertheless, the regional social indicators have been improving lately and the social welfare network has been relatively efficient in mitigating the worst effects of droughts that, in the past, used to generate endemic famines and massive migrations, in this "land of passage" (Josué de Castro 1946).

In view of the above discussion and scenarios it seems that climate change may have significant impact on the Brazilian semiarid region, accelerating desertification of the *Caatinga* biome (Oyama and Nobre 2003) and making it difficult to improve the lives of ordinary people.

A study on climate change in the state of Pernambuco showed an increase of 4°C in maximum temperature, and a decrease in 275 mm of yearly rainfall between 1961–2009. Besides rainfall decrease, drought periods have increased from twenty days to thirty-five days, contributing to more evaporation and water loss (Lacerda and Nobre 2010; Lacerda et al. 2010). All predictions point out negative impacts on the Northeastern climate, particularly in the semiarid region, where drought intensity and frequency is expected to increase, together with the replacement of *Caatinga* by arid vegetation, the decrease of arable land (in terms of soil and water) and the decrease of water availability in general (Nobre 2011: 29).

In the absence of strong preventive measures the rainfed farming will become unviable with "impacts on the northeastern society, particularly on subsistence farmers in the semiarid first, and then to everyone else" (Nobre 2011: 29). It is not possible to ignore the 1.6 million families with small tenures throughout the semiarid region. They already live in poverty and, are most affected by environmental degradation and climate change.

Based on climate models from the National Institute for Spatial Research (INPE), a recent study (CEDEPLAR/UFMg and Fiocruz 2008) has mapped and analyzed the most important social and economic consequences of climate change in the Northeast, with particular emphasis on the semiarid region. The study developed a climate change vulnerability indicator—the General Vulnerability Index (GVI)—which integrates population growth estimates, climate change effects on productive activities, different scenarios for temperature variation, effect of greenhouse effect gases and migration due to climate change, and their impact on population's health. The GVI is a combination of four indicators of the most important dimensions of the impacts of climate change: (i) Health Vulnerability Index; (ii) Desertification Vulnerability Index (IVD); (iii) Economic-demographic Vulnerability Index (IVED); (iv) Healthcare System Cost Vulnerability Index (IVC).

Several important predictions point out to a 11.4 percent decrease in Northeastern GDP growth rate; a reduction in arable land by 72 percent (79 percent in the State of Ceará, 70.1 percent in Piauí, 66.6 percent in Paraíba and 64.9 percent in Pernambuco); a 24 percent increase in migration rate between 2030 and 2050; child vulnerability to malnutrition and diarrhea deaths, chronic and degenerative disease recrudescence in the elderly population, with costs for the public healthcare system.

The effects of climate change are expected to be particularly severe on the economy, mostly on arable land areas, which will affect more intensely small family farms dependent on the commercialization of their crops and livestock products. Family income is estimated to drop by 11.9 percent until 2050. The effects are expected to influence not only consumption, but also family capacity to adapt to these changes. Poverty and low education levels would also force people to migrate to urban areas for jobs and other economic opportunities.

As mentioned above the region most affected by the climate changes is the semiarid hinterland in the Northeast. With the risk of desertification, the region will become unsuitable for the majority of the crops presently farmed there, especially bean and maize. To contain this process and also offer alternative food supplies for the population and the livestock, numerous researchers have maintained that the solution lies within the *Caatinga* (specific natural vegetation in the semiarid region) crops. The idea is to use the knowledge about more drought-resistant plants and to stimulate a wider range of crops. Various species of the biome are more efficient as forage plants, than the exotic grasses implanted in the region. The problem is that producers keep planting some crops which are not suited to the area instead of planting maize, rice, beans, and soybeans. It is possible to cultivate natural plants like mandacaru (*Cereus jamacaru*), xique-xique (*Pilocereus gounellei*), and sorghum, or some leguminous plants, like catingueira (*Caesalpinia pyramidalis*), jurema (*Mimosa tenuiflora*) and angico (*Mimosa* sp.). Fruits like umbu (*Spondias tuberosa*), siriguela (*S. purpurea*), cajá (*S. mombin*), quixabeira (*Sideroxylon obtusifolium*) and wild passion fruit can also be produced in large scale. There is also a rural species of natural color cotton, similar to the commercial cotton, which could easily compete with the ones in the market. Some of these alternatives have been known for years, but have been ignored due to lack of incentives.

DEFORESTATION IN THE AMAZON AND CLIMATE CHANGE

Deforestation, particularly in the tropical forest, is one of the main concerns in the debate on climate change and global warming, mostly because the consequences of deforestation transcend the simulations by most mathematical models. In fact, not even the most catastrophic models, presented in some tabloid press, are able to clearly indicate the consequences to the human kind of the loss of biodiversity or the emergence of new deserts in the American continent. While nowadays deforestation is not the main global source of greenhouse gases emissions, in the medium to long term the preservation of tropical forests is a strategic priority both for the countries where these forests are located and for the rest of the world.

While in the industrialized countries most greenhouse gas emissions come from burning fossil fuels, in Brazil the main source of emissions is due to a change in land use practices, particularly the conversion of forests into farms. The National Inventory of Anthropogenic Emissions and Removals of Greenhouse Gases, as of 1994, pointed out that 75 percent of Brazil's CO_2 emissions are from "changes in the use of land and forests." In 2010, however, these sources of emissions fell to 61 percent of the total. Although this is an undeniable progress (–14 percent), this is not yet a comfortable situation. Amazon, and Cerrado, deforestations are the main sources to blame for gas emissions.

In December 2009 the Brazilian government presented in Copenhagen the national voluntary commitment to reduce emissions. Towards the implementation of the National Policy on Climate Change, Brazil assumed mitigation actions designed to reduce between 36.1 percent to 38.9 percent greenhouse gases emissions by 2020. The commitment includes the reduction of emissions from deforestation in the Amazon by 80 percent by 2020. This represents an annual deforestation rate of nearly 3000 km^2 (Table 6.2; Figure 6.5; Table 6.3).

Since Copenhagen (December 2009), the major national climate change efforts have been to reduce Amazon deforestation, by consistently monitoring deforestation through use of remote sensing techniques. Since 1996, the National Institute on Spatial Research (INPE) has been in charge of monitoring deforestation. It reports annually on the deforestation rates (www.obt.inpe.br/prodes/index.php). The most recent results show the annual deforestation rate is getting close to the target set at Copenhagen.

The INPE, EMBRAPA and IBAMA completed a study to categorize Amazon deforestation into several classes. It concluded that livestock and animal husbandry occupy most of the deforested areas, and that a more efficient use of land by livestock would be enough to release land already cleared for expansion of crops. The data supports the assumption that the increase in efficiency in livestock farming in the Amazon and the regeneration areas with pasture and secondary vegetation will release 20 million hectares, therefore will not need further deforestation to increase agricultural output. Furthermore, it is necessary to regulate farming and broaden supervision and agricultural technology transfer to avoid further deforestation.

MEGACITIES, CLIMATE CHANGE AND NATURAL DISASTERS IN BRAZIL

We now turn to the impact of climate change on megacities in Latin America and the world in general, and Brazil in particular. The following discussion draws heavily on data gathered by Marengo et al. (2010).

São Paulo/Rio de Janeiro extended metropolitan area, is a huge corridor of a megalopolis. It accounts for a third of Brazilian GDP and a total population of 36 million inhabitants (Tolosa 2005: 125). São Paulo metropolitan area has roughly 20 million people (Marengo et al. 2010) and Rio about 7 million.

Table 6.2. Values of Reducing Emissions of Greenhouse Gases in Brazil by 2020 by Industry

Mitigation Action	*Reduction in Million Tons of CO_2 Eq. in 2020*
Land Use	
Reduction of Amazon Deforestation (80%)	564
Reduction of "Cerrado" deforestation (40%)	104
Agriculture and Livestock	
Restoration of grazing land	83 to 104
Integrated Crop-livestock system	18 to 22
No-till farming	16 to 20
Biological N2 fixation	16 to 20
Energy	
Energy Efficiency	12 to 15
Increase the use of biofuels	48 to 60
Increase in energy supply by Hydroelectric power plants	79 to 99
Alternative energy sources	26 to 33
Industry	
Iron & Steel (replace coal from deforestation with coal from planted forest)	8 to 10
Total	975 to 1052

Source: www.inpe.br.

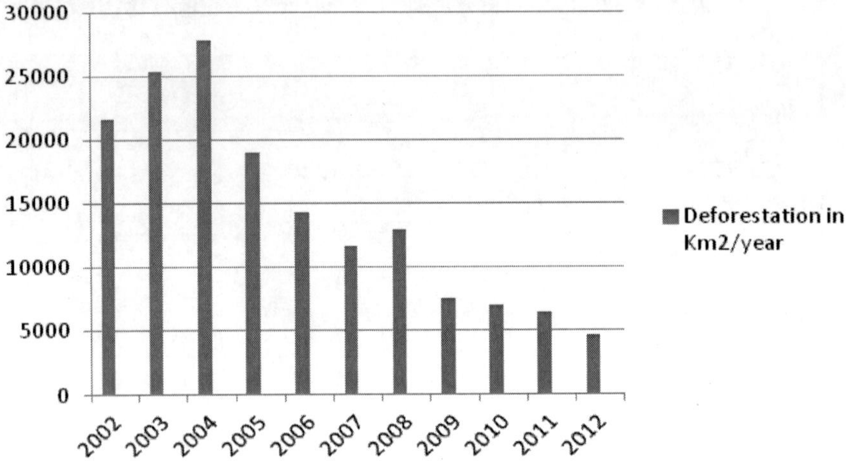

Figure 6.5. Absolute values of reducing deforestation in the Amazon in km² since 2002
Source: www.inpe.br.

Major climate change is due to the impacts of industry and especially human activities, like commuting. In São Paulo traffic jams have become un-surmountable, even with the system of car plate rotations (*mutirão*), that is even plate days for some cars, and uneven plate cars for others. Human ingenuity has found a simpler way to beat the system by acquiring a second car with different plates!

This problem of megacities in Latin America is not exclusive to Brazil. Similar general situation exists in Mexico's capital (Graizbord et al. 2005, 7) in Bogotá, Columbia (Gilbert 2005: 165) and the Tijuana-San Diego trans-boundary urban development between Mexico and the United States (Bae 2005: 181). The urban development and the "explosion" of human concentration in large cities are also seen in the developed world—Los Angeles (Richardson and Gordon 2005: 197); Tokyo (Sorensen 2005: 225), and New York City (Marcotullio et al. 2005: 289). Of course in the emerging world, the process of copy-cat development has led to major urban imbalances—such as Shangai, China (Cai and Sit 2005: 239) and Karachi, Pakistan (Qutub 2005: 251).

As discussed above global warming is already creating new challenges. Brazil is a very important stakeholder and key player in meeting climate change induced global warming challenges. It is part of the global economy, has a large population, and one of the most crucial Amazon forests. It is cleaner and a recycler of the pollution, and thus pollution abatement.

The news in Brazil in the summer of 2010 was quite appalling on sustainable urban development and human security risks. There were major floods, landslides, and destruction around the extended metropolitan Sao

Table 6.3. Reduction of Deforestation by Land Use

CLASS	TOTAL (Km2)
Cultivated Pasture	335,714
Secondary Vegetation	150,815
Shrub	62,823
Regeneration with pasture	48,027
Area Not Observed	45,406
Annual Agriculture	34,927
Mosaic Occupations	24,416
Urban Areas	3,818
Mining	730
Pasture with Bare Soil	594
Others	477
Deforestation in 2008	11,458
TOTAL (Km^2)	719,210

Source: http://www.inpe.br/cra/projetos_pesquisas/sumario_executivo_terraclass_2008.pdf.

Paulo area. It was alluded that this was due to abnormal climate change (Marengo et al. 2010).

Recent news of disasters in Brazil's urban slums (*favelas*) have witnessed increased population concentrations. Landslides have become quite frequent due to the rise in temperatures and relaxation of climate rules especially in *favelas* in Rio posing a threat to Brazilian security. These favelas have become major hubs for the narco-traffic and other crimes in Brazil, and also in Latin America. However, it should be noted that not all favelados are drug centers. Most of them are simply poor and have been excluded from mainstream socioeconomic development activities, and if given a second chance they could be easily rehabilitated. Besides the security risk of narco-traffic, we have a multiplier effect of an additional risk from landslides triggered by climate change. The informal rule of the law of the *favelas* cannot cope with climate change. Landslides do not respect hierarchies. They just hit people indiscriminately.

Marengo *et al.* (2010) suggest that the best way to cope with a sustainable policy regarding megacitites is "to promote technical innovations in urban infrastructure to cope with local conditions; to promote innovation in decision policies (new environmental policies and governance); and to devise new partnerships between the federal, state, and local government bodies to face climate change." A similar extended study has also been conducted for the Rio megalopolis. It shows similar trend as Sao Paulo's (Nobre et al. 2011).

CLIMATE CHANGE AND BRAZILIAN PUBLIC POLICY

Brazilian concerns over climate change, both by the government and by the society are not recent. The Brazilian scientific community has been a participant in the IPCC since its creation in 1988. In a response to the Earth Summit (Rio 92), a presidential decree in 1994 created the Interministerial Commission on Sustainable Development—CIDES in Portuguese—assigned to indicate priorities, strategies, nationwide policies, instruments, and the methodologies required implement the "Agenda XXI" activities, especially their inclusion into the Federal Administration's global and budgetary planning activities. The same Presidential decree created several permanent political and administrative units, including the Climate Change Coordination unit, integrated in the Ministry of Science and Technology. The unit's primary goal was to coordinate the implementation of commitments under the United Nations Framework Convention on Climate Change (UNFCC). The International Affairs and Biodiversity Coordination units were also created.

In July, 1999, a new Presidential decree established the Interministerial Commission on Global Climate Change—CIMGC in Portuguese—designed to articulate the government's actions under the UNFCC. The Commission was updated in 2006 and includes the participation of eleven ministries, empowered to use financial and human resources needed to implement government decisions related to climate change commitments. In 2001, at the "National Science, Technology and Innovation Conference: Challenges to the Brazilian society," the subject of climate change came for discussion between government, scientific societies, and the business community. It became definitely a part of the priorities of the Ministry of Science, Technology, and Innovation, that coordinates the CIMGC.

INSTITUTIONAL FRAMEWORK AND GOVERNMENT ACTIONS ON CLIMATE CHANGE

In November 2007, the Presidential decree (N. 6.263/2007) created the Interministerial Committee on Climate Change (CIMS), to elaborate the National Policy on Climate Change, and the National Plan on Climate Change. The Civil House of the Presidency coordinates the CIM, which includes seventeen federal institutions, including the Brazilian Forum on Climate Change, and other organizations.

The Ministry of Science, Technology and Innovation has been supporting several projects and actions, such as the clean development mechanisms, the national inventory on greenhouse gases, the definition of policies of mitigation and adaptation, national assessment of greenhouse gases emission, among others. Some important publications resulted from these actions, such as: *The Second National Communication of Brazil to the United Nations Framework Convention on Climate Change, A Brazilian Implementation Guide: The Clean Development Mechanism (CDM)* (2009), *First Brazilian Inventory of Anthropogenic Greenhouse Gas Emissions* (2009), and other reports available in Portuguese (www.mct.gov.br/index.php/content/view/327850.html).

The Foreign Affairs Ministry has continued its efforts in the international arena towards supporting the effective application of international environment and climate agreements. From a national point of view, the environmental issues have two major agendas: The Green Agenda, and the Brown Agenda. The first includes the Biological Diversity Convention and Convention to Combat Desertification, among other water and forests issues. The Brown Agenda includes items related to the UNFCC, the Montreal Protocol on Substances depleting the Ozone Layer, and other chemical substances themes.

The Ministry of Environment has been promoting significant actions through the Secretary of Climate Change and Environmental Quality. This department defines the strategies and policies towards mitigation and adaptation on climate change and also coordinates the Executive Group (GEX) of the Interministerial Committee on Climate Change, which prepares the National Plan on Climate Change. During the last seven years this group has coordinated a number of reports on the theme, including: National Plan on Climate Change; Global Climate Change and its Effects on Biodiversity, climate characterization and definition of climate changes for the Brazilian territories throughout the twenty-first century. It is also the coordinator of the ten sector plans defined by decree 7.390 of December 10th 2010 (www.mma.gov.br/clima/politica-nacional-sobre-mudanca-do-clima/planos-setoriais-de-mitigacao-e-adaptacao).

Two other important actions were initiated by the federal government: the Climate Network (Rede Clima—see Box 6.2) and the Brazilian Panel on Climate Change (PBMC—see Box 6.3).

The Climate Network, under the Ministry of Science, Technology and Innovation, is a research network acting on ten subjects: renewable energy, agriculture, health, biodiversity, and ecosystems, water resources, regional development, cities, modeling, economics of climate change and coastal areas.

The PBMC Panel is a common effort of the Ministries of Science, Technology and Innovation and the Environment. It coordinates activities of researchers from public and private institutes and analyzes the scientific output on climate change in all its dimensions. The Panel is also in charge of the diffusion of that knowledge through reports that can be assessed by the UNFCCC, governments and every institution or individual interested in them.

The National Plan on Climate Change results from the combined actions of the PBMC and the Climate Network since 1988. It is expected that in a short-term Brazilian policies can meet and solve the relevant issues on climate change (Box 6.4).

The execution of the Plan is supported by economic, technical, and institutional mechanisms, acting in four areas: (i) promoting scientific and technological development towards industry adaptation to new environmental requisites; (ii) raise social awareness about present environmental problems and the need to create new development standards towards a fairer, more fraternal society; (iii) value forestry and forest conservation as activities at least as economically attractive as deforestation and forest clearing—this requires technological knowledge generation; (iv) support the adoption of regional policies, adapted to the different conditions of each region and state, in order to identify the best opportunities to reduce and remove carbon emissions.

Since 2010 the federal government has invested nearly US$ 2.5 billion per year to carry out the National Plan on Climate Change, most resources are allocated to reducing deforestation in the Amazon and to the Low Carbon Emission Agriculture plan.

CONCLUSION

Above discussion has analyzed climate change issues in Brazil focusing on four topics: the impact on agriculture, assessing the role of climate change through climate change models developed by EMBRAPA/UNICAMP; secondly, the impact of this climate change on the semi-arid Northeast region, one of the most affected in the world by droughts; thirdly the importance of the devastation of the Amazon on Brazil and also the world; and fourthly the impact of climate change on Brazil's megacities. Brazil has emerged not only a major international actor in economic sphere but has taken its role very seriously in cooperating and providing leadership to face challenges in human security affairs such as climate change. ". . . At a time when expectations for progress on climate change are failing, Brazil has given the world a glimmer of hope. In many ways, the hard work is just beginning, but the results so far more than justify continuing the experiment" (Tollefson 2013: 151).

BOXES

BOX 6.1.

The Brazilian semiarid region is characterized by large social inequalities and massive poverty. According to Buainain et al. (2012), 58 percent of the Brazilian poor live in the semiarid region and more than 70 percent of the semiarid region population lives in poverty or extreme poverty. Among 1133 municipalities, 1012 have a low Human Development Index, and only 131 reach national average values of HDI.

The Articulação no Semiárido Brasileiro (ASA) is an organization formed by one thousand NGOs to promote sustainable development in this region. ASA has made a list of social indicators, based on national statistics: 9 million children (almost 68 percent) live in poverty; 82 percent of the municipalities have a low HDI, (0.65), half of the population has no income source or lives on social welfare transfers—this affects mostly women; increased rural exodus between 2000 and 2010 (rural population has decreased 5.7 percent) land tenure is concentrated, irregular, and fragmented; 2.3 percent of farms occupy 46 percent of the land; 27 percent of the farms have precarious tenure and most of them are smallholdings, too small to ensure the livelihood of a family. In 2010, 67 percent of the families had no regular access to water and suffer from periodic drought.

BOX 6.2.
COMPLETED MITIGATION AND ADAPTATION PLANS IN BRAZIL

Action Plan for the Prevention and Control of Deforestation in the Amazon—PPCDAM—www.mma.gov.br/florestas/controle-e-preven%C3%A7%C3%A3o-do-desmatamento/plano-de-a%C3%A7%C3%A3o-para-amaz%C3%B4nia-ppcdam;
Action Plan for the Prevention and Control of Deforestation in the Cerrado—PPCerrado—www.mma.gov.br/florestas/controle-e-preven%C3%A7%C3%A3o-do-desmatamento/plano-de-a%C3%A7%C3%A3o-para-cerrado-%E2%80%93-ppcerrado;
Ten Year Energy Plan—PDE—www.mme.gov.br/mme/menu/todas_publicacoes.html
Low Carbon Agriculture Plan—Plano ABC—www.mma.gov.br/images/2012/80076/Plano_ABC_VERSAO_FINAL_13jan2012.pdf;
Climate Change Mitigation Plan for the Consolidation of an Economy of Low Carbon in the Manufacturing Industry;
Low Carbon Mining Plan—PMBC;
Transportation and Urban Mobility Plan for Mitigation of Climate Change;
Health Sector Plan for Mitigation and Adaptation to Climate Change.

BOX 6.3.
OBJECTIVES OF THE CLIMATE NETWORK

Generate and disseminate knowledge and technologies to help Brazil meet the challenges posed by the causes and effects of global climate change;
Produce data and information necessary for the support of Brazilian diplomacy in the negotiations on the international regime on climate change;
Conduct studies on the impacts of global and regional climate change in Brazil, with emphasis on the country's vulnerability to climate change;
Study climate change adaptation alternatives for social, economic and natural systems in Brazil;
Research the effects of changes in land use and social, economic and natural systems on the Brazilian gas emissions and its effects on global climate change;
Contribute to the creation and monitoring of public policies on global climate change within the Brazilian territory.

BOX 6.4.
OBJECTIVES OF THE NATIONAL PLAN ON CLIMATE CHANGE

Encourage increased industrial performance efficiency in order to reach the best possible practices;
Search maintaining high share of renewable energy in the electricity matrix, preserving a prominent position that Brazil has always occupied in the international arena;
Fostering sustainable increase in the share of biofuels in the national transport matrix and also act towards structuring an international market for sustainable biofuels;
Seek the sustained reduction of deforestation rates in its four-year average in all biomes, until it reaches zero illegal deforestation.
Eliminate the net loss of forest covered area in Brazil by 2015.
Undertake the identification of environmental impacts of climate change and foster the development of scientific research toward a strategy of minimizing national social and economic costs of adaptation to change.

REFERENCES

Affholder, F.; G.C. Rodrigues; E.D. Assad, 1997. "Modelo Agroclimático para avaliação do comportamento do milho na região dos cerrados." *Pesquisa Agropecuária Brasileira*, v. 32, n. 10, 993–1002.

AIE—Agência Internacional de Energia. 2005. Energy Balances of non-OECD countries 2002–2003, Paris.

Assad, E.D.; Luchiari Jr., 1989 "A future scenario and agricultural strategies against climatic changes: the case of tropical savannas." In: *Mudanças Climáticas e Estratégias Futuras*. USP. 30–31 October 1989. Brazil: São Paulo. SP.

Assad, E.D.; Hilton Silveira Pinto; J. Zullo Junior, F.R. Marin, G.Q. Pellegrino. 2008. "Mudanças Climáticas e a Produção de Grãos no Brasil: Avaliação dos Possíveis Impactos." *Revista Plenarium*, V. 5, p. 96–117.

Assad, E. D.; Pinto, H. S.; Zullo Junior, J. e Avila, A. M. H. de. 2004. Impacto das Mudanças Climáticas no Zoneamento Agroclimático do Café no Brasil. Pesq. Agropec. Bras., v. 39, n. 11 pp: 1057–1064. Brasília, DF.

Bae, Chang-Hee C. 2005. "Tijuana-San Diego: Globalization and the transborder metropolis," in *Globalization and Urban Development*. Edited by H.W. Richardson and C.-H.C. Bae. Berlin: Springer, 181–197.

Bastide, R. 1957. Brésil, terre des contrastes. Paris. Le Harmattan.

Brasil, Ministério da Agricultura Pecuária e Abastecimento. Zoneamento Agrícola. 2013. www.agricultura.gov.br/politica-agricola/zoneamento-agricola consulted in 2013/01.

Brasil, Ministério do Meio Ambiente. 2004. *Programa de ação nacional de combate à desertificação e mitigação dos efeitos da seca*—PAN-Brasil, Brasília.

Brasil, Ministério do Meio Ambiente. Politica Nacional de Mudanças Climáticas. 2013.

www.mma.gov.br/clima/politica-nacional-sobre-mudanca-do-clima consulted in 2013/01.

Brasil, Ministério do Meio Ambiente. Plano Nacional sobre Mudanças Climáticas. 2013 www.mma.gov.br/clima/politica-nacional-sobre-mudanca-do-clima/plano-nacional-sobre-mudanca-do-clima consulted in 2013/01.

Brasil. 2009. Ministério do Planejamento, Orçamento e Gestão. Secretaria de Planejamento e Investimentos Estratégicos—SPI. Relatório de Avaliação do Plano Plurianual 2008–2011: exercício 2009—ano base 2008/Ministério do Planejamento, Orçamento e Gestão. Secretaria de Planejamento e Investimentos Estratégicos. Brasília: MP, 2009. 155 p.: il.

Buainain, Antonio Marcio e Junior Ruiz Garcia. 2013. "Pobreza rural e desenvolvimento do Semiárido Nordestino: resistência, reprodução e transformação." In *A Nova Cara da Pobreza Rural: desenvolvimento e questão regional*. Brasilia, IICA, 217–285. Série Desenvolvimento Rural Sustentável, volume 17.

Buainain, António Márcio, Claudio Dedecca, and Henrique Neder. 2012. *A Nova Cara da Pobreza Rural no Brasil: desafios para as políticas públicas*. Brasilia, IICA, Série Desenvolvimento Rural Sustentável, volume 16.

Cai, Jianming and Victor F. Sit. 2005. "Measuring world city formation—The case of Shangai," in *Globalization and Urban Development*. Edited by H.W. Richardson and C.-H.C. Bae. Berlin: Springer, 239–250.

Castro, Josué de. 1946. *A Geografia da Fome: o Dilema brasileiro: pão ou aço*, Brazil: Rio de Janeiro, Antares ed. 1980.

CEDPLAR/UFMG and Fiocruz .2008. *Mudanças Climáticas, migrações e Saúde: Cenários para o Nordeste Brasileiro, 2000–2050*. Belo Horizonte, 250 p.

EMBRAPA Solos. Org.: Celso Vainer Manzatto, Eduardo Delgado Assad, Jesus Fernando Mansilla, Maria José Zaroni, Sandro Eduardo Marshhausen Pereira. 2009. *Zoneamento Agroecológico da Cana-de- Açúcar*. Brazil: Rio de Janeiro, Embrapa Solos: 55 p.: IL.-(Documents/Embrapa Solos, ISSN 1517–2627; 110). www.cnps.embrapa.br/zoneamento_cana_de_acucar/ZonCana.pdf.

Fernandez et al. (2012) World Bank report Impacts of Climate Change on Brazilian Agriculture [P118037].

Gasques, J.G.; E.T. Bastos; C. Valdes; M.R.P. Bacch. 2012. *Produtividade da Agricultura Brasileira e os Efeitos de Algumas Políticas. Revista de Política Agrícola*, v. n. 3, p. 83, 2012.

Gilbert, Alan. 2005. "Globalization and Latin America: Understanding the links of Colombia's capital" in *Globalization and Urban Development*. Edited by H.W. Richardson and C.-H.C. Bae. Berlin: Springer, 165–180.

Graizbord, Boris, Allison Rowland and Adrian Guillermo Aguillar. 2005. "Mexico city as a peripheral global player: The two sides of a coin" in *Globalization and Urban Development*. Edited by H.W. Richardson and C.-H.C. Bae. Berlin: Springer, 147–164.

IPCC 2001. Climate Change 2001: The Scientific Basis. Contribution of Working Group I to the Third Assessment Report of the Intergovernmental Panel on Climate Change, Cambridge University Press. Cambridge, UK.

IPCC. Intergovernmental Panel on Climate Change. 2007. Climate Change 2007: Impacts, Adaptation and Vulnerability. Working Group II Report. www.ipcc.ch/ipccreports/ar4-wg2.htm

IPCC. Intergovernmental Panel on Climate Change. 2007. Climate Change 2007: Mitigation of Climate Change. Working Group III Report. http://www.ipcc.ch/ipccreports/ar4-wg3.htm.

Lacerda, F.F.; H. Dias; E.D. Assad; J.M.O. Assis; M.S.B. Moura. 2010. "Extremos e Variabilidade Climática no Nordeste Brasileiro e em Pernambuco" In: II Workshop de Mudanças Climáticas e Recursos Hídricos de Pernambuco, Recife. Mudanças Climáticas e Recursos Hídricos de Pernambuco.

Lacerda, F. F.; Paulo Nobre. 2010. Aquecimento Global: conceituação e repercussões sobre o Brasil. *Revista Brasileira de Geografia Física*, v. 3, p. 14–17,

Lima, R.C.C., Cavalcante, A.M.B., e Marin, A.M.P. (editors). 2011. *Desertificação e Mudanças Climáticas no Semiárido Brasileiro*. Campina Grande, INSA, 2011.

Manzatto, C. V.; Assad, E. D.; Bacca, J. F. M.; Zaroni, M. J.; Pereira, S. E. M. (Organizadores). 2009. Zoneamento Agroecológico da Cana-de-Açúcar—Expandir a Produção, preserver a vida, garantir o future. Ministério da Agricultura, Pecuária e Abastecimento. Documento 110. Embrapa. CNPS. Brasília, DF. 58 pp.

Marcotullio, Peter J., Sarah Rothemberg and Miri Nakahara. 2005. "Globalization and urban environmental transitions: Comparison of New York's and Tokyo's experiences" in *Globalization and Urban Development*. Edited by H.W. Richardson and C.-H.C. Bae. Berlin: Springer, 289–310.

Marengo, J.A. 2006, *Mudanças Climáticas Globais e seus Efeitos Sobre a Biodiversidade: Caracterização do Clima Atual e Definição das Alterações Climáticas para o Território Brasileiro ao Longo do Século XXI*; Brasília: MMA, 2006; 201 p.

Marengo, J.A. 2007: *Caracterização do clima no Século XX e Cenários Climáticos no Brasil e na América do Sul para o Século XXI derivados dos Modelos Globais de Clima do IPCC*, Relatório 1, MINISTÉRIO DO MEIO AMBIENTE-MMA, SECRETARIA DE BIODIVERSIDADE E FLORESTAS—SBF, DIRETORIA DE CONSERVAÇÃO DA BIODIVERSIDADE—DCBio Mudanças Climáticas Globais e Efeitos sobre a Biodiversidade—Sub projeto: Caracterização do clima atual e definição das alterações climáticas para o território brasileiro ao longo do Século XXI. Brasília, Fevereiro 2007.

Marengo, J.A., L. Alves, M. Valverde, R. Rocha, R. Laborbe, 2007: *Eventos extremos em cenários regionalizados de clima no Brasil e América do Sul para o Século XXI: Projeções de clima futuro usando três modelos regionais.* Relatório 5, MINISTÉRIO DO MEIO AMBIENTE—MMA, SECRETARIA DEBIODIVERSIDADE E FLORESTAS—SBF, DIRETORIA DE CONSERVAÇÃODA BIODIVERSIDADE—DC Bio Mudanças Climáticas Globais e Efeitos sobre a Biodiversidade—Sub projeto: Caracterização do clima atual e definição das alterações climáticas para o território brasileiro ao longo do Século XXI.Brasília, Fevereiro 2007.

Marengo, J.A. et al. 2010. in Nobre; Carlos; Hogan, Paulo (coord) Report *Megacidades, vulnerabilidades e mudanças climáticas: região metropolitano de S.Paulo;* Centro de Ciências do Sistema Terrestre do Instituto nacional de Pesquisas Espaciais

(CST/INPE); Núcleo de Estudos de População da universidade de Campinas (NEPO/UNICAMP); Brazil.

Marin, F.R.; J.W. Jones, A. Singels, F. Royce, E.D. Assad, G.Q. Pellegrino; F. Justino, 2012. Climate change impacts on sugarcane attainable yield in southern Brazil. *Climatic Change*, v. 1, p. 1, 2012.

Marin, Fábio Ricardo; Maria Leonor Lopes-Assad, E.D. Assad; C.E. Vian, M.C. Santos, 2008. Sugar cane crop efficiency in two growing seasons in São Paulo State, Brazil. *Pesquisa Agropecuária Brasileira*, v. 43, p. 1449–1455.

Margulis, S.; C.B.S. Dubeux; J. Marcovitch. 2010 *Economia da Mudança do Clima no Brasil: Custos e Oportunidades*. São Paulo: IBEP, 2010. 82 p.

MCT—Ministério da Ciência e Tecnologia—First Brazilian Inventory of Anthropogenic Greenhouse Gas Emissions—Background Reports. 2002. www.mcti.gov.br/index.php/content/view/17341.html.

MCT—Ministério da Ciência e Tecnologia—Second Brazilian Inventory of Anthropogenic Greenhouse Gas Emissions—Background Reports. 2010. www.mcti.gov.br/index.php/content/view/330034/Relatorios_de_Referencia_2010.html.

MMA. 2004 *Programa de Ação Nacional de Combate à Desertificação e Mitigação dos Efeitos da Seca (PAN-Brasil)*. Brasilia, Ministério do Meio Ambiente, 220 p.

Nepstad, D.; B Soares-Filho, F. Merry; P. Moutinho; M. Bowman; S. Schwartzman; O. Alemida; S. Rivero. 2007. *The costs and benefits of reducing carbon emissions from deforestation and forest degradation in the Brazilian Amazon*. Woods Hole Research Center, IPAM, 2007. 28 p.

Nobre, P.M., D.F. Malagutti, R.A.F. de Almeida Urbano and E. Giarolla, 2009. "Amazon deforestation and climate change in a coupled model simulation." *J. Climate*, 22, 5686–5697.

Nobre, P. 2011. "Mudanças Climáticas e desertificação: os desafios para o Estado Brasileiro." In Lima, Cavalcante e Marin. *Desertificação e Mudanças Climáticas no Semiárido Brasileiro*. Campina Grande, INSA/MCTI, 211 p.

Nobre; Carlos; Young, Andrea (coords) et al. (2011); Report *Megacidades, vulnerabilidades e mudanças climáticas: região metropolitano do Rio de Janeiro;* Centro de Ciências do Sistema Terrestre do Instituto nacional de Pesquisas Espaciais (CST/INPE); Núcleo de Estudos de População da universidade de Campinas (NEPO/UNICAMP); Brazil.

OECD/FAO. 2012, *OECD/FAO Agricultural Outlook 2012–2021*, OECD Publishing and FAO dx.doi.org/101787/agr_outlook-2012-en.

Oyama, M.D., C.A. Nobre, 2003. "A new climate-vegetation equilibrium state for Tropical South America." *Geophysical Research Letter*, 30, 23, 2199.

Qutub, Syed Ayub. 2005. "Karachi—a case of asymmetric inclusion in the current globalization?" in *Globalization and Urban Development*. Edited by H.W. Richardson and C.-H.C. Bae. Berlin: Springer, 239–250.

Pinto, H.S., E.D. Assad, 2008. Global warming and the new geography of agricultural production in Brazil. EMBRAPA e UNICAMP. Brasilia, British Embassy, 2008. 82 p: www.cpa.unicamp.br/aquecimento_agricola_en.html. Consulted in 2013/01 www.agritempo.gov.br/climaeagricultura/ consulted in 2013/01.

Plassman, K., I.M. Harris, R.T. Edwards, G.A.S. Day, A.D. Tomos, S.J. Cowell, D.L. Jones. 2008. "Testing the assertion that 'local food is best': the challenges of an evidence-based approach." *Trends in Food Science and Technology*, 19, 265–274.

Richardson, Harry W. and Peter Gordon. 2005. "Globalization and Los Angeles" in *Globalization and Urban Development*. Edited by H.W. Richardson and C.-H.C. Bae. Berlin: Springer, 197–210

Rodriguez-Puebla, C., S.M. Ayuso, M.D. Frias, L.A. Garcia-Casado, L.A. 2007. "Effects of climate variation on winter cereal production in Spain." *Climate Research* 34, 223–232.

Santana, C.A.M.; D.A.P. Torres, R.C.M. Guiducci; M. Alves; F.L. Garagorry; G.S. Souza, E.D. Assad, G.Q. Pelligrino; L.G. Barioni, M.O. Souza, H. Chaib Filho; R. Marra, Mierson Mota. 2010. *Productive Capacity of Brazilian Agriculture: A Long-Term Perspective Regional Study: Brazil. Foresight Food and Farming Futures Project* Version date: 11 June 2010. London, UK.

Sorensen, André. 2005. "Building world city Tokyo: Globalization and conflict over urban space" in *Globalization and Urban Development*. Edited by H.W. Richardson and C.-H.C. Bae. Berlin: Springer, 225–238.

Tolosa, Hamilton. 2005. "The Rio/São Paulo Extended Metropolitan Area" in *Globalization and Urban Development*. Edited by H.W. Richardson and C.-H.C. Bae. Berlin: Springer, 125–146.

Tollefson, Jeff 2013 "A Light in the Forest," *Foreign Affairs* March–April, 141–151.

Underdal, A.; J. Hovi, S. Kabbelken, T. Skodvin. 2012. "Can conditional commitments break the climate change negotiation deadlock?," *International Political Science Review*, vol. 33, n. 4, September, 475–493.

Wu, W.B., R. Shibasaki, P. Yang, G.X. Tan, K.I. Matsumura, K. Sugimoto. 2007. "Global-scale modelling of future changes in sown areas of major crops." *Ecological Modelling* 208, 378–390.

Zullo Jr., J., H.S. Pinto, E. D. Assad, A.M.H. de Avila, A. M. H. 2011. "Potential for growing Arabica coffee in the extreme south of Brazil in a warmer world." *Climatic Change*, v. 1, p. 1573–1480, 2011.

Zullo Jr., Jurandir; Hilton Silveira Pinto, E. D. Assad, S. R. Evangelista 2008. "Potential Economic Impacts of Global Warming on Two Brazilian Commodities, According to IPCC Prognostics." *Terrae (Campinas)*, v. 3, p. 28.

7

Climate Change Vulnerabilities and Human Insecurities

Understanding the Impact on India

Roopinder Oberoi and M. P. Singh

Science is now unequivocal and undeniable to the reality of climate change. The disconcerting investigation of the human impact of climate change cannot be uncared for at the negotiating tables of world leaders and climatologists. Climate change is a multiplier of human impacts and risks. Until recently, most scientific assessments of such risks focused on the anatomy of conceivable environmental changes themselves, while devoting relatively little attention to the ecosystems and societies the changes might endanger. Recently, however, questions about the *vulnerability* of social and ecological systems are emerging as a central focus of policy-driven assessments of global environmental risks. Climate change as acutely affecting hundreds of millions arose from the nexus of two seemingly different areas of study: the human dimensions of environmental change and the reconceptualisation of security of people at present and in the fact that in next twenty years those affected are likely to double making it the greatest emerging humanitarian challenge of our time. Those seriously affected are in need of instantaneous assistance either following a weather-related disaster, or because their livelihoods have been severely compromised by climate change.

The impacts of weather and climate extremes are largely determined by exposure and vulnerability. This is occurring in a context where all three components—exposure, vulnerability, and climate—are highly dynamic and subject to continuous change. Climate change if contextualized within existing poverty its adverse impacts become more striking because of the fragile geophysical, high dependence on natural resources, and the limited capacity of vulnerable to adapt to a changing climate. Within these poor countries, the poorest, with the least resources and the least capacity to adapt, are most vulnerable (IPCC 2001a; 18). Between 1970 and 2008, over 95 percent of deaths from natural disasters occurred in developing countries (UNISDR, 2009e, 2011; 25–28). Projected changes in the incidence, frequency, intensity, and duration of climate extremes as well as gradual modification in the average climate will particularly threaten already marginalized livelihoods—further escalating inequities and consequent conflicts and struggles.

Given this focus, the analytic emphasis of this chapter is on people and systems that may be adversely affected by climate change, particularly where impacts could have serious and/or irreversible consequences. The chapter analyzes vulnerability "hot spots" and seeks to understand and quantify the vulnerability of India. Furthermore, it has to be acknowledged that the impacts of future climate change will occur in the context of an evolving socioeconomic baseline. Jairam Ramesh while releasing an Assessment Report in 2010 said 'No country in the world is as vulnerable, on so many dimensions, to climate change as India. Whether it is our long coastline of 7000 kms, our Himalayas with their vast glaciers, our almost 70 million hectares of forests we are exposed to climate change on multiple fronts'. The purpose of this discussion is to review the literature on climate change vulnerability of India and explore and synthesize conceptual and analytical aspects considered fundamental in a vulnerability assessment in climate change. It would specifically focus on the issues viz. agriculture and food security, water stress and water insecurity, rising sea levels, biodiversity and human health, which have enormous relevance from the perspective of developing countries in general and India in particular.

THE IMPACT OF CLIMATE CHANGE ON HUMAN SECURITY

The unprecedented economic growth has significantly increased societal vulnerability to the vagaries of climate change, land-use impacts, natural hazards and other environmental events. Future changes in exposure, vulnerability, and climate extremes resulting from natural climate variability, anthropogenic climate change, and socioeconomic development can alter the impacts of climate extremes on natural and human systems. While specific, local outcomes of climate change are uncertain, recent assessments project alteration in the frequency, intensity, spatial extent, or duration of weather and climate extremes, including climate and hydro-meteorological events such as heat waves, heavy precipitation events, drought, and tropical cyclones. Such change, in a context of increasing vulnerability, will lead to increased stress on human and natural systems and a propensity for serious adverse effects in many places around the world (Clark 1998; 68).

Although climate change affects everyone regardless of race, caste, ethnicity, sex, and level of income, its impacts are more profoundly felt by poor nations, communities, and people, as climate change magnifies existing inequalities. For the world's poor the impact will be disproportionately harsh, affecting their livelihoods and security. The discourse on human security goes beyond its mere definition. As Gasper (2005; 228) argues, it "includes normative claims that what matters is the content of individuals' lives, including a reasonable degree of stability." It stands apart from other discourses (e.g., human needs and human development) by disaggregating down to the level of the individual, taking a human rights stance that basic requirements of no individual are to be sacrificed. This reflects John Rawls (1971) theory of justice, and human security and has a strong link to notions of justice, equity, and fairness. But human security is also about freedom from threats and risks—risks that are increasingly likely to become global in their scale (Beck 1992). Although climate change raises a suite of equity-related issues, differential vulnerabilities to climate change are perhaps the most pressing issue from the standpoint of human security. When assessing the impacts of climate change on diverse regions it is imperative to look at the interaction of various dimensions from a human security perspective. Climate change has been described as a threat multiplier; the double vulnerability (poverty compounded by climate change); and one of the greatest threats to development.

To put it simply vulnerability to climate change is the degree (measure) to which our systems are susceptible and predisposed to, and incapable to cope with adverse impacts. (Füssel and Klein, 2006; 312–325). Pattern of vulnerability to environmental and socioeconomic changes, referred to as "archetypes," describes the impacts of these changes on human well-being. Any parley about "dangerous interference with the climate system" and "key vulnerabilities" are structured around thresholds or critical limits (Patwardhan et al., 2003; Izrael 2004: 783–788). Key vulnerabilities must be linked to systemic thresholds where nonlinear processes cause a system to alter from one major state to another associated with multifarious climate sensitive systems, including, for example, food supply, infrastructure, health, water resources, coastal systems, ecosystems, global biogeochemical cycles, ice sheets, and modes of oceanic and atmospheric circulation. The concept of risk, which coalesces with the magnitude of the impact along with the probability of its occurrence, captures the uncertainty and insecurity in the underlying processes of climate change, exposure, impacts, and adaptation.

High exposure and vulnerability are generally the outcome of skewed development processes such as those associated with environmental degradation, rapid and unplanned urbanization in hazardous areas, failures of governance, and the scarcity of livelihood options for the poor. Inequalities influence local coping and adaptive capacity, and pose disaster risk management and adaptation challenges from the local to national levels. These inequalities reflect socioeconomic, demographic, and health-related differences and differences in governance, access to livelihoods, entitlements, and other factors. Inequalities also exist across countries: developed countries are often better equipped financially and institutionally to adopt explicit measures to effectively respond and adapt to projected changes in exposure, vulnerability, and climate extremes than are developing countries. Calculations of vulnerability at the national scale may show comparatively low vulnerability levels despite there being some highly vulnerable sub-populations or individuals in that nation. Nonetheless, all countries face challenges in assessing, understanding, and responding to such projected changes.

As the World Commission on Environment and Development (Brundtland Commission) stated in *Our Common Future*, "A more careful and sensitive consideration of their (vulnerable groups) interests is a touchstone of sustainable development policy" (WCED 1987). The United Nations Conference on Environment and Development, which adopted Agenda 21, proclaimed that "Human beings are at the center of concerns for sustainable development. They are entitled to a healthy and productive life in harmony with nature" (United Nations 1999).The point is that climate exposure, and vulnerability is predisposed by extensive criterions, including anthropogenic climate change, natural climate variability, and socioeconomic development (Figure 7.1). Adaptation to climate change focuses on reducing exposure and vulnerability and increasing resilience to adverse impacts of climate extremes even though hazards cannot be eradicated completely. The figure indicates schematically key concepts involved in disaster risk management and climate change adaptation, and the interaction of these with sustainable development.

Table 7.2 points out the causes of climate change, when correlated to the consequential physical change, impacts humans. The framework gives an idea about how an increased emission physically alters the environment in a way that has grave human penalty or threat.

Links between climate change and the threat of violence and armed conflict are particularly significant because manifold destabilizing circumstances are embellished simultaneously. Climate change intensifies

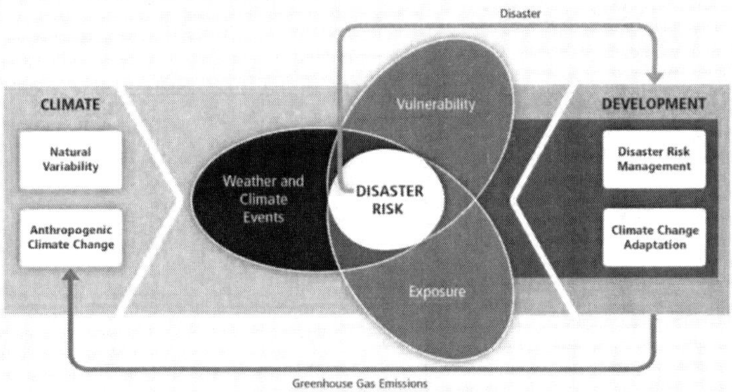

Figure 7.1. Climate Change, Vulnerability, and Impact on Development

Source: Lavell, A., M. Oppenheimer, C. Diop, J. Hess, R. Lempert, J. Li, R. Muir-Wood, and S. Myeong, 2012: Climate change: new dimensions in disaster risk, exposure, vulnerability, and resilience. In: Managing the Risks of Extreme Events and Disasters to Advance Climate Change Adaptation.

Table 7.1. Potential Indicators for Assessing Human Vulnerability to Environmental Change

Human vulnerability	Environmental causes	Indicators
Health	- Urban air pollution - Water pollution/sanitation - Toxic chemicals/food contaminants	- Number of people affected by environmental diseases (pollutants, chemicals), microbial infection, diarrhoea, chronic lung diseases - Number of people having access to safe drinking water and sanitation - Loss of DALY (Disability Adjusted Life Year)
Economic losses/gains	- Environmental diseases - Soil erosion - Deforestation - Siltation	- Amount spent on treating environmental diseases - Amount spent on environmental clean up - Food productivity loss due to soil erosion, deforestation, etc. - Loss due to siltation of dams
Poverty	Depletion of natural resource base to meet the basic needs of food, fibre, firewood , income and employment	- Different income categories affected by natural resource degradation - Different income categories affected by air pollution and sea level rise - Different income categories affected by water contamination and lack of sanitation
Food security	- Loss of natural vegetation and biological diversity. - Soil erosion - Surface and groundwater depletion - Rainfall distribution	- Percentage of natural vegetation cover - Percentage of people directly dependent upon land resources - Extent and distribution of degraded land - Freshwater availability -Rainfall variability

Source: The GEO-4 conceptual framework (UNEP 2007).

harmful environmental trends like desertification, soil salinization and water scarcity; adding to resource scarceness. Essentially it trims down the supply of food, fresh water for people and livestock, agricultural produce, and farmer livelihoods UNEP (2003; 63–65) Thywissen (2006; 24–28). These conditions can trigger increased struggle for food, land and water, producing propensity to conflict. This occurs particularly in areas where governments are not able to provide support or alternative sources of income to already vulnerable societies or marginalized groups. These linkages are becoming increasingly visible at global and local level. The links between the human impacts of climate change contributes to social tensions and instabilities that raise the risk of violence or armed conflict. There are two broad ways in which conflict might be stimulated by climate change. First, conflict could come about through changes in the political economy of energy resources due to mitigative action to reduce emissions from fossil fuels. The second issue is the prospect of conflict stimulated by changes in social systems driven by actual or perceived climate impacts. The outlook for the future is not encouraging, with more frequent, more severe, and more prolonged weather-related disasters on the horizon. Linear projections suggest that by 2030, the number of weather-related disasters recorded in a single year will be approximately three times higher than the average occurrence rate during the 1975–2008 time span (Global Humanitarian Report 2008) The findings of the report indicate that every year climate change leaves over 300,000 people dead, 325 million people seriously affected, and economic losses of US$125 billion. Four billion people are vulnerable, and 500 million people are at extreme risk.

INDIA'S VULNERABILITY TO CLIMATE CHANGE

Determining and projecting climate change induced hazards and perturbations and sensitivity refers to the nature of expected climate change and how a range of possibilities will impact different sectors. The range of impacts is discernible at different scales (e.g., national, subnational, or community level), depending on the relevance and types of management decisions that need to be made and information required to determine impacts. The impact of climate change on the lives and livelihoods of people in India is now widely recognized. Yet, there is neither a consensus on the definition of vulnerability to climate change or a full, regionally-nuanced mapping of the variable impact of such a change. It is only when a better understanding of what constitutes vulnerability to climate change and what its region-specific impact would be is available that proper adaptation strategies can be worked out.

Vulnerable Land, Vulnerable Billions

India is a developing country of subcontinental proportions—home to 1.1 billion people or 17 percent of the world's population. While India is on high economic growth trajectory, it is still grappling with paradoxes of inequity and a large human development backlog. Around 28.6 percent of the population, some 320 million people continue to live in India below the poverty line (UNDP 2007a; 33–36). Climate variability and change can dawdle down the pace of development either through adverse impacts on natural ecosystems or erosion of the adaptive capacity of people and society. The knowledge and understanding of implications of climate change on India is somewhat inadequate and fragmentary. India has a unique topography and climate system with an economy closely tied to its natural-resource-base and climate-sensitive sectors such as agriculture, water, and forestry. India's large population primarily depends on these sectors for livelihood.

India is projected to be severely impacted by changes in climate extremes. India is more vulnerable to climate change than the United States, China, Russia, and indeed most other parts of the world with the exception of Africa. The fatalities related to climatic disasters would be particularly severe, probably calamitous (Table 7.2), if contingencies such as drying up of north Indian rivers and disruption of Monsoon rains came to pass. Consequently, India has a strong national interest in helping to secure a robust climate deal.

Economic development has been the watchword in India's march into the twenty-first century, but a conservative estimate of environmental damage put the figure at more than US$10 billion a year, or 4.5 percent of GDP, in 1992. A breakdown of the estimated costs shows that urban air pollution costs India US$1.3 billion a year; and water degradation has associated with health costs of US$5.7 billion a year, nearly three-fifths of total environmental costs. Land degradation causes productivity losses of around US$2.4 billion and deforestation leads to annual losses of US$214 million. (Suchak 2002; 10)

Besides, Natural catastrophes pose a grave threat in the development of the country. Twenty-two of India's thirty-one states are regarded as

Table 7.2. Number of Reported Natural Disaster Events and Fatalities in Selected Countries 1990–1999

Country	Number of reported Events	Persons killed
USA	242	3 418
India	114	50 777
Bangladesh	86	150 242

Source: CRED 2000 Dataset.

particularly prone to natural disasters: 55 percent of its land is vulnerable to earthquake, 8 percent is vulnerable to cyclone and 5 percent is vulnerable to floods. On average, direct natural disasters losses amount to up to 2 percent of India's GDP. The population of India is projected to reach 1.4 thousand million by 2020 with urban population outpacing the rural population. The coastal mega cities of Chennai (2005: population 6.9 million), Calcutta (14.3 million) and Mumbai (18.2 million) are particularly vulnerable they are situated only a few meters above sea level (German Advisory Council on Global Change WBGU 2008; 5–15). The high population pressure combined with increasing cyclone intensity and sea-level rise as a result of climate change will affect millions of people with storm and flood disasters. Predicted increase in frequency and intensity of floods and droughts are likely to have unfavourable impacts on the occupational structure, food security, health, social infrastructure etc. of the hotspots (Roy et al., 2005; 239–259).

Nearly 700 million rural people in India directly depend on climate-sensitive sectors (agriculture, forests, and fisheries) and natural resources (water, biodiversity, mangroves, coastal zones, and grasslands) for their subsistence and livelihood. Under changing climate, food security of the country might come under threat. India has the highest number of stunted children in the world. According to the Hunger and Malnutrition Report (2012) almost 42 percent of the children in the country numbering over 61 million are malnourished and stunted.[1] In addition, the adaptive capacity of dry-land farmers, forest and coastal communities is low. Climate change is likely to impact all the natural ecosystems as well as health and socioeconomic systems, as indicated by the Indian Governments' National Communication to UNFCCC on Climate Change.

The linkages between climate change and national security works both indirectly and directly. In India, climate change can cause profound impact on internally displaced persons and migrants aggravating pre-existing conflicts. At the same time, it might engender fresh stresses, tensions and conflicts. Climate change may emerge as a key concern in India's bilateral relations with its neighbors. For instance, the influx of environmental refugees can alter the demographic balance in the Indian states. Water sharing issues could arise between India and China, India and Bangladesh, India and Pakistan. The adverse impacts of climate change would threaten India's human security and livelihood. The rising temperature, variations in rainfall, disappearance of glaciers, increased frequency and severity of extreme events of floods in the plains as well as in the coastal areas, droughts, storms are likely to impact more particularly the vulnerable section of the economy.

The impact of vulnerability is not merely determined by the extent of climate change, but also by the robustness of the developmental progression in the economy. It is the quality of development that provides an indemnity against the impacts of climate change and augments the adaptive capacity of the vulnerable. Superimposing the incremental risk of climate change on huge development deficit confronts with the goal of 'inclusive growth.'

TEMPERATURE, PRECIPITATION AND OCCURRENCES OF EXTREME EVENTS

Heat waves can directly impact ecosystems by constraining carbon and nitrogen cycling and reducing water availability, with the result of potentially decreasing production or even causing species mortality. As per the Indian Institute of Tropical Meteorology (IITM), Pune simulation, annual temperature rises from 3–5°C under A2 and 2.5–4°C under B2 scenarios of IPCC, with warming more pronounced in the northern parts of India by the end of century. In 1998, India experienced its worst hot spell in fifty years, which took a toll of over 3,000 lives; tropical cyclone of Orissa in 1999 took a toll of about 10,000 lives. Indian annual mean temperature showed a significant warming trend of 0.51°C per hundred years, during the period 1901–2007 (Kothawale et al., 2010; 89–104).

Among the most significant potential impacts of climate change on India are changes in the monsoon pattern. Several studies have shown that in general, the mean monsoon intensity and variability is expected to increase (Ashrit et al. 2001; Chung et al. 2006; Kumar et al. 2006). The summer monsoon intensity may increase beginning from 2040 and by 10 percent by 2100 under A2 scenario of IPCC. In addition, the variations in the distribution of rainfall are also being observed in the range of plus-12 percent to minus-8 percent of the normal over the last hundred years. The analyses of daily rainfall data set have shown (i) a rising trend in the frequency of heavy rain events and (ii) a significant decrease in the frequency of moderate events over central India between 1951 to 2000 (Goswami et al, 2006; 15–17). The daily rainfall data for 1954–2003 show vulnerability of Kerala to increasing water scarcity in the pre-monsoon time, and a delaying monsoon onset (Pal and Tabbaa, 2009; 47) Snow, ice, and glaciers in the Himalayan region are approximately equivalent to about 1,400 km of ice. The melting of snow will lead to flood disasters in Himalayan catchments (IPCC, 2001). Mahanadi, Brahmani, Godavari and Cauvery basins are projected to have increasing precipitation due to increase in evapotranspiration on account of increased temperature, further intensifying flooding conditions. Sabarmati and Luni basins will be experiencing drastically decreased precipitation which will further deteriorate drought condition (Shukla: 2003).

Agriculture

The economies of many developing countries rely heavily on agriculture, dominated by small-scale and subsistence farming. People's livelihoods in this sector are especially exposed to weather extremes. According to the "Agricultural Statistics at a Glance" (2010), 46.1 percent of the reported area for land use estimation is under agriculture. Agriculture in India makes up roughly 17 percent of GDP (GOI). The latest data from greenhouse gases (GHG) emissions inventory of 2007 indicates that agriculture sector contributes 28 percent of the total GHG emissions from India. The emissions are primarily due to methane emission from rice paddies, enteric fermentation in ruminant animals, and nitrous oxides from application of manures and fertilizers to agricultural soils.

Out of the total 329 million hectares of geographical area 174 million hectares or (53 percent) of the total land area in India suffers from serious degradation. Water and wind erosion amounts to 144 million hectares. Other problems like ravines, salinity, and water logging further contribute to degradation of another 30 million hectares (Koty Reddy T. 2010; 28). According to Sinha and Swaminathan (1991), an increase of 20°C in temperature could decrease the rice yield by about 0.75 ton per hectare in the high yield areas, and 0.50°C increase in winter temperature would reduce wheat yield by 0.45 tons/ha. Rai and Shina (1994) showed that wheat yields could decrease between 28–68 percent without considering the CO_2 fertilization effects. Aggarwal and Sinha (1993) point out that a 20°C temperature rise would decrease wheat yields in most places and for every one degree rise in temperature the decline in rice yield would be about 6 percent.

In India, while the wheat crop is vulnerable to an increase in maximum temperature, the rice crop is sensitive to an increase in minimum temperature. Yet global warming poses serious threats to the weather system, which can potentially affect millions of small, marginal, and poor farmers and all those who depend on agriculture for their livelihood (Mitra Amit, 2009: 13). Acute water shortage, combined with temperature stress negatively affects both wheat and rice productivity in north-west India. Coconut yields are likely to be affected by global climate change. Plains of Karnataka, Eastern Tamilnadu, coastal Andhra Pradesh, Pondicherry, West Bengal, and Assam are 'hot spots' as per HadCM3 model scenarios of climate change; No change in productivity was projected due to climate change in coastal Karnataka and Kerala.

Preliminary estimates indicate that global warming is likely to lead to a loss of 1.5–2.0 million tons in milk production by 2020 and 15 million tons by 2050. Increasing sea and river water temperatures are likely to affect fish breeding, their migration and other harvests. By 2040 coral reefs in the Indian seas are predicted to decline. A rise in temperature as small as 1°C could have important and rapid effects on the mortality of fish and their geographical distributions.

Food Systems and Food Security

Food systems and food security can be affected by extreme events that impair food production and food storage and delivery systems (food logistics).[2] An increase in the price of food can be especially challenging for the urban poor in developing countries (FAO: 2008). Global food price increases are borne disproportionally by low-income countries, where people spend more of their income on food (OECD-FAO, 2008).

Food security is directly or indirectly related to climate change. Any alteration in climatic parameters such as temperature and humidity which govern crops growth will have a direct impact on the quality of food produced. Indirect linkages pertain to catastrophic events such as floods and droughts which are projected to multiply as a consequence of climate change leading to huge crop loss and leaving large patches of arable land unfit for cultivation, and hence threatening food security (Chaudhry, Aggarwal, 2007; 56). Studies by Kavi kumar and Parikh (2001: 147–150) point out that the interannual variability in rainfall will have major impact on food grain production in India and therefore the economy of the country as a whole. Despite fast economic growth and piling food stocks in the government storage facilities, India is home to the largest number of hungry and deprived people in the world—to be precise 360 million undernourished and 300 million poor people. Sustaining supply of food itself is emerging as a critical issue. Growth in food grain production is slow, rather decreasing over the last few decades. During 1996–2008 it increased by just 1.2 percent per annum: from 199 to 230 million tons, as against an annual rate of growth of 3.5 percent achieved during the 1980s (UNDP, 2009).

The Indian Agricultural Research Institute (IARI) examined the vulnerability of agricultural production to climate change, to determine the differences in climate change impacts on agriculture by region and by crops. The study shows the growth of major crops and their per hectare yield in India between 1989–2000 and 2001/2007–2008. During 1989–2000 the production of rice declined from 17.96 percent to 3.45 percent in the year 2007–2008. During the same period the production of wheat declined from 34.37 percent to 3.51 percent. This clearly indicates that food security of India may be at risk in the future due to the threat of climate change. Studies by Indian Agricultural Research Institute (IARI) indicate every 1°C rise in temperature will also significantly affect the quality of fruits, vegetables, tea, coffee, aromatic and medicinal plants,

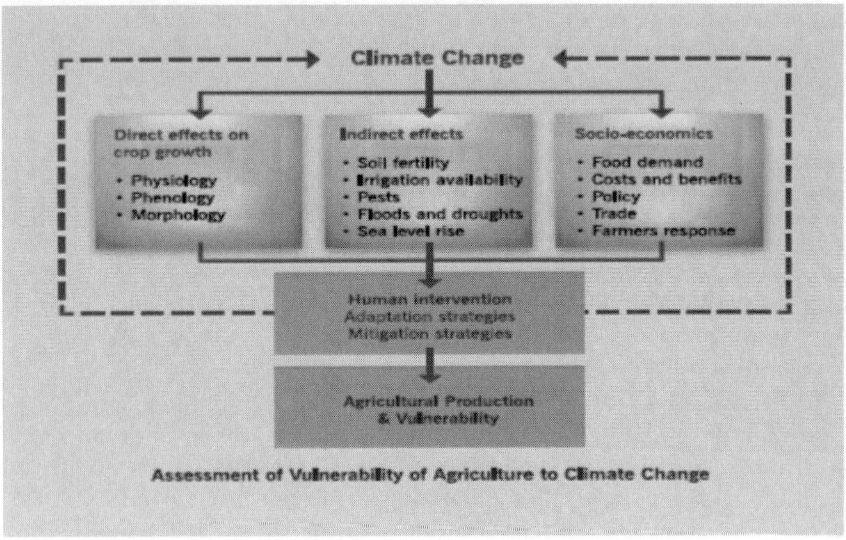

Figure 7.2. Assessment of Vulnerability of Agriculture to Climate Change
Source: IIPC 2007.

and basmati rice. Global reports indicate a loss of 10–40 percent in crop production by 2100. By the end of the next century, it can cause annual damages in farm income between 4 percent and 26 percent in India (Sanghi and Mendelsohn, 2008).

If the current production practices continue, India will face a serious food shortage in the near future. Sixty-five percent of arable land in India is rain fed and the increasing demand for food and feed has to be met with the increased production in the rain fed areas, as there is very little scope for expansion of cultivable area or irrigation facilities. At present approximately 68.35 million hectare of land is Wasteland, of this approximately 50 percent is non-forest land which can be made fertile again if treated properly. Rajasthan has the highest component of degraded land, followed by Madhya Pradesh, Maharashtra, Uttar Pradesh, Gujarat, Andhra Pradesh and Karnataka. Where land has been subjected to light or moderate degradation, the same level of inputs will give lower outputs. (Ahmad, Alam and Haseen 2011)

IMPACT ON FORESTS AND OTHER NATURAL ECOSYSTEMS

As per State of Forest Report 2007 (*Forest* Survey of India (FSI) 2009) assessment, the forest cover of India is estimated to be 69.07 million hectare (Mha), which is 21.02 percent of the total geographical area of the country. Forests in India are already subjected to multiple stresses including—over extraction, insect outbreaks, livestock grazing, forest fires and other anthropogenic pressures. The Fourth Assessment Report of IPCC (IPCC 2007) concluded that forest ecosystems could be seriously impacted by future climate change, even with moderate global warming of 1°C to 2°C. Forests in India play a crucial socioeconomic role, contributing 0.7 percent to GDP in 2008 (GOI, 2009). Nearly 55 million people in India depend upon non-timber forest products. The forest covers is 23.60 percent in 2008, and there are plans to extend then to 33 percent by 2012 (MoEF, 2008). Impacts of climate change on forests has severe implications for the people who depend on forest resources for their livelihoods. With nearly 173,000 villages classified as forest villages, depletion and degradation of these forest resources could be very stressful and socially and economically destabilizing.

Approximately 47 percent and 42 percent of tropical dry deciduous grids are projected to undergo shifts under A2 and B2 scenarios respectively, as opposed to less than 16 percent grids comprising of tropical wet evergreen forests. Similarly, the tropical thorny scrub forest is projected to undergo shifts in majority of forested grids under A2 (more than 80 percent) as well as B2 scenarios (50 percent of grids). States such as Chhattisgarh, Karnataka, and Andhra Pradesh are projected to experience a change of 73 percent, 67 percent, and 62 percent of the forested grids while in Madhya Pradesh it is about 50 percent under the A2 scenario. According to another study (Gopalakrishnan et al, 2011), projected impacts of climate change, using a moderate A1B scenario and IBIS vegetation model for the period of 2030s and 2080s, are projected in India.

Impact on Coastal Areas

Coasts are highly dynamic and geomorphologically complex systems, which respond in various ways to extreme weather events. Coastal floods are regarded as among the most dangerous and harmful of natural disasters. India has a low-lying densely populated coastline extending to about 6500 km and has been identified as one of the most vulnerable to sea level rise (UNEP, 1989). Most of the coastal regions are agriculturally fertile, with paddy fields that are highly vulnerable to inundation and salinization.

The rise in sea level in the north Indian Ocean has been observed to be in the range of 1.06–1.75 millimeters per year in the past century. There is a threat of coastal inundation in some of the low lying coastal areas. On the other hand, the glaciers show a mixed behavior. While some of them are receding, others are growing. Indian coast line is prone to undergo severe impact due to increased frequency of climatically induced extreme events like cyclones, storm surge, high-tides resulting in flooding of vast stretches of area all-along the coast line. The rise in sea level in the north Indian Ocean has been observed to be in the range of 1.06–1.75 millimeters per year in the past century. The IPCC (2007) has projected a mean sea-level rise of 0.18 to 0.59m by 2090, relative to the 1980–1999 level. If these predictions are proved correct, it can impact India's coastal population due to increased flooding of low-lying areas and loss of crop yields from inundation and salinization of fresh water sources. An estimated 6 percent of people in India are highly vulnerable to cyclonic hazards. The most vulnerable, approximately 53 million, are in Andhra Pradesh, Orissa, Tamil Nadu and West Bengal. The coastlines of Andaman and Nicobar Islands in the Bay of Bengal and Lakshadweep Islands in the Arabian Sea is 7517 km of which 81 percent (6100 km) is along the Indian mainland surrounded by the Arabian Sea are the west, Bay of Bengal in the east, and Indian Ocean in the south. More than 40 million people reside along this coastline.

There are thirteen Coastal States and union territories susceptible to sea level rise in the country, with about eighty-four coastal districts affected by tropical cyclones. Four States (Tamil Nadu, Andhra Pradesh, Orissa, and West Bengal) and one Union Territory (Pondicherry) on the East Coast and one of the States on the West Coast, Gujarat, are the States that are the most affected by cyclonic activities. The mega cities of Mumbai, Chennai, and Kolkata lie along this coastline, and dotted with several major ports such as Kandla, Mumbai, Navasheva, Mangalore, Cochin, Chennai, Tuticorin, Vishakhapatnam, and Paradip. Ranger et al. (2011; 79–83) have observed that the risk from heavy rainfall in the city of Mumbai, direct and indirect losses could triple by the 2080s compared with the present cost from US$ 700 to 2,305 million). Only that adaptation could help reduce future damages.

A large portion of the population along the coastline is dependent on climate activities such as marine fisheries and agriculture. Sea level changes and occurrence of extreme events such as cyclones and storm surges are of considerable significance for India as these adversely impact on human populations living in coastal regions and on islands as well as the sensitive ecosystems such as the mangroves in the Sundarbans. The flooding would result in loss of coastal habitat and displacement of humans.

It is projected that the total frequency of cyclonic storms that form over the north Indian Ocean does not show any significant trends during the period 1891–2008, but a slightly decreasing trend. An analysis of past tide gauge records of long duration for the Indian coastline regions give an estimate of average sea-level-rise trend as 1.30 mm/year which is consistent with the values reported elsewhere. However, in the Indo-

Gangetic deltaic region, for instance at Diamond Harbour (Kolkata), the record shows a trend of 5.74 mm/year, which is partly attributed to the subsidence in the delta.

Impact on Water Sources

Past and future changes in exposure and vulnerability to climate extremes in the water sector are driven by both changes in the volume, timing, and quality of available water and changes in the property, lives, and systems that use the water resource or that are exposed to water related hazards (Aggarwal and Singh, 2010; 389–393). India faces a number of water related challenges, including increasing water scarcity and competition for water between different sectors and states. Climate change will affect the water balance, and particularly the amount of runoff and recharge, which in turn determines the water resources available for human and ecosystem uses. The amount of water available per person in India is decreasing steadily—from 3450 cm in 1951, to 1250 cm in 1999 and further to 760 cm per person in 2050. Accordingly, the per capita availability of water for the country as a whole decreased from 5177 m3/year in 1951 to 1654 m3/year in 2007. "There will be constant competition over water, between farming families and urban dwellers, environmental conservationists and industrialists, minorities living off natural resources and entrepreneurs seeking to commodify the resources base for commercial gain" (Progress on Drinking Water and Sanitation: 2012 Update UNICEF 2012).

It has been predicted that the drought-affected areas will likely create more stress on already stressed ecosystems of India. Such a situation, in conjunction with manmade interventions, can cause disruptions in river systems. By the year 2050, the average annual runoff in the river Brahmaputra will decline by 14 percent (TERI, *The Energy and Resources Institute* 2004). The Himalayan river systems draining into the Ganga basin are gradually dying out (Tangri, 2003; 213).

There are also some areas predicted to experience extreme precipitation events, with increased frequency and intensity, thus causing enhanced flood risk, especially in the north eastern systems of Mahanadi and Baitarni rivers (Gosain et. al, 2006; 25). Increased frequency and severity of floods and droughts will have implications on the functioning of the ecosystems.

Water yield (which is a function of precipitation, total surface run off, evapotranspiration and soil properties), is projected to increase in the Himalayan region in 2030s by 5–20 percent, however, water yields are likely to be variable across the North Eastern region, Western Ghats, and Coastal region. In some of these regions, it is projected to increase and in others it is projected to decrease. The daily rainfall data for 1954–2003 show vulnerability of Kerala to increasing water scarcity in the pre-monsoon period and a delayed monsoon onset (Pal and Tabbaa, 2009; 62–69).

Water volumes stored in glaciers and snow cover are very likely to decline, reducing summer and autumn flows in the Himalayan river systems in the long run. This shall be a major impact on the breadbasket of India since the Himalayan glaciers feed many major systems of India.

HUMAN HEALTH

Global climate change and associated increases in climate variability will have severe implications for human health with disproportionate effects on countries such as India, which already faces significant public health and health care delivery challenges including resource constraints, high rates of endemic infectious disease, and substantial inequalities in health-care access. ICMR has identified four areas of risks arising from climate change, such as (a) Climate Change and Vector Bornc diseases, (b) Aerosols and Respiratory Diseases, (c) UV-A and UV-B and Corneal Damage and Cataract and (d) Environment and Heart Diseases. Indirect health impacts are therefore a potentially large but under-examined outcome of extreme weather events that lead to a substantial underestimation of the total health burden.

Impact of climate change on dengue also reveals increase in transmission with 2°C rise in temperature in northern India as reviewed in 2010. According to the World Bank, India ranks at number 2 after Bangladesh where about 47 percent of children exhibit a degree of malnutrition. Diseases like diarrhoea, malaria, typhoid, and pneumonia further add misery to the problem. It is estimated that around 37.7 million Indians are affected by waterborne diseases annually, 1.5 million children are estimated to die of diarrhoea alone and 73 million working days are lost due to waterborne diseases each year. The resulting economic burden is estimated at $600 million a year.

Spatial and temporal distribution of vector-borne diseases like malaria, dengue and chikungunya are transmitted by mosquitoes. Their life cycle and development of pathogen in their body are likely to be affected at varying temperature and relative humidity. Studies undertaken in India with A2 scenario on malaria reveal that the transmission window in Punjab, Haryana, Jammu, and Kashmir and north eastern states are likely to extend temporally by 2–3 months and in Odisha, Andhra Pradesh and Tamil Nadu there may be reduction in transmission windows. The link between reemergence of kala-azar in northern parts of India and reappearance of chikungunya mainly in southern India appears to be due to changing climatic conditions which needs to be elucidated.

Rising temperatures are also likely to cause increase in eye diseases like cataract, dry eyes, pterygium, and vernal kerato conjunctivitis and skin diseases. Warmer air temperatures can influence the concentration of regional air pollutants and aeroallergens. Changes in the climate also affect diseases like chronic obstructive pulmonary disease, pneumothorax, and respiratory infections in children.

In view of the observed changes, India has reasons to be concerned about climate change. Its large population depends upon climate-sensitive sectors like agriculture and forestry for its livelihood. Any adverse impact on water availability due to recession of glaciers, decreases in rainfall and increased flooding in certain pockets would threaten food security, cause dieback of natural ecosystems including species that sustain the livelihood of rural households, and adversely impact the coastal system due to sea-level rise and increased extreme events. It will also retard vital national development goals such as habitats, health, energy demand and infrastructure investments. In view of the existing and emerging challenges at the domestic and international level in the area of climate change, it is necessary, therefore, that India should develop a concerted response strategy to minimize its likely adverse impacts on livelihoods, natural ecosystems and growth potential of the country.

INDIA'S POLICY RESPONSE TO CLIMATE CHANGE CHALLENGES

Quite often the systems and policies in developing countries are not equipped to handle even the present climate related stress and climate variability. Income disparities and population growth further constrain the opportunities and equitable access to existing social infrastructure. Projected climate change could further accentuate these conditions. The challenge then is to identify opportunities that facilitate sustainable use of existing resources. It entails considerations that make climate-sensitive systems, sectors, and communities more resilient to current climate variability and to pave the way to enhance their adaptive capacity to future climate change. Faster economic development with more equitable income distribution, improved disaster management, sustainable sectoral policies, and careful planning of capital-intensive and climate-sensitive long-life infrastructure assets are some measures that could assist in ameliorating India's vulnerability to climate change.

India insists for equity and differentiates responsibility in global negotiations on climate change. India is committed to a path of sustainable development. Amongst the key actions needed to achieve this objective are (i) establishing a framework of policy and institutional arrangements and provision of financial resources for implementation of national Missions under the National Action Plan on Climate Change (NAPCC), (ii) preparation of a mitigation strategy and institution of an implementation mechanism for a lower carbon inclusive growth, (iii) building and strengthening the institutional capacity for scientific research into climate change including observation of ecological changes and assessment of climate change, and (iv) involving States in implementing Action Plans on Climate Change at the state level.

India, with 17 percent of the world's population, contributes only 4 percent of the total global greenhouse gas emissions against 30 percent approx. of the United States and 25 percent of the EU countries. In terms of per capita GHG emissions, the divergence in the status becomes blatantly evident when seen against the backdrop of the fact that around 55 percent of India's population still does not have access to commercial energy.

The Ministry of Environment and Forests is the nodal agency for climate change issues in India. India has initiated several climate-friendly measures, particularly in the area of renewable energy. It has one of the most active renewable energy programs besides having perhaps, the only dedicated Ministry for non-conventional energy sources in the world (Ministry of New and Renewable Energy). India adopted the National Environment Policy 2006 which provides for several measures and policy initiatives, to create responsiveness about climate change and help capacity building for adaptation and mitigation measures. Government of India spent 2.6 percent of its GDP during 2006–2007 on adaptation related activities to the existing climate variability. Some of the major schemes/policies significantly addressing adaptation objectives are as follows:· Swarnajayanti Gram Swarozgar Yojana (rural self-employment program);· Sampoorna Grameen Rozgar Yojana (comprehensive rural employment scheme);· Pradhan Mantri Gram Sadak Yojana (Prime Minister's rural roads program);· National Rural Health Mission; Accelerated Rural Water Supply Programme; Desert Development Programme; Major and Medium Irrigation Sustainability of Dryland/Rainfed Farming System; Disaster Management.

On 30 June 2008 India also unveiled its National Action Plan on Climate Change (NAPCC) with a view to lay down the priorities and future actions of the Government for addressing climate change and updating India's national program relevant to addressing climate change. The National Action Plan identifies measures that encourage development objectives while also yielding co-benefits for addressing climate change efficiently. Eight national missions (solar mission, energy efficiency, sustainable habitat, water, Himalayan ecosystem, green India, Eco-green agriculture and knowledge) have been exclusively delineated to simultaneously press forward India's development and climate change related objectives of adaptation and GHG mitigation. Further, the Government has put up an "Expert Committee on Impacts of Climate Change" 2007 to assess the impacts of anthropogenic climate change and to recognize the instruments that government needs to embark on in the prospect to addressing vulnerability to anthropogenic climate change impacts.

Moreover, a Council has also been set up under the Chairmanship of the Prime Minister of India in 2007 consisting of experts to develop a coordinated response to issues relating to climate change at the National level and provide oversight for the formulation of action plans in the area of assessment. "Green India" program envisages undertaking massive afforestation of degraded forests land in the country has been initiated. Monetary support for these programs include funds available under "Compensatory Afforestation Fund Management and Planning Authority (CAMPA)," mobilising funds from the market, developing partner associations, and income from tree felling at ecologically appropriate intervals. The "Green India" program proposes to cover about six million hectare in the country in about ten years.

The existing policies and programs under the Eleventh Plan have made some headway in addressing climate change, for example, building adaptive capacity through ongoing development programs, successful commencement of a National Action Plan on Climate Change, and reducing the energy and emissions intensity of the economy and in the society. However, there is still a substantial potential to enhance the effectiveness of policies and programs to accelerate sustainable development and guide the country along a more proactive path in addressing climate change question. The National Forest Policy also envisages active measures for expanding carbon sinks through increase in forest and tree cover to 25 percent by 2007 and 33 percent by 2012. A major afforestation programme covering 6 mn hectares has been launched under the Eleventh Plan for same purpose.

For the first time, the country is integrating low carbon growth into its 12th Five Year Plan (2012 to 2017). If India effectively implements its current climate change initiatives, by 2020 the country could shrink its emissions intensity by 25 percent over 2005 levels with a "determined-effort" scenario, and by a 33 to 35 percent with an "aggressive effort" scenario. India is making major strides in its efforts to reduce its reliance on fossil fuels and increase its production and use of clean, sustainable energy. The country is already pitched as one of the global leader in wind energy. As the world's fifth largest wind energy producer, the country has onshore wind power potential of up to 3,000 gigawatts (GW). Total grid interactive renewable energy installations reached 25,000 MW in the first quarter of 2012. India's installed solar power jumped to more than 1 GW in 2012, from just 17 MW in 2010. Under the National Solar Mission, India allocated projects totaling 1,000 MW during Phase I. As part of Phase II, which begins in 2013, the Mission plans to develop 20 GW of solar power capacity by 2022, focusing on both grid-connected and decentralized solar projects.

The Indian government also announced plans to double the amount of clean power generated by 2017 under the twelfth Five Year Plan, and levied an innovative coal tax of 50 rupees (approximately $1) per metric ton to create a National Clean Energy Fund. $90 million of the current funds accrued are dedicated to clean energy development, including transmission lines for renewable energy. In 2009, India had created the Indian Network of Climate Change Assessment (INCCA), a comprehensive scientific program that includes 120 research institutions. INCCA has published an emissions inventory for India that assesses the impacts of climate change in the country by 2030, and launched a Black Carbon Research Initiative on the emission sources and impacts of black carbon. In October 2011, India became one of the first developing nations to launch a dedicated satellite, SRMSAT, to monitor its emissions.

The active engagement of States in implementing the NAPCC through preparation of state level action plans is another vital consideration in formulating a cross sectoral strategy. State Governments are preparing State level Action Plans for climate change (SAPCC) and adequate provisions will have to be made either through the developmental plans of the concerned State Governments or the nodal Ministries to support the SAPCCs.

In conclusion, the principle of common but differentiated responsibility in negotiations has to be endorsed to take forward the concept of equalizing per capita emissions of countries, conserving the use of fossil fuels and reducing GHG emissions. It is highly desirable for all, including the governments, to understand and internalize the seriousness of climate change and its varied impacts on the environment and human security. The ways in which society and the economy take corrective measures or adapt to climate change may be as momentous as the direct impacts. While all possible mitigation and adaptation measures should be considered, it necessitates that the climate change issue is not overplayed as cautioned HDR Report 2007/2008 of UNDP that warned that "climate change may undermine international efforts to combat poverty. Climate change is hampering efforts to deliver the MDG promise. Looking to the future, the danger is that it may stall and then reverse progress built up over generations not just in cutting extreme poverty but also in health, nutrition, education, and other areas." This danger needs to be guarded against especially by India as she needs to continue the balancing act of sustainable and inclusive development.

NOTES

1. The IPCC special report "Regional Vulnerabilities to Global Climate Change" explores the potential consequences of changes in climate for ten continental or subcontinental-scale regions. Because of the uncertainties associated with regional projections of climate change, the report assesses the sensitivities and vulnerabilities of each region, rather than attempting to provide quantitative predictions of the impacts of climate change. The report defines vulnerability as the extent to which climate change may damage or harm a system; it is a function of both sensitivity to climate and ability to adapt to new conditions.

2. Definitions of vulnerability range from a focus on physical exposure (Mitchell 1989; Schneider and Chen 1980; Barth and Titus 1984), through measures of socioeconomic status and access to resources (Susman, O'Keefe, and Wisner 1983; Timmerman 1981; Cannon 1994) and sociological investigations of the differential ability of groups to resist harm and to recover afterwards (Drabek 1986; Bolin 1982; Quarentelli 1991), to discussions of how the 'hazard of place' is linked to social profiles (Dow 1992; Cutter 1996). Generally speaking, therefore, vulnerability is the manifestation of social, economic, and political structures, and environmental setting. Vulnerability can be seen as made up of two elements: exposure to hazard and coping capability. People having more capability to cope with extreme events are naturally also less vulnerable to risk.

REFERENCES

Aggarwal, P.K. and A.K. Singh, (2010): Implications of global climatic change on water and food security. In: *Global Change: Impacts on Water and Food Security* [Ringler, C., A.K. Biswas, and S. Kline (eds.)]. Springer-Verlag, Berlin and Heidelberg, Germany, pp. 49–63.

Aggarwal, P.K., and Sinha, S.K., (1993) "Effect of Probable Increase In Carbon Dioxide and Temperature on Productivity of Wheat In India" *Journal of Agricultural Metrology.*

agricoop.nic.in/Climatechange/ccr/india-climate-6-agriculture.pdf

Agricultural Statistics at a Glance 2010 Directorate of Economics and Statistics ; Department of Agriculture and Cooperation; Ministry of Agriculture; Government of India eands.dacnet.nic.in/latest_2006.htm.

Ahmad, Alam and Haseen et al. (2011) Impact of Climate Change on Agriculture and Food Security in India *International Journal of Agricultural, Environment, and Biotech. Vol. 4, No. 2*: 129–137

Ashrit, R.G., K. Rupa Kumar and K. Krishna Kumar (2001): "ENSO-Monsoon Relationships in a Greenhouse Warming Scenario," *Geophysical Research Letters,* 28(9): 1727–30.

Beck, Ulrich. (1992) *Risk Society: Towards a New Modernity*. London: Sage Publications.

Bhadwal S. et al. (2003) *Coping with global change: vulnerability and adaptation in India agriculture.* TERI Report, Venema, H. et al. (eds) IG Printers, New Delhi.

Chaudhry, Anita and Aggarwal, P.K., (2007), "Climate Changes and Food Security in India," Indian Agriculture Research Institute, New Delhi.

Chennai 2005 German Advisory Council on Global Change; Summary for policymakers World in Transition: Climate Change as a Security Risk.

Chung, C.E., and V. Ramanathan. (2006). Weakening of North Indian SST Gradients and the Monsoon Rainfall in India and the Sahel. Journal of Climate 19: 2036–45.

Clark, J.M., Orr, H.G., Freer, J., House, J.I., Smith, P., Freeman, C. (2010) Assessment of projected changes in upland environments Using simple climatic indices. 45: 87–104.

Clark. Moser, Schwarz et al. (1998) Assessing The Vulnerability of Coastal Communities to Extreme Storms: The Case of Revere, MA, USA. *Mitigation and Adaptation Strategies for Global Change*, 3, 59–82.

Cohen, M.J. et al. (2008*) Impact of climate change and bio energy on nutrition*. International Food Policy Research Institute.

Cruz, R.V. et al. (2007) Asia, Climate Change 2007: *Impacts, Adaptation and Vulnerability Contribution of Working Group 2 to the 4th Assessment Report of the Intergovernmental Panel on Climate.*

Change. Parry M.L. (eds), Cambridge University Press; Cambridge.

FAO, 2008, "Climate Change and Food Security: A Framework Document." Food and Agriculture Organization Interdepartmental working group on climate change. www.fao.org/clim/docs/climatechange_foodsecurity.pdf.

Food and Agriculture Organization of the United Nations; 2009 The State of Food Insecurity in the World 2008. Web site: www.fao.org/icatalog/inter-e.htm.

Gasper, Des. 2005. Securing Humanity: Situating "Human Security as Concept and Discourse" *Journal of Human Development* 7(2):221–245.

Global Humanitarian Forum; 2008 *Human Impact Report* Geneva.

Gosain, A.K. et al. (2006): "Climate Change Impact Assessment on Hydrology of Indian River Basins," *Current Science,* 90(3): 346–53.

Gopalakrishnan, R., Jayaraman, M., Bala, G., Ravindranath, N.H., 2011. Climate change and Indian forests. Current Science 101, 348–355.

Goswami, B. N. (2005), *Interdecadal variability, in The Asian Monsoon,* Edited by B. Wang, Springer, New York. pp. 295–328.

Goswami B.N., Venugopal V., Sengupta D., Madhusoodanam M.S., Xavier Prince K., (2006), Increasing Trend of Extreme Rain Events Over India in a Warming Environment, Science, Vol. 314 no. 5804, 1442–1445.

Government of India; Ministry of Environment and Forests: CDM Government of India Interim Approval Criteria. envfor.nic.in:80/divisions/ccd/cdm_iac.html.

Government of India (2008); Climate change & 12th five year Plan Report of subgroup on Climate change; Planning commission; New Delhi.

Government of India (2008), Economic Survey 2007-08, New Delhi.

Government of India (2008); "*National Action Plan on Climate Change.*" Prime Minister's Council on Climate Change: New Delhi.

Government of India (2011) *Climate Change and 12th Five Year Plan Report of Sub Group on Climate Change*; Planning Commission, New Delhi.

Government of India. (2009) "*The Road to Copenhagen: India's position on Climate Change issues,*" Ministry of External Affairs: Public Diplomacy division New Delhi.

H.-M. Füssel and R.J.T. Klein. (2006) Climate Change Vulnerability Assessments: An Evolution of Conceptual Thinking. *Climatic Change*, 75:301–329, 2006.

Hunger and Malnutrition Report (2012). www.siasat.com/english/news/pm-42-malnutrition-childrens-national-shame.

IASC (2009). *Climate Change, Food Insecurity and Hunger, Technical Paper for IASC Task Force on Climate Change*. Submitted by WFP, FAO, IFRC, OXFAM.

ICAR (Indian Council of Agricultural Research), (2009) *DARE/ICAR Annual Report 2008–2009*. Department of Agricultural Research and Education, Ministry of Agriculture, Government of India; New Delhi.

Indian Network for Climate Change Assessment (INCCA) 2010. *Climate Change and India: A 4 × 4 Assessment—A sectoral and regional analysis for 2030s*, Government of India.

Indrani Pala, Abir Al-tabbaab Regional changes in extreme monsoon rainfall deficit and Excess in India 2009 Elsevier, 62–69.

Intergovernmental Panel on Climate Change (IPCC), 2007. *Climate Change 2007: Impacts, Adaptation and Vulnerability: Contribution of Working Group II to the Fourth Assessment Report*. Cambridge University Press, Cambridge, UK 469–506.

IPCC (2001). Climate Change 2001: *Impacts, Adaptation, and Vulnerability*. Cambridge: Cambridge University Press. www.ipcc.ch/ipccreports-/tar/wg1/index.htm.

IPCC (2001a). Climate Change 2001: The Scientific Basis. A Contribution of Working Group I to the Third Assessment Report of the Intergovernmental Panel on Climate Change Cambridge University Press, Cambridge, United Kingdom and New York, NY, USA, pp. 881.

IPCC (2007a). Climate Change 2007: *Climate Impacts, Adaptation and Vulnerability. Working Group II to the Intergovernmental Panel on Climate Change Fourth Assessment Report,* DRAFT Technical Summary 2006, Geneva: Intergovernmental Panel on Climate Change.

IPCC (2007b) Climate Change 2007: *The Physical Science Basis: Summary for Policymakers'* Working Group I to the Intergovernmental Panel on Climate Change Fourth Assessment Report. Geneva: Intergovernmental Panel on Climate Change.

IPCC, *Climate Change 2007: Synthesis Report*, Geneva, 2007.

Izrael Y.A. (2004). On the concept of dangerous anthropogenic interference with the climate system and capacities of the biosphere (in Russian).Meteorol.Hydrol., 4, 30-37.

Kavi Kumar, K.S. and Parikh, Jyoti (2001). "Indian Agriculture and Climate Sensitivity" Global Environmental Change, 11, pp: 147–154

Kelkar Ulka, Narula Kapil Kumar, Sharma Ved Prakash, and Chandna Usha (2008), Vulnerability and Adaptation to Climate Variability and Water Stress in Uttarakhand State, India, *Global Environmental Change*, 18, 564–574.

Kothawale D.R., Munot A.A., Krishna Kumar K. (2010) Surface Air temperature variability over India during 1901–2007 and its association with ENSO. 42: 89–104.

Koty Reddy T., 2010, "Impact of Environment on Poverty in India" *The Indian Economy Review*, Vol. Vii, March Issue, P. 28.

Kumar, K. and J. Parikh (1998) "Climate Change Impacts on Indian Agriculture: The Ricardian Approach" in A. Dinar, R. Mendelsohn et al. (ed.), *Measuring the*

Impact of Climate Change on Indian Agriculture, World Bank Technical Paper No. 402 (Washington DC: The World Bank).

Kumar, K.K., B. Rajagopalan et al. (2006): "Unravelling the Mystery of Indian Monsoon Failures during El Nino," *Science,* 314: 115–19.

Kumar, R., K.K. Kumar et al. (2003): "Future Climate Scenario" in P.R. Shukla, S.K. Sharma et al. (ed.*), Climate Change and India: Vulnerability Assessment and Adaptation;* Orient Longman Hyderabad; 69–127.

Lal, M. (2003) "*Global climate change: India's monsoon and its variability,*" Journal of Environmental Studies and Policy 6: 1–34.

Lavell, A., M. Oppenheimer et al 2012: Climate change: new dimensions in disaster risk, exposure, vulnerability, and resilience. In: Managing the Risks of Extreme Events and Disasters to Advance Climate Change Adaptation Cambridge University Press, Cambridge, UK, and New York, NY, USA, pp. 25–64.

Ministry of Agriculture MoA, (2010) Annual Report, 2009–10. Government of India, Planning Commission, 2007. 11th five-year plan—Vol I–III, New Delhi.

Ministry of Environment and Forests (2009) *State of Environment Report* 2009. MoEF, Government of India, New Delhi.

Ministry of Environment and Forests (2010) *India: Greenhouse Gas Emissions 2007.* Ministry of Environment and Forests, Government of India, New Delhi.

Ministry of Environment and Forests (2010) *India: Taking on Climate Change—24 Recent Initiatives Related to Climate Change.* MoEF, Government of India, New Delhi.

Mitra Amit, 2009, "Climate changes: Adaptation Activities in India," Gorakhpur Environmental Action Group, Gorakhpur U.P.

MNRE (2009), *Renewable Energy at a Glance in India, Renewable Energy-Akshay Urja,* Ministry of New and Renewable Energy, Government of India, 2(5), 48.

MoEF (2008), *Annual Report 2007–08,* New Delhi: Ministry of Environment and Forests, Government of India.

NAPCC (2009), *National Action Plan on Climate Change,* Government of India, Prime Minister's Council on Climate Change, New Delhi.

NATCOM (2004), *India's Initial National Communication to the United Nations Framework Convention on Climate Change,* Ministry of Environment and Forests, Government of India, New Delhi. Available at: http://www.natcomindia.org/natcomreport.htm.

O'Brien, Karen L. and Robin M. Leichenko. 2006. Climate Change, Equity and Human Security. *Die Erde* 137: 165–179.

OECD-FAO (2008) *Agricultural Outlook 2008–2017* OECD-FAO.

Pachauri, R.K. (2009) *Climate Change and its implications for India's fragile ecosystems; The Human Impact of Climate Change and Development*; OECD; New Delhi.

Pal and Tabbaa (2009). Trends in seasonal precipitation extremes—An indicator of 'climate change' in Kerala, India Journal of Hydrology 367 (2009) 62–69.

Parikh Jyoti K. and Parikh Kirit, (2002), "*Climate Change: India's perceptions, Positions, Policies and Possibilities,*" Climate Change and Development, Paris: Organization for Economic Cooperation and Development.

Patwardhan, A., S.H. Schneider and S.M. Semenov, (2003) Assessing the science to Addressunfcccarticle 2: A concept paper relating to cross cutting theme number *Assessing key vulnerabilities and the risk from climate change Chapter 19.* IPCC, 13 pp. www.ipcc.ch/activity/cct3.pdf.

Planning Commission (2011), *Interim Report of the Expert Group on Low Carbon Strategies for Inclusive Growth. Planning Commission*, Government of India, New Delhi.

Raghavan, S. and S. Rajesh, (2003): Trends in Tropical Cyclone Impact: A Study in Andhra Pradesh, India. *Bulletin of the American Meteorological Society*, 84, 635–644.

Rajiv K. and N.H. Ravindranath (2011) "Impact of climate change on Indian forests: a dynamic vegetation modeling approach," *Mitigation and Adaptation Strategies for Global Change*, Volume 16, Number 2, pp. 119–142.

Ranger, N., S. Hallegatte et al. (2011) An assessment of the potential impact of climate change on flood risk in Mumbai. *Climatic Change*, 104, 139–167.

Ravindranath N.H, Joshi N.V., Sukumar R. and Saxena A. (2006), Impact of Climate Change on Forests in India, *Current Science*, 90, 354.

Rawls, John. (1971) *A Theory of Justice*. Oxford: Oxford University Press.

Roy J., A. Ghosh, A. Majumdar, P. Roy, A.P. Mitra and C. Sharma, (2005), Socio-economic and physical perspectives of water related vulnerability to climate change: results of field study in India, *Science and Culture* 71, (8), pp. 239–259.

Sanghi, A. and Mendelsohn, R. (2008). The impacts of global warming on farmers in Brazil and India. *Global Environmental Change*, doi:10.1016/j.gloenvcha.2008.06.008.

Shukla P.R., Rana A., Garg A., Kapshe M. and Nair R. (2003) *Climate Change Assessment for India: Applications of Asia Pacific Integrated Model (AIM)*, Universities Press, Hyderabad.

Singh Nityanand and Ranande Ashwini (2009) The History of Climate Change in India, *Geography and You*, March–April, 30–35.

Sinha, S.K. and Swaminathan, M.S. (1991), "Deforestation Climate Change and Sustainable Nutrients Security," *Climate Change* 16, pp. 33–45.

Sinha, S.K. Singh, M. Rai, (1994), "In Decline in Crop Productivity In Haryana and Punjab: Myth or Reality?" *Indian Council of Agricultural Research*, New Delhi, p. 89.

Suchak, Y.K. (2002). *Development and Environment Issues with Special Reference to Gandhian Perspective*. Ghandian Institution, Bombay Sarvodaya Mandal www.mkgandhisarvodaya.org/kavita_suchak.htm [Geo-2-438].

Tangri, A.K. (2003), "Impact of climate change on Himalayan glaciers," Proceedings of V and A Workshop on Water Resources, Coastal Zones, and Human Health, NATCOM, *27–28 June*, Indian Institute of Technology (ITT), New Delhi.

Teri (2004), *India Specific Impacts of Climate Change*, Tata Energy Research Institute, New Delhi.

Thywissen, K. (2006) *Components of Risk. A Comparative Glossary*. UNU-EHS, Source No. 2/2006.

UNDP (1999). Human Development Report 1999, Oxford: Oxford University Press.

UNEP (1989). Criteria for Assessing Vulnerability to Sea Level Rise: a Global Inventory to High-risk Areas, Delft, The Netherlands: Delft Hydraulics.

UNEP (2003), *Assessing Human Vulnerability due to Environmental Change: Concepts, Issues, Methods and Case Studies*. UNEP's Global Environment Outlook (GEO 3) Report, USA.

UNICEF and World Health Organization (2012); *Progress on Drinking Water and Sanitation*: 2012 Update.

UNISDR, 2009a: Adaptation to Climate Change by Reducing Disaster Risk: Country Practices and Lessons. Briefing Note 02, United Nations International Strategy for Disaster Reduction, Geneva, Switzerland.

UNICEF/WHO Joint Monitoring Programme for Water Supply and Sanitation; United States of America.

Venkateshwarlu, B. (2009) *Climate Change and Sustainable Agriculture: Securing the Small and Marginal Farmers in India,* The Human Impact of Climate Change: Policy Notes for Parliamentarians, CLRA; New Delhi.

WBGU (2008) German Advisory Council on Global Change "World in Transition: Climate Change as a Security Risk." London: *Earthscan*. www.wbgu.de/wbgu_jg2007_engl.html. Accessed December 2012.

World Bank (June 2008) *Climate Change Impacts in Drought and Flood Affected Areas: Case Studies,* Report No 43946-IN.

World Bank (2006). *Overcoming Drought: Adaptation Strategies for Andhra Pradesh, India.* Directions in Development: Environment and Sustainable Development. Washington, D.C.: World Bank.

World Bank. (2009) *South Asia: Climate Change Strategy*. Draft Report, World Bank: South Asia Region.

World Commission for Environment and Development WCED (1987) 'Our Common Future' (Oxford: Oxford University Press).

8

Climate Change and Its Impacts

Domestic and International Responses of Japan

Cheng Fang-Ting

In Japan, the climate change and its impacts are a major topic of concern, and they actively support international efforts to face the challenges posed by global warming. Japan's post World War II economic development, rapid industrialization and urbanization have both positive and negative impacts on its life style, consumption habits and dependence on fossil fuels. These factors also impact Japan's environmental health and economic and social security. Following is an analysis of two main aspects of Japan's environmental politics:

- Impact of climate change on its weather, meteorological conditions and overall human security
- How Japanese policymakers and society in general responds to climate change

CLIMATE CHANGE AND NATURAL DISASTERS IN JAPAN

Climate Change in Japan

In Japan, according to the recent data from five observation stations, concentration of greenhouse gases shows similar trends as in the rest of

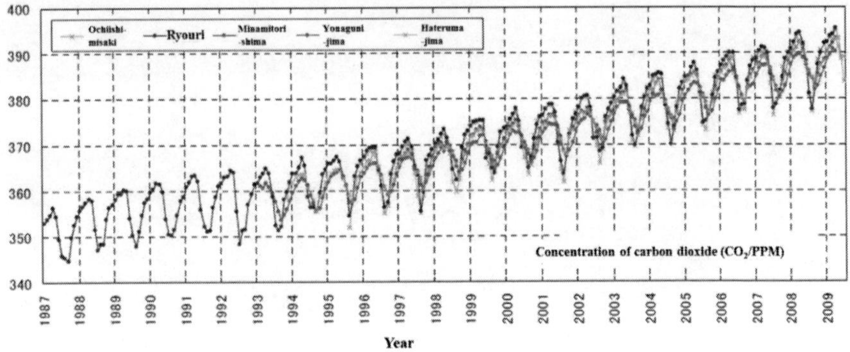

Figure 8.1. Observed results of the concentration of greenhouse gases in Japan, 1987–2009 (CO_2)[1]

Source: "A Unified Report of Observation, Prediction and Impact Assessment on Global Warming: Climate Change in Japan and its Influence," October, 2009

the world. They show an increase in concentration of CO_2, CH_4, and N_2O (Figures 8.1, 8.2, and 8.3). (*the Unified Report*, 2009).

Cold and Heat Wave

It has been observed that increased concentration of greenhouse gases has also contributed to a rise in temperature. Since 1898, the temperature in Japan has increased more than 1.1 degrees Celsius, but since the late 1980s the rise in temperature has been more frequent—more "hot years"

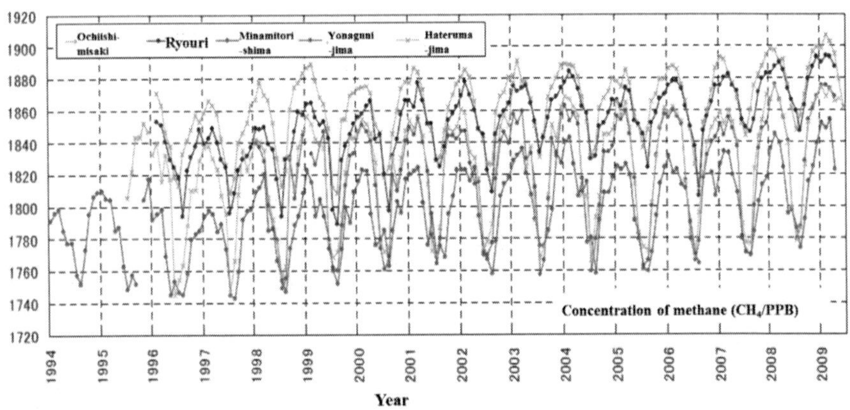

Figure 8.2. Observed results of the concentration of greenhouse gases in Japan, 1994–2009 (CH_4)

Source: "A Unified Report of Observation, Prediction and Impact Assessment on Global Warming: Climate Change in Japan and its Influence," October, 2009

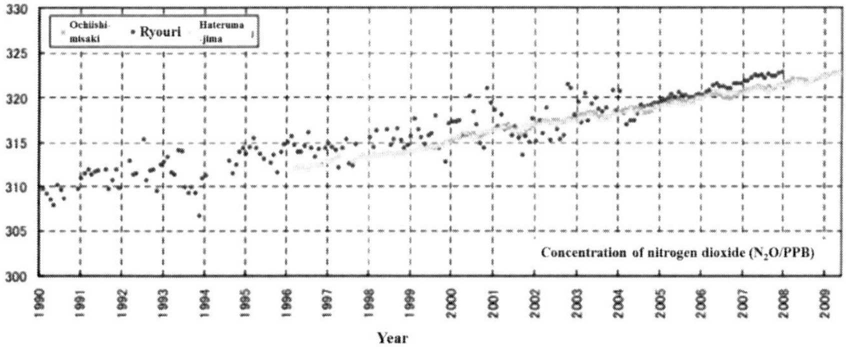

Figure 8.3. Observed results of the concentration of greenhouse gases in Japan, 1994–2009 (N_2O)

Source: "A Unified Report of Observation, Prediction and Impact Assessment on Global Warming: Climate Change in Japan and its Influence," October, 2009

specifically in Tokyo metropolitan and other major cities situated on the seas (Figure 8.4 and 8.5) (the heat island effect) has exceeded the mean temperature in Japan (1.1 degrees Celsius). In the last one hundred years the temperature in Tokyo has risen 3 degrees Celsius, while in Sapporo, Nagoya, Osaka, and Fukuoka the increase has been around 2 degrees Celsius (*the Unified Report*, 2009).

The bar shows the difference between each year and the average years. The dark line shows five-year average of the difference, while the straight line demonstrates a long-term tendency of the change in temperature.

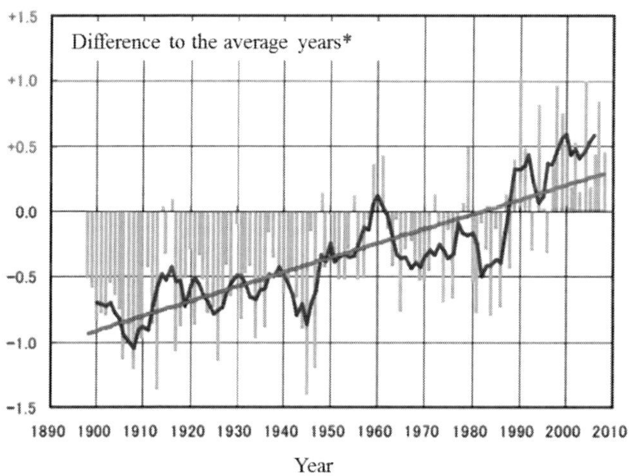

Figure 8.4. Annual Temperature in Japan 1890–2010[2]

Source: "A Unified Report of Observation, Prediction and Impact Assessment on Global Warming: Climate Change in Japan and its Influence," October, 2009

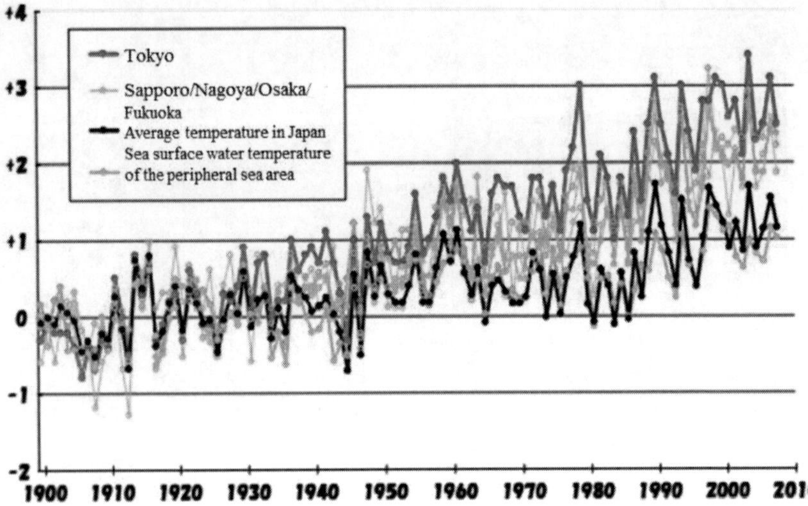

Figure 8.5. Temperature of the largest cities in Japan, the average temperature in Japan, and changes in Sea surface temperature of the peripheral sea area 1900–2010[3]

Source: "A Unified Report of Observation, Prediction and Impact Assessment on Global Warming: Climate Change in Japan and its Influence," October, 2009

It is predicted that by 2100 summers and winters will become warmer, as shown in Figure 8.6; the temperature is predicted to increase 4.0°C, 3.2°C, and 2.1°C (IPCC).

It has been observed that the maximum temperature on "hot days" will exceed 30 degrees Celsius, on "extremely hot days" it will exceed 35 degrees Celsius and on "sultry nights" it will fall below 25 degrees Celsius (Figure 8.7).

EXTREME WEATHER EVENTS

Typhoons

It in generally accepted that climate change will result in extreme weather such as typhoons, floods, and major ice storms. Between 1980 and 2010, 35.6 typhoons reached Japan's vicinity. While 11.4 typhoons touched Japan only 2.7 typhoons reached its coastlines every year on average (Figure 8.8). From Figure 8.8, however it is hard to assert that typhoons increased in number during the past sixty years. In addition, the power of typhoons does not show an obvious trend of strengthening (Figure 8.9). The typhoons accompanied by heavy rainfalls took a high toll on human life despite improved protective facilities to manage these disasters (Table

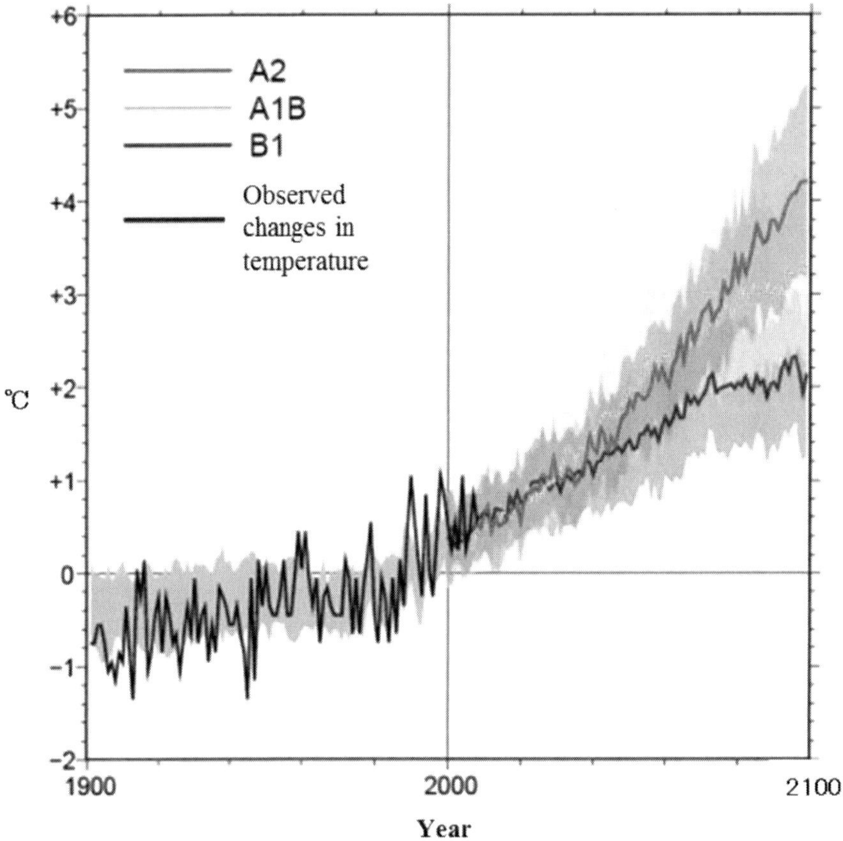

Figure 8.6. Observation and Projections of Changes in Temperature in Japan 1900–2010[4]

Source: "A Unified Report of Observation, Prediction and Impact Assessment on Global Warming: Climate Change in Japan and its Influence," October, 2009

8.1). For instance, in 2004 Japan was hit by typhoon number twenty-three causing ninety-eight deaths and lost people. In late summer 2011 Kinki area and Shikoku Island were hit by typhoon number twelve causing a considerable loss of human life and property.

Floods and Water-related Disasters

Floods and water-related disasters are other serious problems faced by Japan. Heavy rainfalls have caused large-scale loss and dislocation of human life. In the last three decades, heavy rainfalls have resulted in 635 deaths. The number of dead goes up to 1,522 if typhoons and snow damages are included (Table 8.1). And despite advanced technologies and

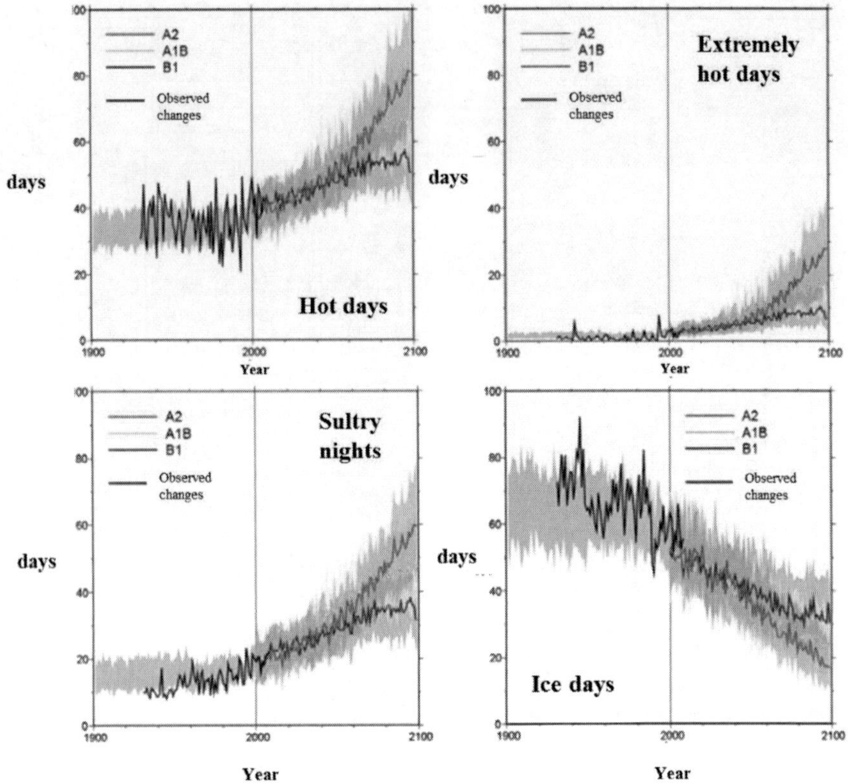

Figure 8.7. Observations and Projections of hot days, extremely hot days, sultry nights, icy days during summer and winter in Japan 1900–2100[5]

Source: "A Unified Report of Observation, Prediction, and Impact Assessment on Global Warming: Climate Change in Japan and its Influence," October, 2009

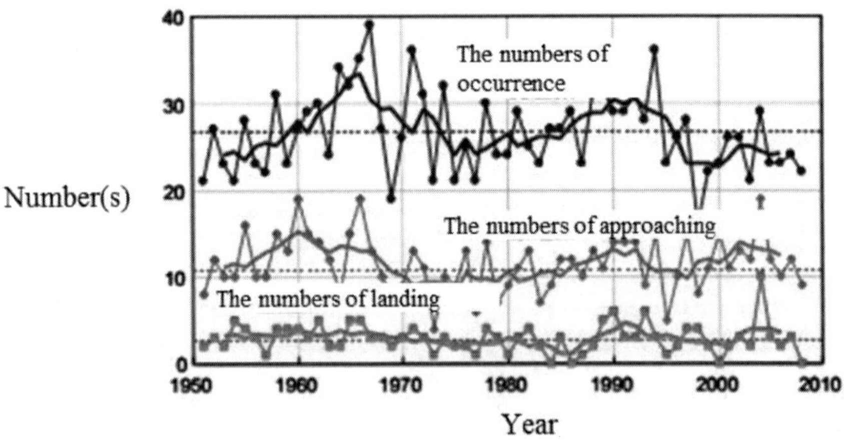

Figure 8.8. The number of occurrences of typhoons, the number of typhoons that reached near Japan, and the number of typhoons that landed in Japan, 1950–2010

Source: "A Unified Report of Observation, Prediction and Impact Assessment on Global Warming: Climate Change in Japan and its Influence," October, 2009

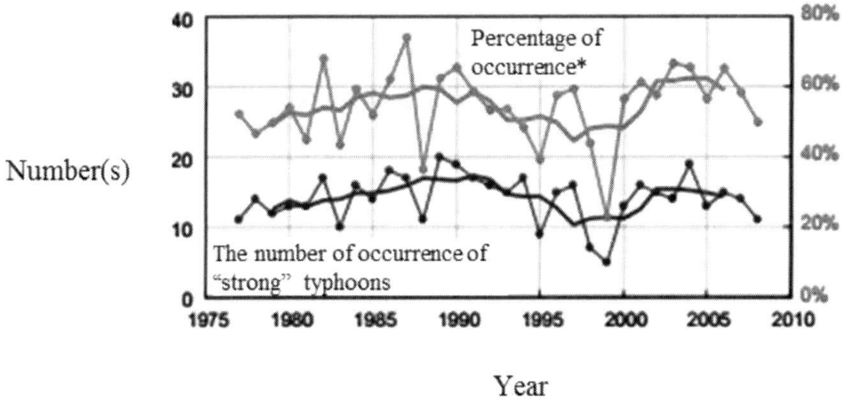

Figure 8.9. Number of occurrences of "strong" typhoons and their percentage to the total number of typhoons 1975–2010[6]

Source: "A Unified Report of Observation, Prediction and Impact Assessment on Global Warming: Climate Change in Japan and its Influence," October, 2009

improved managements have failed to prevent loss of human life. Since the 1970s the number of heavy rainy days have increased (Figure 8.10), and it is projected that the number of these days will increase by the end of the twenty-first century.

According to several IPCC scenarios, annual average precipitation in Japan will increase 5 percent by the end of the twenty-first century compared to the twentieth century. Although there is still uncertainty in simulation models (since the observed precipitation varied much wider than what had been predicted), global warming, along with an increase in the amount of water vapor would continue to cause heavy rainfall (daily rainfall that exceeds 100mm) resulting in floods or landslides (Figure 8.11). In addition, in the models of climate prediction, the Baiu (the long spell of rainy weather in early summer) front tends to be stagnant around Japan due to the weakened force approaching from the north; delaying the end of rainy season.

Sea Level Rise

Besides typhoons, heavy rainfalls and water-related disasters, the third dimension of impacts of climate change is loss of sandy beaches and expansion of inundated areas due to the rise in sea level. According to the Model for Interdisciplinary Research on *Climate (MIROC)*, the increase in days of heavy rain and in precipitation are predicted to bring about storm surge, river and flood damages, sediment-related disasters as well as damages accompanied by strong typhoons in Japan. Based on the above model, the inundated areas in western Japan will expand by 2100 *(The Unified Report,* 2009).

Table 8.1. Deaths and Missing Persons Due to Disaster Related to Weather Events, 1945–2011

Date	Disaster	Location	Death & Missing
1945.9.17 ~ 18	Typhoon Makurazaki	West Japan (Hiroshima)	3,756
1947.9.14 ~ 15	Typhoon Kathleen	North of Tokai area	1,930
1948.9.15 ~ 17	Typhoon Ione	Shikoku to Tohoku area (Iwate)	838
1950.9.2 ~ 4	Typhoon *Jane*	North of Shikoku island(Osaka)	539
1951.10.13 ~ 15	Typhoon Ruth	Whole country (Yamaguchi)	943
1953.6.25 ~ 29	Heavy rain	Kyushu island, Shikoku island, Chugoku area (Kitakyushu)	1,013
1953.7.16 ~ 24	Nanki Heavy rain	West of Tohoku area (Wakayama)	1,124
1954.5.8 ~ 12	Wind damage	North Japan, Kinki area	670
1954.9.25 ~ 27	Typhoon Touyamaru	Whole country (Hokkaido, Shikoku island)	1,761
1957.7.25 ~ 28	Isahaya Heavy rain	Kyushu island (Isahaya area)	722
1958.9.26 ~ 28	Typhoon Kanogawa	East of Kinki area (Shizuoka)	1,269
1959.9.26 ~ 27	Typhoon Isewan	Whole country (Aichi, besides Kyushu island)	5,098
1963.1.	January Heavy snowfall	Hokuriku area, Sanin area, Yamagata, Shiga, Gifu	231
1965.9.10 ~ 18	Typhoon #23, 24, 25	Whole country(Tokushima, Hyogo, Hukui)	181
1966.9.23 ~ 25	Typhoon #24, 26	Chubu area, Kanto area, Tohoku area (Shizuoka, Yamanashi)	317
1967.7. ~ 8.	Heavy rain (July, August)	West of Chubu area, South of Tohoku area	256
1972.7.3 ~ 15	Typhoon #6, 7, 9, Heavy rain (July)	Whole country (Kitakyushu, Shimane, Hiroshima)	447
1976.9.8 ~ 14	Typhoon #17, Heavy rain (September)	Whole country (Kagawa, Okayama)	171
1977.1.	Snow damage	Tohoku area, (North of Kinki area, Hokuriku area)	101
1979.10.17 ~ 20	Typhoon #20	Whole country (Tokai area, Kanto area, Tohoku area)	115
1980.12. ~ 81.3.	Snow damage	Tohoku area	152
1982.7. ~ 8.	Heavy rain (July, August, Typhoon #10	Whole country (Nagasaki, Kumamoto, Mie)	439
1983.7.20 ~ 29	Heavy rain (front)	East of Sanin area (Shimane)	117
1983.12. ~ 84.3.	Snow damage	Tohoku area, Hokuriku area (Niigata, Toyama)	131
1993.7.31 ~ 8.7	Heavy rain (August)	Whole country	79
2004.10.20 ~ 21	Typhoon #23	Whole country	98
2005.12. ~ 06.3.	Heavy snowfall	Hokuriku area, the Sea of Japan side	152
2010.12. ~ 11.3.	Snow damage	North Japan to West Japan, the Sea of Japan side	128
2011.8.29 ~ 9.7	Typhoon #12	Kinki area, Shikoku island	94
2011.11. ~ 2012.3.	Heavy snowfall (November)	North Japan to West Japan, the Sea of Japan side	132

Source: White Paper on Disaster Management 2011, White Paper on Disaster Management 2012, the Cabinet Office, Government of Japan, http://www.bousai.go.jp/hakusho/h24/bousai2012/html/honbun/4b_8s_17_00.htm (downloaded on October 25, 2012).

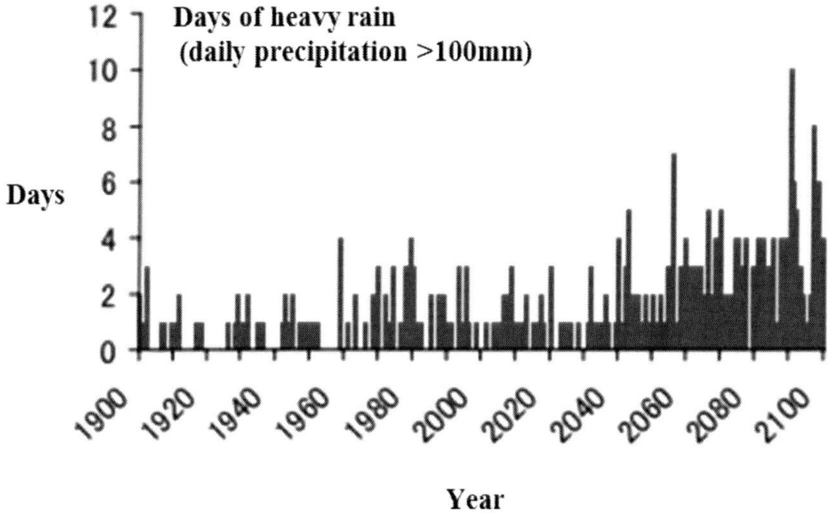

Figure 8.10. Observations and Projections of changes in days of heavy rains (100mm) during summer in Japan 1900–2100

Source: "A Unified Report of Observation, Prediction and Impact Assessment on Global Warming: Climate Change in Japan and its Influence," October, 2009

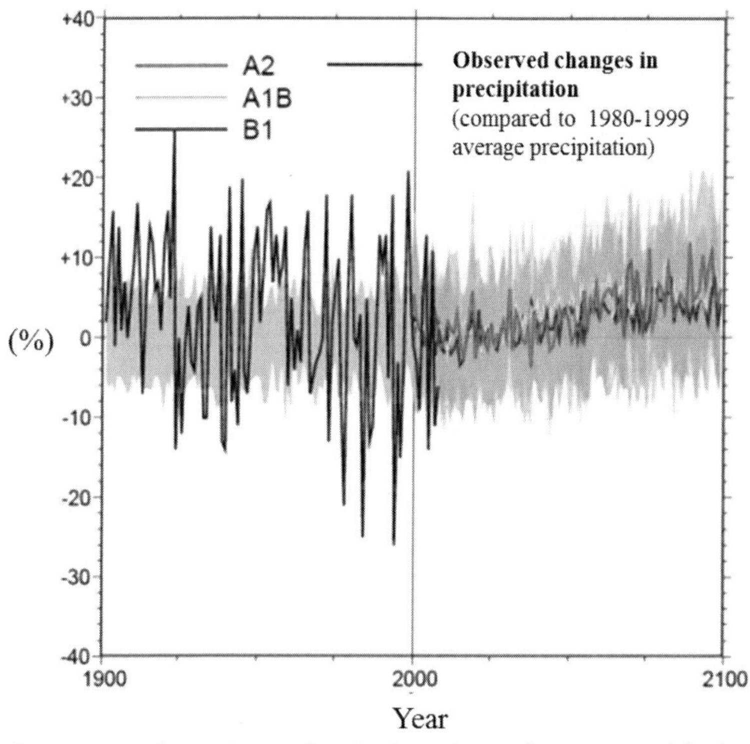

Figure 8.11. Observations and Projections of Annual average precipitation in Japan 1900–2100[7]

Source: "A Unified Report of Observation, Prediction and Impact Assessment on Global Warming: Climate Change in Japan and its Influence," October, 2009

THE IMPACTS OF CLIMATE CHANGE TO HUMAN SECURITY IN JAPAN AND ITS RESPONSE

As discussed above Japan has experienced numerous natural disasters such as typhoons, heavy rainfall's and other meteorological events causing huge loss of property and human life. Even though the capabilities and facilities of disaster management have improved considerably, climate change related impacts have been producing harmful influences to the whole country. Under the Kyoto Protocol Japan has made efforts to reduce greenhouse gas emission by 6 percent from 1990 level by making use of flexible mechanisms such as emission trade (ET) and clean development mechanism (CDM). However, these efforts do not necessarily contribute to the mitigation of the ongoing climate change owing to weakness of the Protocol (Keohane and Raustiala 2008; Keohane and Victor 2010; Victor 2004, 2011). In order to tackle the climate change impacts, Japanese government and municipalities have been taking measures to adapt or prevent adverse influences from causing more serious damages to the country. Following is a discussion of three cases:

Observation, Modeling and Assessment: the Establishment of "the Analysis and Assessment Group on Extreme Weather" and Important Research Projects

The Analysis and Assessment Group on Extreme Weather, along with its supportive body "Working Group on Analysis of Extreme Weather" were founded on June 12, and on October 1, 2007 respectively under Japan Meteorological Agency (JMA) after the huge damage caused by the heavy snow fall in 2006. The purpose of the group is to analyze and assess the factors responsible for extreme weather. The Group functions in cooperation with other experts and academic research institutes (Japan Meteorological Agency, 2007). The group analyzes past extreme weather events in or around Japan mainly due to abnormal atmospheric circulation, and lasted for a relatively long period of time (around two weeks). It studied the extremely hot summer of 2004, heavy snowfall of 2006 and the warm winters of 2006 and 2007. It functions as an advisory body on short-term events such as typhoons, heavy rainfalls, and squalls, and analyzes the relationship between weather events and climate change. As a specialized body it provides a variety of data, assessments and analyses to JMA. Since August 2008, the Agency has released eleven reports on weather events such as extreme high or low temperature, heavy or lack of rainfall, remarkable cold wave, lack of sunshine and heavy snow (JMA, 2007–2012). In recent years these efforts have helped to understand the relationship among ongoing climate change, characteristics of atmospheric circulation and unusual weather events (*the Unified Report*, 2009).

For Japan, as well as other countries, one of the most urgent tasks is to understand climate change and its influences by taking advantage of latest knowledge and technologies. In January 2009 Japan launched the world's first Greenhouse gases Observing Satellite (GOSAT), named *IBUKI*, to precisely observe global distribution and temporal variations of methane and CO_2 concentration. Complemented by ground and civil air observation, GOSAT project is expected to accurately estimate regional absorption and emission of greenhouse gases (GOSAT, JAXA).

By pursuing effective modeling of climate change and greenhouse gases absorption/emission, Japan has developed its climate change prediction models (*"Innovative Program of Climate Change Projection for the 21st Century"* or *"KAKUSHIN Program," 2002–2012)1* under Ministry of Education, Culture, Sports, Science, and Technology (MEXT) to contribute to IPCC assessment reports including the coming 5th Assessment Report. In order to improve the modeling's accuracy, the program aims to further take advantage of the super computer *Earth Simulator* for long-tern (up to the year 2300), near future (in twenty to thirty years) about extreme weather event, typhoons, heavy rainfalls/snowfalls, projections (*Executive Report of KAKUSHIN Program*, 2012).[8] To monitor climate change's real impacts by understanding its long-term changes continues to be an extremely important task for Japan in the near future. Data from the research program will be broadly used for the assessment of vulnerability and adaptation to climate change, and for the quantification of damages related to adverse climate effects (*the Unified Report*, 2009).

Enhancement of Preventive Measures: Summer Heat Stroke and Corresponding Actions

According to current projection, the possible impacts brought by climate change may cause significant adverse influence on a wide range of fields that closely relate to people's daily lives. It is estimated that by the end of the twenty-first century, the loss of property will approximately reach to 2,040 billion USD ($1=100 JPY, at present value) because of floods, sediment disasters, vanishing of suitable habitats for beech forest, loss of sandy beaches, storm surge damages in western Japan, and higher human mortality due to heat stroke. The data shows that heat stroke during summer in Japan have caused tremendous damages to human health. For instance, in 2008 the extremely high temperature caused 23,071 casualties (including 47 deaths). The number went up to 53,843 (including 167 deaths) in 2010 due to the increase in the number of extremely hot days and nights (Figure 8.12). Although heat stroke

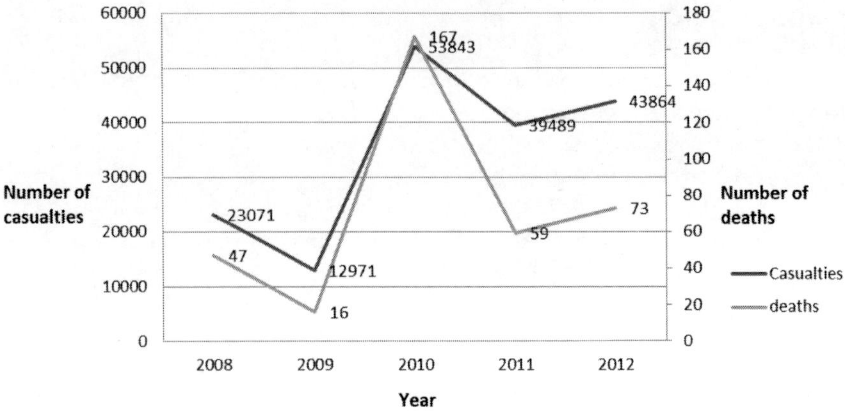

Figure 8.12. Number of casualties and deaths caused by Heat strokes during the summer (July–September 2008–2012)

Source: Information of Heat Stroke (2008–2012), Fire and Disaster Management Agency (FDMA), http://www.fdma.go.jp/neuter/topics/fieldList9_2.html

is not considered as a serious type of disaster and has never been included into *the White Paper on Disaster Management*, the damages can be comparable to natural hazards such as typhoons, floods, and rainfalls/snowfalls (*the White Paper on Disaster Management*, 2012).

In Japan the Liaison Conference of Related Ministries and Agencies on Heat Stroke, is the Fire and Disaster Management Agency (FDMA), MEXT, the Ministry of Health, Labour and Welfare (MHLW), the Ministry of Agriculture, Forest and Fisheries (MAFF), JMA and the MOE, had its first meeting on December 21, 2007. To integrate efforts of government's compartmentalized public administration dealing with heat stroke damages. The MOE serves as the supervisory authority to heat stroke, provides prompt information on the heat index (WBGT)[9], risk and real-time casualties of heat stroke, and suggests approaches to adapt to high temperature during day and night time (MOE Website of Information for Preventing Heat Stroke). Besides government's actions, information related to heat stroke is delivered to municipal, medical educational institutions and media (TV and any other forms of public broadcasts, radio, and so on.) to raise citizens' awareness in coping with extreme high temperatures. Despite various levels of governmental efforts to raise citizen awareness on climate change.

The social consciousness has been moderate on the heat stroke issue specifically among highly vulnerable populations such as elder people and children. In 2012, 19,848 casualties among elderly people, which accounted for 45.2 percent of the total casualties. To prevent vulnerable populations, especially the elderly (above sixty-five-years-old), children, young students and workers from being harmed by heat related illness, the preventive methods play important roles to mitigate possible dangers. In recent years governments at all levels have taken preventive actions against heat stroke.

During the summer of 2012, the governmental activities of combating heat stroke involved various ministries and agencies. The JMA provides observed data of temperature and related meteorological information. In addition, besides the publication of WBGT index, MOE works with MHLW and JMA to take precautionary measures against harmful heat stroke in daily life, especially for the elderly. Furthermore, MEXT, MHLW and MAFF respectively provide preventive and counter measures against heat stroke in schools, workplaces and agricultural fields. Meanwhile, FDMA is in charge of ambulance services, as well as providing statistical information on casualties and related damages, and heat stroke countermeasures. The Ministry of Economy, Trade and Industry (METI) and MOE are responsible for publicizing accurate power saving behaviors (Website for Information of Heat Stoke, MOE 2012). After the Tōhoku earthquake and Fukushima accident in March 2011, governmental administrative offices, educational institutions, commercial buildings, corporations, stores and private houses in eastern Japan area have been encouraged to reduce power consumption and to refrain from using electricity extravagantly. However, the concern of excessive power saving emerges as a new challenge of dealing with heat stroke because the attempts of power saving may cause dangerous heat illness during extremely hot days by not using air conditioning. Therefore METI and MOE started a new initiative in 2012 to call for appropriate use of electricity to avoid possible damages. Indeed, there is a need to educate users to appropriately take advantages of cooling facilities in order to prevent severe negative effects from high temperatures. These agencies have organized various workshops and lecture classes to train public health nurses, staffs, district welfare officers and event organizers of local governments to enhance preventive and counter measures in accordance with different characteristics of heat stroke in each region. In addition to this, both of the ministries have promoted researches and surveys on heat stroke/heat stress and its relationship with global warming to understand the realities of adverse influence of the problem, and further establish early warning systems/networks and take measures urgently needed to be implemented in the future (Japanese Government, June 2012).

Adaptation: Adverse Impacts on Agriculture, and Local Responses

The negative influences of climate change on Japanese agriculture include production and growth conditions of paddy rice, wheat, soybean, vegetables, flowers and ornamental plants, forage and crops, and livestock.

According to the survey conducted by the National Agriculture and Food Research Organization (NARO), the negative effect of global warming to fruit growing can be witnessed all over the country. The negative effects of global warming to vegetables, flowers and ornamental plants are reported to be around 90 percent in all prefectures (Sugiura, 2007). In recent years, the negative influences to paddy rice production have become serious in at least 70 percent of all prefectures (Figure 8.13). It is to be noted that paddy rice is one of the most important staple foods to Japanese people.

Compared to paddy rice cultivation, fruit crops in general have lower adaptive capability against weather and temperature. Hence fruit production can be profoundly affected by changing weather and temperature. Low temperature triggers poor germination in spring because buds need longer period to be dormant in winter. The shortened winter reduces the time to be dormant and flowering and fruition. Global warming weakens the level of fruit's freezing tolerance which eventually makes fruit more vulnerable to snow/frost damages. Another problem of fruit production in Japan is the variation of suitable growing areas. As a result of the global warming and climate change suitable areas of to grow satsuma mandarins (Unshu-mikan) will shift gradually

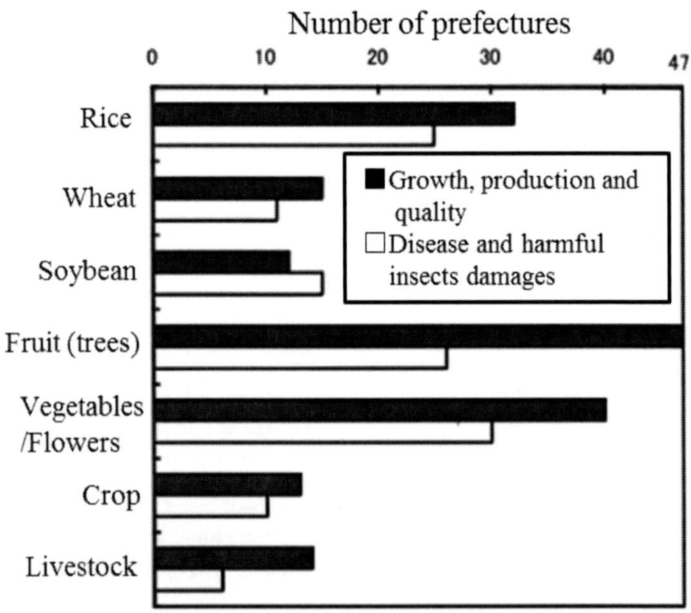

Figure 8.13. Number of prefectures with at least one problem caused by global warming

Source: Sugiura, 2007: 23.

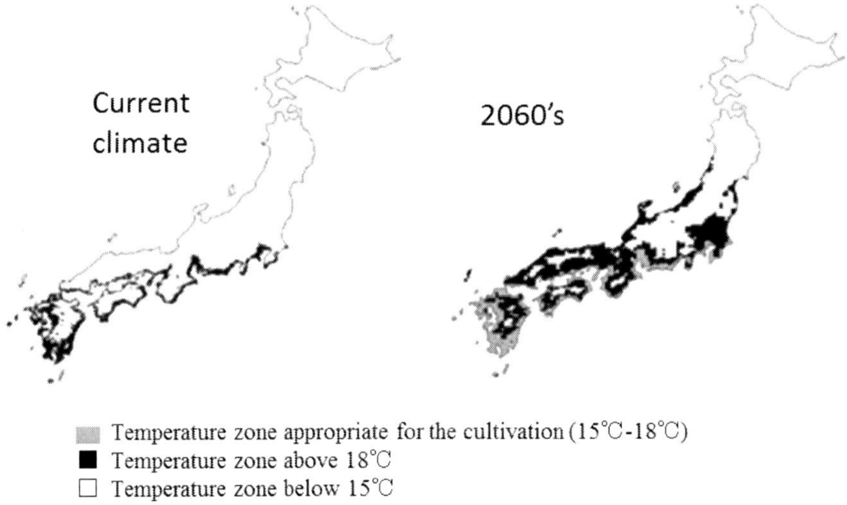

Map 8.1. Spatial distribution of favorable regions for satsuma mandarin production, assuming that the annual mean temperature appropriate for satsuma manjorin ranges 15–18°C

Source: Sugiura, Yokozawa, 2004, p.76.

to the north and inner lands (Map 8.1) (The ideal annual mean temperature to grow Satsuma mandarins is between 15 to 18 degrees Celsius). Therefore, by 2060 it will be quite difficult to grow these mandarins in coastal areas of the Seto Inland Sea, and Kyushu Island.

Similar but different problem will be faced for cultivation of paddy rice (Map 8.2). It is estimated that paddy rice production will increase in the northern Japan while decrease slightly in western Japan between 2046 and 2065. On the other hand, as the influence of climate change strengthens and expands in central Japan and Kyushu Islands it has been observed that deterioration of aquatic resources' quantity and the increase of pests damages may lower down the rice production between 2081 and 2100 (*the Unified Report*, 2009).

Other problems include the outbreak of red mold disease, declining productions due to the shortened ripening period, a decline in quality, and frost damage to wheat cultivation. It has been pointed out that there is an increase of pest harmful to soybean production. The production of vegetables and flowers, has also suffered in recent years do to a shortened growing period and stagnation of growth during summer (Sugiura 2007). The drop in production during summer and increase during winter can affect the balance between domestic supply and demand, further adversely affecting the market of meat and processed meat products. And for the livestock, it is reported that deaths due to heat stroke, increase of

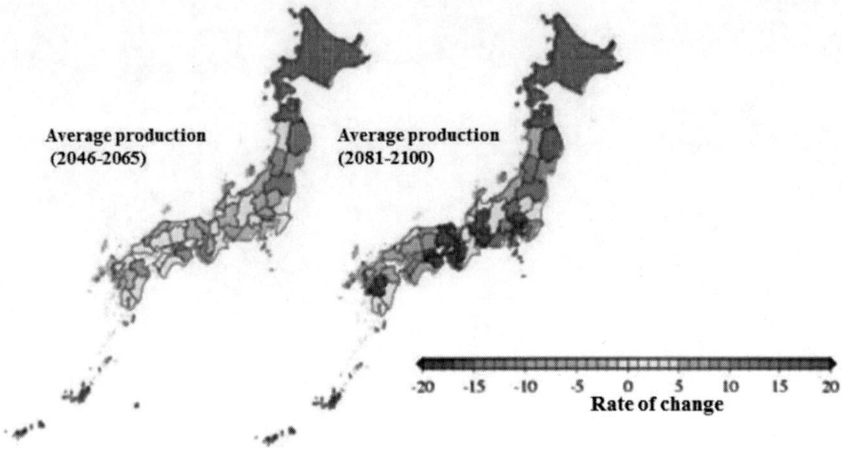

Map 8.2. Projections of Changes in the yield of Paddy rice in Japan, 2046–2065; 2081–2100

Source: "A Unified Report of Observation, Prediction and Impact Assessment on Global Warming: Climate Change in Japan and its Influence," October, 2009

reproductive disorders, decrease in milk production, decrease in intake of forage and increase of pests have taken place.

Last but not least, the reduced quality and quantity of paddy rice production can be another problem for Japanese agriculture. One of the most serious problems includes "chalky grains" in paddy rice. Chalky grain means kernels of grain which have some portion as opaque or milky white in color. According to a survey, twenty-two prefectures out of twenty-seven reported increased of chalky grains due to global warming. Besides the chalky grain issue, cracked rice, increase of sterility due to high temperature and frequent occurrence of spotted rice, are also reported in many prefectures. The increase of pests and related damages to soybean, vegetables, and livestock's has also adversely affected the yield of paddy rice in Japan adversely (Nishimura, 2006). Various preventive measures are being taken by farmers. Three pillars constitute these adaptive measures: 1) Adjusting the timing of transplantation, 2) Efficient water management, and 3) Control of annual production (Sugiura, 2007); (Table 8.2).

Several new approaches have been introduced to satsuma mandarin production to improve the quality and various other problems. Reflection sheets are often used for improving the efficiency of photosynthesis, to solve the problem of poor coloration. In addition, the application of agricultural chemicals (Ethychlozate) to improve physiological disorder, simple storage with Tyvek, the DuPont's brand of high-density polyethylene fibers are also being used (Table 8.3).

Table 8.2. Measures for the Occurrence of Chalky Grains

Purpose	Methods
Optimization of timing of transplantation	■ Early cultivation ■ Later cultivation ■ Decentration of the seasons for transplantation
Response by water management	■ Intermittent irrigation ■ Constant flowing irrigation ■ Prevention of early drainage ■ Waterlogging treatment ■ Water saturation management ■ Evening water conduction ■ Deep ponding irrigation
Control of annual production	■ Appropriate application of panicle fertilizers ■ Sparse planting ■ Reduction of use of basal fertilizer ■ Ditching and midseason drainage ■ Prevention of excessive branching
Others	■ Selection and implementation of corresponding varieties ■ Enhancement of soil development ■ Optimization of selection and adjustment work ■ Cultivation by direct seeding

Source: Sugiura, 2007.

Table 8.3. Measures for the Satsuma Mandarin Adversely Affected by Global Warming

Damages caused by global warming		Examples of implemented measures
Fruit	Poor coloration	· Use of reflection sheets
	Physiological disorder	· Application of agricultural chemicals (Ethychlozate) · Application of calcium · Anti-glare planting
	Drop of storability	· Simple storage with Tyvek · Early harvest · Application of calcium
	Biennial fruiting	· Alternate fruiting cultivation · Fruit thinning (upper part of tree canopy)
Flower bud	Lack of spontaneous dormancy and diapause termination	· Temperate cultivation in open field · Temperature and humidity management · Review of the timing of artificial heating season
Body of tree	Frost damage	· Straw winding
	Leaf scorch and sunburn of trees' body	· Application of whiten powder (calcium) on the body

Source: Sugiura, 2006 & 2007.

However, results of these measures are not easy to assess. Also as global warming and climate change become prominent and serious, these preventive measures in agriculture might have to be reevaluated. Application of new technology along with adaptable varieties of crops and other agricultural products are constantly needed. Also changing climate and warming temperatures are not always negative to Japanese agriculture (Sugiura, 2007; Nishimura, 2006).

CONCLUSION AND PROSPECT

For several decades, measures/suggestions such as mitigation, adaptation and integration have been made, Japan has leaned more towards facilitating adaptation. To understand the causal relationship between the extreme weather events and climate change, Japan has used advance technology such as satellites and computers to achieve information on climate change and analyze the adverse effects of these events on human security.

In this chapter, three cases were introduced from different perspectives: In the first observation, modeling and assessment of climate change. The Japanese government has established "the Analysis and Assessment Group on Extreme Weather" under MOE to provide prompt, transparent and accurate information and analysis to the public. Due to the scientific uncertainty of climate change, the effort of improving models and data set continues to be crucial to our quest to analyze the impacts of climate change and changing metrological systems.

Measures adopted by Japanese administration and society as a response to the rising temperatures. The awareness of adapting to extremely high temperature has risen among citizens due to efforts made by all levels of Japanese government and civil society. As an aging society, Japan is under pressure to construct social networks with strong abilities to adapt to changing natural environment. Combined with mitigation actions, strengthened efforts of adaptation will encourage promotion of comprehensive measures such as early warning systems, appropriate use of air conditioning, improved education and knowledge in dealing with relevant impacts of global warming and climate change.

Local governments and agricultural workers in Japan widely recognize the adverse influences caused by global warming and climate change on agriculture. The impacts of changing weather and temperature is different from one region to another hence various types of products demand different preventive and adaptive methods.

As mentioned above, Japan is facing a problem of aging population. It is urgent for stakeholders to combine preventive and adaptive measures in order to respond the risk of changing environment. For instance, in dealing with the occurrence of extreme weather events, serious floods, sea level rise, sediment and natural disasters, implementations of policies for recovery and restoration are no longer considered sufficient. Preventive approaches such as introduction of early-warning/alert system, provision of precautionary information, risk assessment and management are regarded as a necessity that needs to be further addressed in coming years. These are not only to reduce the level of damages that large-scale disasters can cause to the society, but to respond to the uncertainty that exists in the scientific knowledge of contemporary era.

NOTES

1. The Observation stations used for Figure 8.1, Figure 8.2, and Figure 8.3 are located at Ochiishimisaki (Hokkaido), Ryouri (Iwate), Minamitorishima (Tokyo), Yonagunijima (Okinawa), and Haterumajima (Okinawa).
2. The average years are set from 1970 to 2000.
3. The average years are set as twenty years, from 1901 to 1930.
4. The average years are set as twenty years, from 1980 to 1999.
5. The average years are set as ten years, from 1990 to 1999.
6. The percentage of occurrences of strong typhoons to the total number of typhoons.
7. The average years are set as twenty years, from 1980 to 1999.
8. The KAKUSHIN Program was developed upon "the Human, Nature, Earth, Co-Exist Project," which was conducted from March 2002 to March 2006.
9. WBGT (Wet Bulb Global Temperature) is reference value that that incorporates humidity, radiant heat, and temperature, which substantially affect heat balance of human body. WBGT is calculated by using dry-bulb temperature, wet bulb globe temperature and black-bulb temperature. WBGT is widely used for assessing heat stress as well as risk if heat stroke.

REFERENCES

Keohane, O. Robert, Kal Raustiala (2008). "Toward a Post-Kyoto Climate Change Architecture: A Political Analysis," *Law-Econ Research Paper*, No. 08–14, pp.1–26.

Keohane, O. Robert, David, G. Victor (2010). "the Regime Complex of Climate Change," Discussion Paper 10–33, the Harvard Project on International Climate Agreements, January 2010, pp. 1–30.

Nishimura, Motoki (2006). "The Impact of Global Warming to Agriculture and Food security," *Global Environmental Research*, Vol. 11, No. 1, pp. 35–42

Sugiura, Toshihiko (2007). "The Problems that Caused by Global Warming to Japanese Agricultural Production and Adaptation," lecture paper at the symposium of National Institute for Agro-Environmental Sciences, pp. 23–26.

Sugiura, Toshihiko (2006). "The Current Situation of Fruit Growing Influenced by Global Warming and Countermeasures," *Journal of Agricultural Science*, Vol. 61, No. 3, pp. 107–122.

Sugiura, Toshihiko, Yokozawa Masayuki 2004 "Impact of Global Warming on the Environments for Apple and Satsuma Mandarin Production Estimated from Changes of the Annual Mean Temperature," *Journal of the Japanese Society for Horitulcultural Science,* No. 73, pp. 72–78.

Victor, G. David (2011). *Global Warming Gridlock: Creating More Effective Strategies for Protecting the Planet,* Cambridge University Press: New York.

Victor, G. David (2004). *The Collapse of the Kyoto Protocol and the Struggle to Slow Global Warming,* Princeton University Press: New Jersey.

"The Executive Report of the Innovative Program of Climate Change Projection for the 21st Century (KAKUSHIN project)," 2012 www.jamstec.go.jp/kakushin21/jp/reports/general%20report-j.pdf (downloaded on November 25, 2012).

"Governmental Response to Heat Stroke in 2012 Summer," Ministry of Environment, June 11, 2012, www.env.go.jp/chemi/heat_stroke/ic_rma/torikumi.pdf.

"About the High Temperature in North and East Japan from Mid-August to Mid-September in 2012," Press Release, September 21, 2012, Japan Meteorological Agency.

"The Characteristics of the Flow of Atmosphere and Weather in winter in 2012," Press Release, February 27, 2012, Japan Meteorological Agency.

"About the Low Temperature and Heavy Snow from Late December in 2011 to Early February in 2012," Press Release, February 3, 2012, Japan Meteorological Agency.

"Analysis of the Factors that Led to the Extreme Temperatures of summer in 2010," Press Release, September 3, 2010, Japan Meteorological Agency.

"About the High Temperature of the Northern Hemisphere Mid-latitudes in July 2010," Press Release, August 11, 2010, Japan Meteorological Agency.

"About the Lack of Sunshine and Variation in Temperature between March and April in 2010," Press Release, April 23, 2010, Japan Meteorological Agency.

"About the Atmospheric Flow that Brought a Remarkable Cold Wave in the Northern Hemisphere Mid-latitude Zone," Press Release, March 3, 2010, Japan Meteorological Agency.

"Analysis of the Factors that Led to Unseasonable Weather in July 2009," Press Release, August 3, 2009, Japan Meteorological Agency.

"Characteristics of Circulation and Weather in Japan during 2008/2009 winter," Press Release, February 23, 2009, Japan Meteorological Agency.

"The Flow of the Atmosphere that Led to the Heavy Rain in the End of August in 2008," Press Release, September 12, 2008, Japan Meteorological Agency.

"About the High Temperature and Lack of Rainfall in West Japan after July 2008," Press Release, August 8, 2008, Japan Meteorological Agency.

"The Establishment of the Analysis and Assessment Group on Extreme Weather," Press Release, June 12, 2007, Japan Meteorological Agency.

"A Unified Report of Observation, Prediction and Impact Assessment on Global Warming: Climate Change in Japan and its Influence," MEXT, Japan Meteorological Agency and MOE, October, 2009.

"About the Guidance for Heat Stroke Occurred during Farming," Press Release, July 4, 2011, The Ministry of Agriculture, Forest and Fisheries (MAFF), www.maff.go.jp/j/press/seisan/sien/110704.html.

Information of Heat Stroke, Ministry of Environment (MOE), www.env.go.jp/chemi/heat_stroke/index.html.

Japan Aerospace Exploration Agency (JAXA), www.jaxa.jp/projects/sat/gosat/topics_j.html.

Japan Meteorological Agency (JMA), www.jma.go.jp/jma/index.html.

The Fire and Disaster Management Agency (FDMA)—information of heat stroke, www.fdma.go.jp/neuter/topics/fieldList9_2.html.

The Ministry of Foreign Affairs of Japan (MOFA)—climate change policy, www.mofa.go.jp/mofaj/gaiko/kankyo/kiko/index.html.

The Ministry of Education, Culture, Sports, Science and Technology in Japan (MEXT), www.mext.go.jp/b_menu/houdou/21/10/1285613.htm.

The Cabinet Office, Government of Japan *(the White Paper on Disaster Management, 2001–2012)*, www.bousai.go.jp/hakusho/hakusho.html.

The Ministry of Health, Labour and Welfare (MHLW)—Heat Stroke relevant information, www.mhlw.go.jp/seisakunitsuite/bunya/kenkou_iryou/kenkou/nettyuu/.

9

Domestic Debates on Climate Change in Russia

Leonid Grigoryev, Igor Makarov, and Alla Salmina

The Russian Federation plays an important part in mitigation of the global climate change. The dramatic economic crisis of 1990–1999 during the transition from "planned/command" to more liberal—open economy created severe economic and social stress for Russian people. Many not "so urgent" issues such as the environment did not receive priority. Most of the population is still trying to cope with a myriad of other social and political issues. Development of social life in post-Soviet period has not been favorable to make Russia a leader in coping with climate change, however, as Russia emerges from its economic—social woes, and as its civil society gains greater visibility and strength it is bound to play a more active role in global environmental issues.

CLIMATE CHANGE IN RUSSIA

Climate change problem in Russia is quite acute. Over the past one hundred years, the average surface temperature in Russia has increased by about 1.3°C (Otsenochnyj doklad ob izmenenijah klimata, 2008: 9) compared to less than 0.8°C in the world as a whole. The overwhelming proportion of this warming has occurred in the last few decades. Between 1976 and 2011 there has been a pronounced trend in the increase of surface temperature estimated as 0.44°C (Doklad ob osobennostyah klimata na territorii Rossijskoj Federatsii, 2012: 17). According to most climate models this trend will continue in the future (Katsov and Porfiriev 2011: 23).

Compared with many other countries, Russia has several unique features related to climate issues. First of all, it has a huge territory (nine time zones) and significant diversity in its natural conditions (Table 9.1). It's European territory has about 79 percent of the country's population

Table 9.1. Climate Characteristics of Russian Regions

Region	City / town	Average temperature, °C	January average temperature, °C	July average temperature, °C	Precipitations, mm per year	Precipitations, max mm per year
European part	Moscow	5.8	-6.5	19.2	707	882
	Sochi	14.2	6.0	23.3	1016	2835
Western Siberia	Omsk	2.1	-16.3	19.6	415	589
Central Siberia	Dudinka	-9.4	-26.8	13.8	522	689
Cisbaikalia and Transbaikal	Irkutsk	0.9	-17.9	18.2	477	797
Amur River region and Primorye territory	Vladivostok	4.9	-12,3	17.6	818	1272
Eastern Siberia	Yakutsk	-8.8	-38.6	19.5	237	330

Source: www.pogoda.ru.net

with different climate issues than central continental territories of Siberia (close to Central Asia), the Arctic covered by permafrost, and the coastal zone along the Pacific Ocean. Due to these differences climate related challenges are not uniform (Table 9.2). The most intense warming takes place in its European territory and Eastern Siberia.

Different climatic conditions of life, from landscape properties to seasonal features, including temperature and humidity, lead to different perceptions of environmental conditions by regions.

Rainfall patterns in Russia are also changing, although these patterns are not as obvious as temperature increase. In recent decades, the average annual rainfall over almost the entire territory of Russia has increased. It is 0.7 mm per ten years and accounts only for 20 percent of interannual variability (Doklad ob osobennostyah klimata na territorii

Table 9.2. Estimates of the Linear Trends in Average Annual and Seasonal Temperature Rise 1976–2011

Region	Year		Winter		Spring		Summer		Fall	
	B	D	B	D	B	D	B	D	B	D
The Russian Federation	0.44	36	0.20	2	0.57	29	0.43	56	0.53	25
European part	0.53	35	0.45	5	0.38	13	0.58	32	0.60	25
Western Siberia	0.28	10	-0.12	0	0.64	19	0.09	1	0.45	9
Central Siberia	0.46	21	0.19	1	0.70	22	0.43	31	0.42	7
Cisbaikalia and Transbaikal	0.38	24	0.13	1	0.58	19	0.56	44	0.28	6
Amur River region and Primorye territory	0.42	42	0.54	14	0.29	8	0.34	26	0.51	30
Eastern Siberia	0.51	39	0.00	0	0.79	32	0.50	46	0.77	37

Source: (Doklad ob osobennostyah klimata na territorii Rossijskoj Federatsii 2012: 17).

Rossijskoj Federatsii, 2012: 26). More active precipitations and stronger winter thaws lead to increasing river runoff, especially in the Volga basin, and the rivers flowing into the Arctic Ocean. On one hand, it increases available water resources but on the other hand, it makes floods and inundations more frequent and intene (Katsov and Porfiriev 2011: 19). It is estimated that in the future rainfall volume and river runoff across Russia except for the North Caucasus, where the highest possible aridity growth in the country is anticipated (Otsenochnyj doklad ob izmenenijah klimata, 2008: 15–17) will increase.

These developments increase in frequency and intensity of severe natural hydro-meteorological phenomena are of serious (Figure 9.1) concern. Russia is also very vulnerable to heat waves. The heat wave in the European part of Russia in 2001, resulted in more than a thousand deaths (Revich 2010: 140–150). 2010 saw another natural disaster in Russia, it was considered the most destructive one in the Northern Hemisphere. Severe heat wave lasted more than fifty days, and led to large economic damage and resulted tens of thousands of deaths (Katsov and Porfiriev 2011: 18).

It is interesting to note that several international scientific studies believe that Russia is one of the few countries that would even benefit from relatively minor climate changes. Indeed, its northern location offers some advantages due to climate change (Stern 2006, chapter 5: 5), conditions not available to other countries with similar levels of development. Sustainable reduction of ice-covered area (in 2012, its historic minimum was recorded (WMO Provisional Statement, 2012)) in the Arctic, and expanding navigation period of the Northern Sea Route allow more intensive use of its transport both for domestic transportation and for transit of goods from the Pacific area to Europe (Khon et al. 2010: 757–768). Total savings for European corporations, using the Northern Sea Route instead

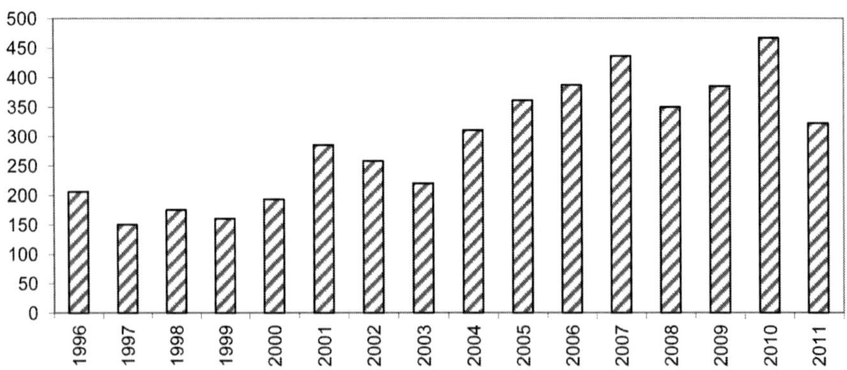

Figure 9.1. The number of hydro-meteorological disasters causing significant damage in Russia between 1996–2011

Source: (Doklad ob osobennostyah klimata na territorii Rossijskoj Federatsii, 2012: 62).

of the traditional route through the Suez Canal, could exceed $500 thousand per one trip (more than 15 percent of total transport costs). However, availability of these routes requires significant capital investments in the Arctic fleet as well as in port and navigation infrastructures.

Climate change also provides Russia with other positive opportunities, such as:

reduction in heating season throughout the country which will reduce the heating costs
due to sea ice melting, development of hydrocarbon reserves in the Arctic shelf would become less complicated;
as a result of increased river runoff, hydroelectric potential would increase;
due to less frequently extreme winter cold, the health protection systems could work under more normal conditions;
crop-growing seasons would become longer and area available for agriculture could be expanded.

In spite of these advantageous opportunities and positive aggregate impacts of climate change on the Russian economy one should not ignore the potential threats of climate induced changes. One of the negative effects of climate change on the economy and living standards will be caused by permafrost melting. Permafrost currently covers 63 percent of the territory of Russia, but over the next twenty-five to thirty years this area would be reduced by 10–18 percent (Oliphant 2011), adversely affecting buildings, transportation facilities, and infrastructure of extractive industries (metals, oil, and gas).

Other natural disasters may also inflict significant damages to the economy of Russia. The extreme heat in 2010 summer resulted in the loss of 1.2 percent of GDP (Katsov and Porfiriev 2011: 52). Climate change-induced proliferation of disease vectors (e.g., tick-borne encephalitis) and pests (primarily of the locust family) may also occur. More frequent heat waves are expected to exert greater demands on the health care system and the energy system due to increased demand for air conditioning, and will harm various ecosystems.

All these factors should not make us overoptimistic about the rosy positive impact scenarios discussed above. Average surface temperature rise by 2°C, is estimated to a loss of 2–3 percent of GDP (Porfiriev et al. 2011: 62). While this amount may be significantly lower than more vulnerable small island states and poor countries of Africa, Asia, and Central America, but it is more than the climate change-induced damage (at the same level of average temperature rise) for a majority of developed countries. However, losses in Russia will probably occur mainly in the remote regions, inhabited by a relatively small portion of the population.

The specific impact of climate change on the Russian economy is also different. Unlike most regions of the world, climate change wouldn't lead to worsening access to water and food in Russia the only notable exception is the North Caucasus. In fact global warming will contribute to increasing water resources and food production capacity. Russia is also less vulnerable to climate change-related natural disasters than most other large states: total damage from floods, hurricanes and storms for the national economy is minimal, and only heat waves may cause a serious threat to its economic growth. The main risk factor is melting permafrost, that is atypical for most other countries.

In sum climate change could bring more opportunities to Russia than to many countries, however the Russians can benefit from them only if comprehensive adaptation policies are implemented. Efficient management and implementation of these policies require highly skilled administrative personnel and significant financial investment. At present both of these are not easily available.

CLIMATE CHANGE AND HUMAN SECURITY IN RUSSIA

Climate change impact on people and human safety in Russia is significantly underestimated both by society and the government. This has largely been a legacy of the Soviet past, when environmental safety issues were ignored or even suppressed. However, the global warming is a significant threat for the Russian population. Mainly due to possible destruction of facilities as a result of permafrost melting, high vulnerability to indigenous peoples of the North, as well as generally low social and economic security of the population, which is much lower than in developed European countries.

For an objective analysis of the climate change impact on the Russian society, we need to know the dynamics of the environment situation in the country and its impact on human health. It is difficult to separate impacts on deteriorating (or improving) population health in Russia due to such factors as climate change, environment pollution (not directly related to climate change, for example water pollution from industrial wastes), and the deterioration of quality of life in general due to economic crisis of the 1990s and thereafter. However, in assessing the social impact of climate change, we cannot ignore the general state of the environment and human health in Russia.

Reduction of air and water pollution due to temporary decline in industrial activities in the 1990s had partially improved the environment, including certain purification of its rivers. It was hoped that these changes would lead to better health, but in the 1990s Russia saw the opposite trends. From 1990 to 2010 the total number of diseases increased by 1.4 times. Russia ranks second in the world in suicides (thirty suicides per 100 thousand people). While it is not connected to climate issues but it does show the general mental health of the people. Even after decline in pollution in the 1990s, the state of environment remains rather disappointing. Fifteen percent of Russian territory is considered as very polluted (Otchet ekspertnoj gruppy № 11 Strategii-2020, 2011). Mortality rates due to increased air pollution in Russia is about forty thousand people a year, or about 2 percent of the *total* number of *deaths* per year (Revich et al. 2004). According to the latest assessment by the Blacksmith Institute, three of the ten most polluted cities globally are located in Russia (Pickrel 2012) specifically Dzerhinsk, Norilsk, and Roudnaya Pristan. Average life expectancy in Dzerzhinsk in 2007 was forty-two years for men and forty-seven years for women. The average life expectancy in Russia for men is sixty-one, for women is seventy-four. These health problems are mainly due to air pollution and polluted drinking water. According to the Institute of Human Ecology and Environmental Health, about 50 percent of the population in Russia uses drinking water which does not meet health and safety standards. Economic conditions of daily lives for people in different strata of the Russian society in many regions are so severe that issues related to climate change seem very distant and uncertain. But the fact remains that climate change in Russia does impact its population and its health both directly due to higher maximum temperatures and an increasing number of natural disasters, and indirectly as a result of changes in crop areas, volume and quality of available water resources, and air pollution. A study conducted by Russian scientists in the city of Tver' showed that the increase of maximum daily temperature by 10 degrees led to an increase in the mortality rate by 8 percent (Revich et al. 2005: 20–24). According to the World Health Organization, climate change globally accounts for 0.3 percent of total deaths (Climate Change and Human Health, 2003: 322). It also shows that additional mortality caused by climate change and use of fossil fuels in 2010, as well as forecast for 2030 will increase (Table 9.3).

The increasing number of natural disasters in the past twenty years have been largely caused by climate change (Figure 9.1) contributing to a significant increase in mortality. More floods (especially in Primorye, Yakutia, the Far East and the North Caucasus, Krasnodar and Stavropol Territories) have led to substantial destruction, spread of infectious diseases and additional deaths.

Climate change has also led to a significant increase in diseases in Russia. More cases of cardiovascular and infectious diseases have been reported. These diseases are transmitted by tick-borne encephalitis (Zlobin

Table 9.3. Additional Mortality Due to Climate Change and Carbon Emissions—Annual average

Country	2010	2030
China	1500000	1500000
India	1000000	1500000
Russia	***100000***	***80000***
United States	80000	100000
Brazil	55000	70000
Ukraine	45000	40000
Japan	35000	40000
Poland	15000	15000
Germany	15000	20000
South Africa	15000	20000
United Kingdom	15000	20000
South Korea	10000	15000
France	10000	15000
Spain	10000	10000
Kazakhstan	9000	10000
Saudi Arabia	6000	10000
Canada	4500	6500
Australia	4000	6500

Source: (DARA and the Climate Vulnerable Forum, 2012: 119–120).

et al. 2004: 121–124; Korotkov et al. 1992: 7–10) and malaria (Mironova 2006: 20–25). The geographic range of these diseases is gradually expanding to the north of Russia. Climate change has contributed to growth of tick-borne borellioz. Over the last twenty years the number of morbid events almost doubled (Korotkov 2005: 52–56).

Due to permafrost melting and increased precipitation the problem of climate change-induced diseases will become more acute in the northern regions of the country. All these factors negatively affect health of the Russian population and reduce their natural lifetime. It is the poorest people who suffer particularly badly. These impacts of climate change on the Russian society greatly vary from region to region; therefore reaction to climate change created diseases also varies.

In the southern regions of Russia there is a decrease in precipitation, and increase in average air temperature is expected to reduce the access of the population to high-quality drinking water, and to increase morbidity and mortality from infectious, cardiovascular and respiratory diseases. In some places, higher average temperature will particularly affect food security, as farmers expect losses from droughts and floods. Increase in aridity in the South of Russia, especially in the North Caucasus, will decrease yield. These changes are already visible. Farmers, including those in central Russia, state that drought, lack of rainfall, and drying rivers and reservoirs have become very common. Due to frequent droughts, they have to rebuild the entire system of agricultural management. At the same time for the rural population of northern Russia, climate change is creating new opportunities for agriculture. Due to these regional differences Russia faces unique challenges and thus requires a diverse and coherent policy framework to solve climate change induced problems. A thoughtful agricultural policy is needed, including changing the varieties of cereals grown in the south and developing insurance against losses from droughts.

Increasing aridity in recent years has contributed to more frequent forest fires and peat bogs. They emit massive carbon dioxide and other greenhouse gases detrimental to human health. Severe air pollution in Moscow during the long hot summer of 2010 brought an unbearable heat wave and smog in August. These events occurred due to forest fires and pose many health problems. In the summer of 2010, the number of deaths in Moscow reached eleven thousand (Table 9.4). According to the report of Ministry of Economic Development, due to heat waves, forest fires, and smoke in July and August 2010, the number of the dead in Russia was at fifty-six thousand more than for the same period the previous year.

Climate change is also quite noticeable in the northern regions of Russia, specifically in the Arctic region's population of small indigenous people somewhat similar to northern Alaska, Canada, and Greenland. It is to be noted here that the Russian Arctic population is much larger than

Table 9.4. Deaths in Moscow in Summer 2010

Index	July	August	Total
Days with temperatures at 5°C above long-term average (in July above 23,4°C, in August above 21,4°C)	27	18	45
Additional deaths in 2010 compared to 2009, the absolute number (growth, %)	+4824 (50,7)	+6111 (68,6)	+10935 (59,6)
Including (growth, %) from: diseases of the circulatory system	51,5	66,1	58,8
respiratory diseases	59,1	110,1	84,5
infectious diseases	56,3	66,7	61,5
cancer	58,8	81,6	70,2
external causes	48,0	57,8	52,9

Source: (Revich, Shaposhnikov 2012: 131).

one in the Arctic territories of other countries. There are also different industrial processes (extraction of metals, coal, etc.), radioactive waste disposal sites, sites for the testing of nuclear weapons there threaten the environmental safety of northern people, especially the indigenous people who are more vulnerable to the effects of climate change. The health status of the population is already at a fairly low level due to the quality health care. Between 2005 and 2006, the number of newly reported diseases among the adult population in Nenets Autonomous Area was approximately two times higher than the Russian average. Respiratory diseases are particularly prevalent. Mental disorders are also very high. At particular risk are children in the northern regions of Russia. About 70 percent of children have health problems (Vlijanie global'nyh klimaticheskih izmenenij 2008: 5).

The level of financial security and quality of the health care system in Northern Russia is much worse than the Russian average. Shortage of qualified medical personnel, medical equipment and medicines are acute. Another major problem is long distances between small human settlements to the nearest adequate medical facilities. For example, access to health care is "seven hours per week for Taimyr Autonomous Area, sixty-two hours per week for Evenk Autonomous Area, twenty-eight hours per week for Chukchi Autonomous Area, twenty-fix hours per week for Republic of Sakha (Yakutia), and eighteen hours per week for Yamal-Nenets Autonomous District" (Vlijanie global'nyh klimaticheskih izmenenij 2008: 13–14). It is estimated that the number of infectious and parasitic diseases will also increase in the northern region.

Climate change will negatively influence the traditional way of life and culture of indigenous people of the North. They will be forced to change their centuries-old way of life and adapt to new conditions. Melting permafrost could lead to destruction of housing and other infrastructure and will lead in general to deteriorating quality of life of northern indigenous people. Most of these people are still engaged in traditional farming. Reduction in populations of many species of animals and plants will degrade the quality of food of indigenous people. Traditional activities of northern people, reindeer herding, fishing and hunting, are in danger. Further climate change may even cause the population migration from the most vulnerable areas to other "alien" areas where adoption to new life conditions and environment will be difficult for these people. Of course impacts of climate change on various groups will be different. Results will vary not only by the region of residence (north or south), but also by the level of income and health status of different population groups. In general, climate change will strongly affect the poorest people; due to their fewer material resources (access to quality water, air conditioning). The most affected groups will be, of course, the elderly, children, pensioners in Russia who suffer from poor health and lower level of income, and the farmers.

The problem of climate change impacts on society in Russia is currently recognized neither by the authorities nor the media. Climate change creates new challenges and risks for the Russian society and increases uncertainty for the future. The new "risk society" requires adaptation but presently Russia is hampered by issues of underdevelopment, and by traditions of paternalism in relationship between the state and society. But inaction and hesitation to take bold steps will adversely have negative social and economic consequences.

Climate change impacts on public safety should not be separated from problems of inequalities by income and access to information. However, the problem of climate change and broader problems of environmental safety in Russia are usually not considered in the context of inequality. In Russia very serious social inequalities exist. The contribution of climate change to inequality is one of the most invisible factors compared to others, such as unequal access to education, and corruption. It is difficult to find a direct correlation between climate change to the inequality in Russia, but it is assumed that it exists.

ATTITUDES TOWARD CLIMATE CHANGE

As discussed above the problems of climate change in Russia are already quite acute. Yet Russians are more skeptical about them than most European countries. According to a Pew poll in 2007, global warming was considered as a serious problem by 73 percent of Russians, which would seem quite high but still lower than Ukraine (89 percent) and much lower than in most developed European countries such as in Germany (86 percent) or in France (95 percent) but closer the United States (75 percent).

Russians' attitude towards climate change could be better understood in comparison with their attitude to environmental issues in general. According to a poll by the Levada Center in 2011, 73 percent of the population were concerned about environment in their communities and surrounding areas (including 24 percent of the population being very concerned) (Levada-centr 2011) . At first glance, this proportion of the population is not low. However, attention should be drawn to the fact that during the past two decades the number of skeptical Russians about environmental issues and global warming in particular has greatly increased and continues to grow. During the Soviet period, especially after the Chernobyl accident (1986), and in the early 1990s, public concern with environmental issues was much higher than at present. A survey in 1993 showed that 88 percent of Russians rated environment as poor or very poor, 62 percent believed that environment is a very serious threat for them (Douhovnikoff 2008: 448). One of possible reasons for this decrease may be the economic crisis of the 1990s, which shifted their attention to more urgent problems of survival. DeBardeleben and Heuckroth believe that in the early 1990s environmental issues and economic development were of equal concern for Russians (DeBardeleben and Heuckroth 2001: 49–76). This economic explanation for the behavior of "rational man" is quite credible, but there is an alternative explanation for it. Russians as discussed above that concern about environmental safety had decreased due to improving environmental situation in the country due to a steep decline in industrial production. Russians are even more skeptical about global warming. They are more interested in local environmental issues such as air and water pollution. According to a 2011 poll by the Levada Center most Russians are concerned about the following environmental issues: water pollution (50 percent), high levels of air pollution (48 percent), and poor drinking water (36 percent) (Table 9.5). Only 24 percent of Russians consider climate change as a serious environmental problem. These opinions and attitudes vary from small towns to metropolitan and rural areas. Muscovites are most concerned about climate change (42 percent) that may be due to the simple fact that they are better educated, more aware and have higher economic stakes than their rural counterparts (23 percent) 20 percent of rural residents worry more about shoaling water, expanding deserts, and damaging water regime in comparison to 7 percent of Muscovites. It can be concluded that villagers worry about visible and tangible climate change no less than Muscovites, but they do not always make a connection between their problems and climate change.

In an earlier opinion poll (2008) 90 percent of Russians knew something about global warming, 42 percent were well aware of it (VCIOM 2008) 51 percent of Russians believed that global warming had already begun. Seventeen percent thought that it started recently, and 6 percent were sure that it will never happen (VCIOM 2008). According to a survey of the Fund "Public Opinion" (2008), only 51 percent of Russians noted changes in the average temperature of the region where they lived.

People with higher and some education were best informed about climate change. Members of this group believed that they knew a lot about global warming (41 percent), among the least educated group of Russians having incomplete secondary education the share was only 22 percent (VCIOM 2007). Age was also a factor in awareness; responses. The proportion of young people who were aware of climate change/global warming was higher than older people. They viewed themselves as more pragmatic and realistic. They felt that climate change is a global phenomenon, and is visible in their own regions. Also the Russians living in the coldest parts of the country demonstrated greater awareness (60 percent) of climate change and higher temperatures in their regions—Far Eastern Federal District (cold) and the Southern Federal District (warm) (Table 9.6). The lowest number were in the Siberian Federal District (35 percent) and in the North-West Federal District (31 percent), including the educated population of St. Petersburg. Thus, the share of Russians who recognize that global warming has already begun is higher in those districts where its effects are most noticeable.

It is also interesting to note the responses about "favorable" and "unfavorable" impacts of climate change on different regions. While almost all Russians knew about global warming only a few of them were afraid of its effects. Fifty percent of Russians expected catastrophic consequences, 27 percent expect slightly worsening climate. Only 7 percent believe in improving climate (VCIOM 2007). Most respondents believed that global warming will have a negative impact on Russia especially on the Southern Federal District (70 percent) and the least one (rather oddly) on Far Eastern Federal District (52 percent) (Table 9.7). Some of the most optimistic people were inhabitants of the coldest Siberian Federal District. Only 27 percent of its respondents expect adverse effects (VCIOM 2007).

The most affluent segments of the population expected catastrophic consequences of global warming (51 percent) than the poorest (44 percent) (Table 9.8).

Table 9.5. "Which local environmental problems concern you most?"

Problems that are of Most Concern	Human settlements					
	Total	Moscow	More 500,000	100,000-500,000	up to 100,000	Villages
Pollution of water bodies (rivers, sea)	50	41	50	46	54	54
Levels of air pollution	48	60	65	54	40	31
Poor, polluted drinking water	36	28	30	54	39	25
Unsanitary condition of the territory	30	14	35	29	35	27
Harmful chemicals in food	28	39	31	32	22	24
Climate change	**24**	**42**	**23**	**21**	**23**	**23**
Deforestation	19	26	23	13	18	19
High levels of radiation	13	23	9	18	11	10
Shoaling water, enlarging deserts, and damaging water regime	12	7	16	6	10	20
Disappearance of certain species of birds, fish, animals, plants, insects	10	5	16	9	6	14

Table 9.5. (*continued*)

or change of flora and fauna						
Increased noise	10	27	6	15	7	5
Lack of green spaces	10	15	16	5	7	7
Acid rain	9	11	5	9	9	10
Inadequacy of water for irrigation	3	0	3	3	3	3
Other	1	0	1	2	0	2
Do not know	2	0	0	2	2	5

Source: Levada-centre May, 2011.

Table 9.6. Russian Public Opinion Research Center (VICOM) Poll of April 2007—"Which of the following Opinions about global warming do you most agree to?"[1]

	All Respondents	Federal Districts*						
		CFD	NWFD	SFD	VFD	UFO	SFO	FEFO
Global warming has already begun.	45	46	31	58	43	46	35	60
Global warming is about to begin	17	17	14	11	21	18	23	13
Global warming will not be very soon.	21	21	28	14	19	24	30	6
Global warming will not be at all.	6	7	14	6	4	3	5	4
Hard to say.	11	9	13	11	13	9	7	17

Note (Abbreviations): CFD—Central Federal District, NWFD—North-West Federal District, SFD—Southern Federal District; VFD—Volga Federal District, UFO—Ural Federal District, SFO—Siberian Federal District, FEFO—Far Eastern Federal District.

Source: (VCIOM 2007).

Table 9.7. Russian Public Opinion Research Center (VCIOM) poll of April, 2007: "What consequences of global warming for Russia will be, in your opinion?"

	All respondents	Federal Districts*						
		CFD	NWFD	SFD	VFD	UFO	SFO	FEFO
More likely favorable	18	18	20	6	19	21	29	24
More likely unfavorable	59	56	56	70	59	56	57	52
Hard to say	23	26	24	24	22	23	14	24

Note (*Abbreviations*): CFD—Central Federal District, NWFD—North-West Federal District, SFD—Southern Federal District; VFD—Volga Federal District, UFO—Ural Federal District, SFO—Siberian Federal District, FEFO—Far Eastern Federal District.

Source: VCIOM 2007.

Table 9.8. Russian Public Opinion Research Center , 2007 "What effects will global warming cause?"

	All Respondents	Self-assessment of material well-being		
		Very good, good	Average	Bad, very bad
Global warming will lead to disastrous consequences for all mankind.	45	51	44	44
Global warming will cause some deterioration of climate, but major changes in environment will not happen.	29	25	31	28
Global warming will lead to a better climate in some regions of the earth.	10	9	11	9
Hard to say.	16	15	14	19

Source: VCIOM, April 2007.

Above data demonstrates that the most affluent and young population, the so-called middle class (especially in Moscow), was more aware of dangers of climate change and was anxious about them. They are generally better educated, informed, and engaged in solving everyday essential problems at a quite different level.

More detailed analysis of attitudes of Russians to global warming would be needed not only to show variety of perceptions of the problem in Russia, but to understand whether the public requests to address climate change, as well as to find out what are possible changes in public opinion in Russia on this issue. Several survey data shows that a large number of Russians know about Climate Change, but only very few of them consider it as a threat to their regions. According to a survey by the Fund "Public opinion" in 2008, only 33 percent of Russians believed that global warming was due to human activity. Twenty-five percent responded that it was the result of a combination of natural and human factors, and 8 percent thought that is was due solely to natural processes, 36 percent of Russians believed that humans cannot stop global warming, while 21 percent felt that humans can stop it. It seems that Russians skepticism is largely due to the fact that they do not see a relationship between human activities and changing climate. These attitudes are due to several reasons including ambiguous opinions of Russian on the topic and a weak interest in the media, and the government.

The accident at the Chernobyl nuclear power plant in 1986 had perked up Russian interest on environmental issues but since the mid-1990s up to the present time popularity and activities of movements and organizations specializing in environmental protection declined. According to the World Value Survey, in 1990 only 1.7 percent of the Russian population was members of environmental organizations compared to the average 5.2 percent in 56 other countries. By 1999 these numbers in Russia dropped to 0.7 percent (Douhovnikoff 2008: 450), reflecting a relatively low level of civil society development under crisis conditions. In Russia, we see a strong presence of such international organizations as Greenpeace, the World Wildlife Fund, and a number of other NGOs and movements specializing in certain areas of nature protection, but the number of their members is relatively small. Political parties advocating for environment in Russia are few: Russian Ecological Party, "The Greens," of Anatoly Panfilov; Party "Green Alliance—People's Party" of Gleb Fetisov are in opposition hence exert very little influence in decision-making. They are not well known to the population and have not been active lately. At the local level their meetings are periodically held but they usually do not get much attention from the media and the Russians who are not directly affected by these problems. Very few of them are willing to pay taxes to finance climate change mitigation efforts.

THE CURRENT STATE OF GHG EMISSIONS

According to the UNFCCC Russia is the fourth largest emitter of greenhouse gasses in the world. It trails China, the United States, and India. The Soviet industrialization of the 1930–1980s was accompanied by rapid increase greenhouse emissions. In seventy years its emissions went from 11.2 million tons of carbon in 1922 to 1.1 billion tons in 1988, and prior to its collapse, the volume of its emissions was very close to the United States (Marland et al. 2011).

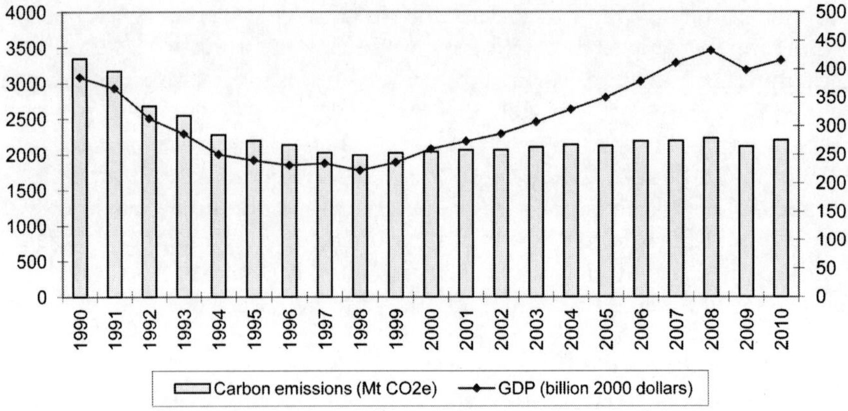

Figure 9.2. Russian GDP (right scale) and volume of carbon emissions (without—LULUCF) (left scale)

Source: World Development Indicators, (Nacional'nyj doklad o kadastre antropogennyh vybrosov 2012: 8).

According to World Development indicators the Russian GDP between 1990 and 1998 dropped by 42.5 percent. Many industrial plants were shut down, hence reduction in greenhouse gas emissions (Figures 9.2 and 3). By 1998, their volume (excluding LULUCF) compared to 1990 decreased by 40.6 percent. Economic recovery began in 1999. In 2008 GDP reached the 1989 level. The recovery did not increase emissions due to a shift in Russian restructuring of their economy in general and industries in particular. The carbon-intensive industries that dominated the Soviet era economy were replaced by the service sector. In the first decade of the twenty-first century, emissions grew only slightly, and during 2010–2012 their total volume was below the 1990 level by about 34 percent (UNFCCC 2012).

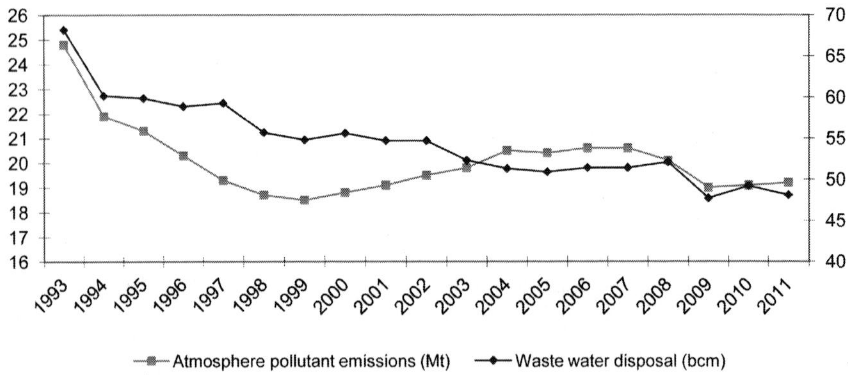

Figure 9.3. Atmosphere (left scale) and water (right scale) Pollution in Russia

Source: Russian statistical service (Rosstat)

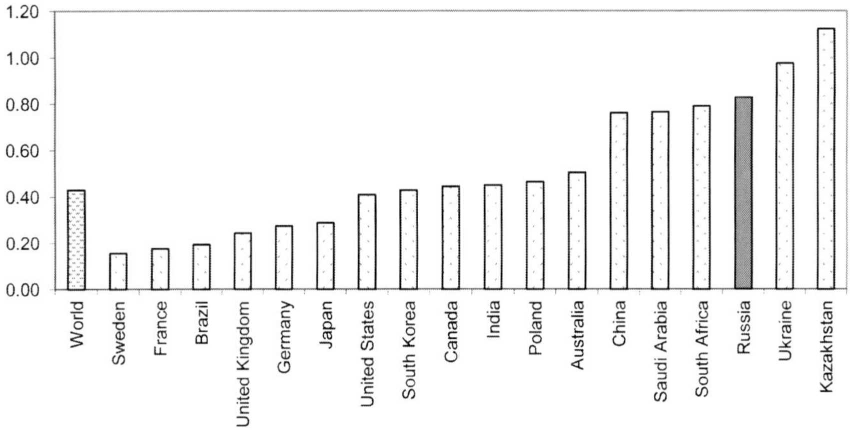

Figure 9.4. Russian Greenhouse gas emissions in comparison to other countries
Source: Enerdata

Despite a significant reduction in greenhouse gas emissions, the Russian economy remains extremely carbon-intensive, by quantity of CO_2, however it is doing better in CO_2 emissions per unit of GDP, than all BRICS countries and even the Gulf States (Figure 9.4).

Most of GHG emissions in Russia emanate from heating its buildings in cold regions. In 2010 it was 82.6 percent in comparison to 81.1 percent in 1990. The emission share of this sector in total emissions is gradually growing (in 1990 it was 81.1 percent). The share of the industrial sector associated with producing metals, paper, and fertilizers remains stable at 7.8 percent, while the agriculture's share is declining (6.2 percent). The only sector experiencing an absolute increase in emissions over the past twenty years is the waste sector. It has almost doubled (3.3 percent) (Figure 9.5). (Nacional'nyj doklad o kadastre antropogennyh vybrosov 2012: 7.

CLIMATE CHANGE AND DOMESTIC POLICY OF THE RUSSIAN FEDERATION

Due to many complicated economic, political and social issues environmental debate in Russia remains on the back burner. Popular awareness and NGOs are quite weak to exert much influence. The fate of Kyoto Protocol is such an example. There was an active discussion in Russia about its ratification. Public and Private discourses were led by two groups (Tynkkynen 2010: 188). The first consisted of the experts and the politicians who understood the scale of the problem and were convinced that only international cooperation could solve it. However,

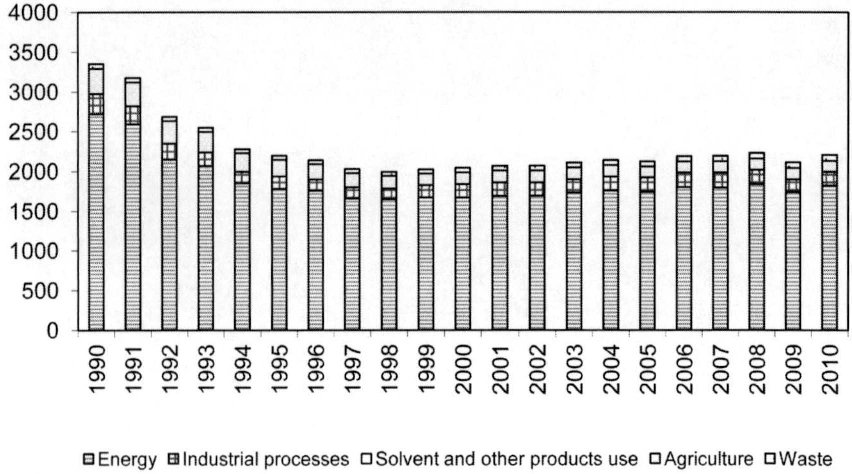

Figure 9.5. Russian Greenhouse gas emissions by sectors (without LULUCF); Mt CO_2e
Source: (Nacional'nyj doklad o kadastre antropogennyh vybrosov, 2012: 8)

many members of this group (Victor Danilov-Danilyan) recognized the weakness of the Kyoto Protocol to address the climate problem, but considered it as a necessary first step on a long road to tackle climate change (Danilov-Danilyan 2003).

The second group included the politicians and public figures who believed that the fight against climate change could become a new mission for Russia. They urged Russia to take a very active part as a global environmental leader, a model for an environmentally-oriented approach. In their view, Russia has enough natural potential (especially forests, which absorb CO_2), and significant record of success in emissions reduction (by that time Russia had reduced them by one third compared to the 1990 levels). Such ideas were promoted for example, by Russian Ecological Party *"Zelyonye"* ("The Greens") established in 2002. The party however failed to be approved by the State Duma (Tynkkynen 2010: 189).

The opponent to the ratification of the Kyoto Protocol included a very heterogeneous people and group—lobbyists related to the energy sector, bureaucrats, and even some scientists. The most influential opponent of the Kyoto Protocol was an economic adviser to President Vladimir Putin, Andrei Illarionov. He labelled the protocol a "fascist" document (Dall'Olio 2004: 9), and a group of academics of the Russian Academy of Sciences led by Yuri Izrael expressed doubts about the viability of the scientific data of the Protocol (Tynkkynen 2010: 189)

The main arguments of the opponents were that man-made nature of climate change has not been proven. Secondly, the Protocol was discriminatory against Russia, because it did not take into account the need for rapid growth of the Russian economy after a long transitional crisis. In 2004, Vladimir Putin had set a goal of doubling the GDP by 2010. The opponents of the Kyoto Protocol expressed concern that ratification could prevent the achievement of this goal.

Some representatives of Russian industry and influential group of scientists including climatologists questioned the basic provisions of scientific reports on climate change presented by the assessment reports of the IPCC or Stern Report. During the economic boom of the 2000s the Russian business, and intellectual elite were split. Weak support of joint implementation projects by businesses and the delay in adopting internal emissions regulation reflect an absence of social support needed to cope with climate change. The political elites of the country failed to play the role of the "enlightened elites" and did not pay due attention to these problems. They were mainly concerned about economic growth and development, unemployment, and the uneven state of the economy. The government couldn't dissociate itself from the idea of "carbon trading" and hardly perceived the evolution of carbon trade rules. As a consequence Russia failed to get any significant compensation for its huge GHG emissions reductions in the 1990s.

Several political issues and differences between the United States, Russia and the European Union also complicated Russian attitude and behavior. Russians were also unhappy that certain developing countries were exempted and were not forced to decrease greenhouse gas emissions. It is necessary to note that constant demands from some EU countries for Russia to take on new commitments to reduce domestic consumption of hydrocarbons was met with suspicion. Due to these ambiguities the content of the internal debate on climate change often changed. Instead of discussing ways to further reduce carbon emissions, the Russians questioned motives of other countries (United States, European Union, China). Internal pressure of the "green" and other public forces in favor of a modern climate policy within the country and outside was not strong enough to make significant progress.

Under these circumstances, the focus of the domestic policy of Russia related to problems of climate change mitigation was set in the "Climate Doctrine" of 2009. The Doctrine recognized the importance of human factor in climate change: "The scientific justification of this Doctrine includes the recognition of the fact that the anthropogenous factor may have an effect on the climate system triggering an important reaction which is adverse and dangerous, first of all, for human beings and environment." (para 4) (Klimaticheskaja Doktrina Rossijskoj Federatsii 2009). "The strategic goal of climate policy is to achieve secure and sustainable development of the Russian Federation including institutional, economic, environmental and social as well as demographic aspects of development in the context of changing climate and emerging challenges." (para 6) "The main climate policy principles are as follows: the global scope of the interests of the Russian Federation concerning climate change and its effects; the priority of national interests in the development and implementation of climate policy; the clarity and informational transparency of climate policy; the recognition of the need for domestic as well as international equal partnership actions of the Russian Federation in the framework of international research programs and projects concerning climate change; the comprehensive consideration of potential losses and advantages related to climate change; the prudential planning and implementation of measures intended to protect human beings, economy and State from the adverse effects of climate change" (para 7).

The Doctrine covers a fairly wide range of issues related to climate. It envisages an active monitoring of climate situation in Russia and participation of scientists in the development of relevant issues, particularly in international programs. It recognizes certain unique characteristics of the Russian situation: "Exceptional (in comparison with other countries) variety and scale of climate change in the regions of the Russian Federation and their consequences for its environment, economy, and the population are a natural result of its immense territory and the diversity of natural conditions" (Para 30).

The Doctrine calls the country to actively participate in initiatives of international community, addressing issues related to climate change. It also provides a rather comprehensive list of tasks of assessing vulnerability of the economy. It widely discusses problems of adaptation, especially in the context of monitoring, evaluating, and developing measures to reduce economic losses. The Doctrine actually gives the government freedom to pursue any necessary pragmatic actions. The plan of the Doctrine implementation consists of nonrigid, flexible instructions to the ministries and agencies engaged in dealing with relevant problems, but does not provide any guidance for specific purposes, levels, or thresholds for the indicators that should be achieved by a certain time. It has no relation to the budget process, the country's other strategies (energy), and no funds are allocated for solving practical problems!

As for the energy sector, the Russian Federation is still guided by the Energy strategy developed in 2008 and adopted in 2009 during the global recession. It is based on the assumption of high energy prices in the world as a reliable financial source for Russian corporations and the national budget.

The Russian energy sector is a huge complex producing 10 percent of global primary energy. Russia consumes approximately half of this energy and exports the other half. It is important to note that about one-third of produced coal and gas and two-thirds of oil (the energy is also used for their production and transportation) are exported. Furthermore, the energy consumed in the country is used for the production of large amounts of exported metals (aluminum alone amounts to four million tons), paper, chemicals, and fertilizers. This accordingly reduces GHG emissions in importing countries. Production of wood pellets in 2011 reached about one million tons, but 90 percent of it was exported. Anually this sector invests about 4.5 percent of GDP (at 21 percent capital formation rate), which is much more than global figures (1–1.2 percent). In its forecasts for the period up to 2035, International Energy Agency anticipates that Russian economy (its share of global GDP is about 2.5 percent) will invest 6.6 percent of investments in global energy industry. The economic crisis of the 1990s ruled out any significant activities in the area of energy efficiency. These issues returned in 1996, when the necessary legal framework was developed, and the program "Save Energy in Russia" was replaced in 2001 by a federal target program "Energy Efficient Economy."

As for energy efficiency, the government took a number of important steps that potentially could solve many of the problems of hydrocarbon fuel saving and emission reduction in Russia. As mentioned earlier that a large sphere of energy is used for heating. Efficiency measures would provide opportunities to save energy (Kurdin 2012). Promoting energy efficiency and introducing energy-efficient technologies are regarded as one of the main goals announced by President Dmitry Medvedev in the modernization strategy of the Russian economy. It is expected that by 2020, energy intensity of gross domestic product would be reduced by 40 percent compared to 2007. The law "On energy saving and energy efficiency" was passed at the end of 2009. The law requires a series of energy audits of buildings and required energy certificates for efficiency, requirements for programs to improve energy efficiency for regions and municipalities, and mechanisms of information support. It also refined a number of other organizational/administrative problems. "Russian Energy Agency" (REA) became an authorized state body to implement this law to realize energy efficiency. A state information system "Energy efficiency" was set up to collect and store data on energy efficiency across Russia, and to formulate programs of energy saving and energy efficiency.

This government program has become a key instrument in the field of practical energy conservation policy. The envisaged measures should ensure the reduction of energy intensity by 13.5 percent. The residue should be provided through structural changes, price factor, and "autonomous technical progress." Total cumulative energy savings to be achieved during the implementation of the State program, between 2011 and 2020, is targeted at 1124.2 million tons of fuel or about 800 million tons of energy conservation is being carried out in seven areas (Figure 9.6).

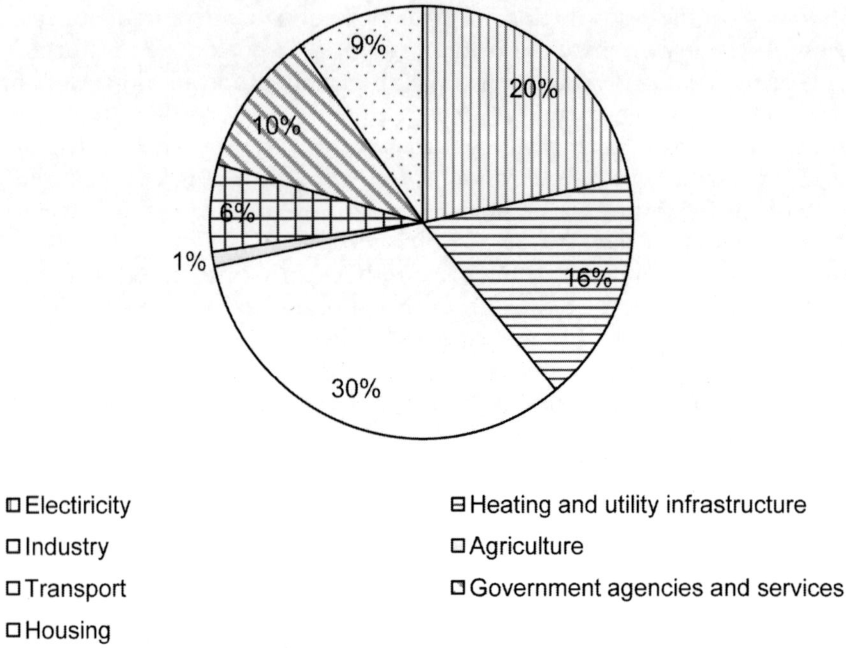

Figure 9.6. Structure of Energy Savings in areas of the state Program in Energy Savings and Efficiency of the period of up to 2020

Source: State program "Energy saving and energy efficiency for the period of up to 2020"

Major areas of energy savings include overcoming physical deterioration of fixed assets, particularly in the energy sector. According to with the State program on energy saving and energy efficiency, more than half of the total expected energy savings expected by 2020 could be from three key related areas: modernization of gas and coal-fired thermal power plants (23.4 percent reduction), extracting and processing oil (15.9 percent, including recycling associated gas—10.4 percent), reduction of losses in thermal and electrical networks (12.4 percent).

Debate on the climate change mitigation measures has not been really intensive in Russia in recent years, despite huge publicity of the problem in the world negotiations in Copenhagen in 2009, Durban in 2011 and Doha in 2012. Due to GDP decline in Russia in 2009 by 8 percent, attention of business and policy makers has been directed to pressing economic problems.

Most raw material producers in Russia prefer to be neutral on climate issues. At the same time, organizations representing business interests in Russia support Russia's participation in international programs of climate change mitigation. Thus, in early October 2012 representatives of the Association of Entrepreneurs "Business Russia," Russian Union of Industrialists and Entrepreneurs (RUIE), and experts participated in the special conference on climate supported development of projects during the second debate on the Kyoto Protocol. This can be seen as an important step for the future. Strong positive role is played by professional meteorologists and scientists working in this area. They have been consistently studying climate issues. For example, the Roshydromet (The Federal Service for Hydrometeorology and Environmental Monitoring of Russia) published two volumes, more than five hundred page assessment report on climate change in Russia containing important material for public discussion in the future (Katsov and Porfiriev 2011).

Non-governmental organizations such as World Wildlife Fund (WWF) Russia continue their work to educate the population mainly its middle class. For example, among the top six programs of the Strategic Plan for WWF-Russia 2012–2017, climate issues occupy a paramount position. Other Green NGOs have also become more active, mainly those which have close ties with global NGOs, and the EU. "WWF Russia will work to achieve adoption and implementation of Russian commitment of 50 percent reduction with transition to 80 percent greenhouse gas emission by 2050" (WWF Russia Strategic Plan 2012–2017, 2012: 17).

In general the civil society, business, and the scientific community have been more active and aware of climate issues and change impact on Russia in recent years. In Russia, programs and actions in this area, in our view, should be linked to development of a new, long-term energy strategy. Could public opinion strongly influence the government in the foreseeable future in formulating modern politics of further reducing emissions? The answer to this question depends on general economic, sociopolitical factors in the country and the international cooperation.

RUSSIA'S PARTICIPATION IN INTERNATIONAL EFFORTS

Russia has been a participant in the present international climate change regime since its inception in the early 1990s. Russia participated in developing both the UN Framework Convention on Climate Change (UNFCCC) and the Kyoto Protocol. Russia accepted obligations under the Kyoto Protocol: to retain greenhouse gas emissions in 2012 at the 1990 levels. Several observers have commented that despite much publicity Russia has not been an actor in global climate efforts but more concerned with its own domestic image and has used international opportunities to improve economic and political relations with the European Union and the United States.

The Russian attitude and role in the international climate regime radically changed after 2001. The American failure to ratify the Kyoto Protocol put Russia in a very interesting position. For the Kyoto Protocol to succeed it was necessary to reduce gas emissions to 55 percent. Without the United States, Russia remained the only country that could close this gap. The fate of the climate agreement depended on Russia (Dall'Olio 2004: 3). It gave Russia an effective leverage for defending its own foreign policy interests, especially in relations with the EU, the main sponsor of the Kyoto Protocol. It has been observed that Russian behavior was driven more by its foreign policy interests. Vladimir Putin finally tilted in favor of Kyoto ratification only when Russia received the EU concessions in negotiations on joining WTO. These concessions were seen as a price that the European Union paid for the ratification of the Kyoto Protocol by Russia (Safonov 2010: 131; Dall'Olio 2004: 8; Vatansever 2012: 7). It is also important to note that the deal, if it was a deal, was not too advantageous for Russia. It finally joined the WTO only eight years later.

With the passage of time and benefit from hindsight it seems that the expected devastating impact of the Kyoto Protocol on the Russian economy was minimal. Despite rapid economic recovery, Russia has not come even close to the limit set by the agreement on greenhouse gas emissions, however it has done better than many other countries. As a reward it received the world's largest reserve of unused quotas—6 billion tons of CO_2-equivalent, which could be sold or used to attract foreign investment through joint implementation projects. As for joint implementation projects, these opportunities were squandered. Russia took a long time to prepare adequate procedures to utilize these opportunities. Finally it appointed Sberbank as the project operator, and a quota at 300 million tons of CO_2-equivalent was set for Russian corporations. The mechanism to realize the goals of these projects have repeatedly undergone significant changes. Until 2011 the system was highly bureaucratized and a competitive mechanism for selecting projects was overly sluggish. As a result the mechanism was cancelled, and changes made. It started to operate in 2012. As of October 2012, 108 projects totaling 311 million tons have been approved, earning Russian corporations about $600 million (Ljutova 2012), and despite losing several years and billions of dollars of carbon investments Russia still occupies a dominant position in the market of joint investment projects. The future is uncertain in this sphere. In 2013 Russia has refused to participate in the second phase of the Kyoto Protocol.

Despite the fact that the first period of the Kyoto Protocol brought Russia significantly more benefits, Russia did not attempt to broaden its involvement in international cooperation on climate change. Its future role is quite uncertain. It has made it clear that it was not ready to take on significant commitments. For example, in June 2009, President Medvedev observed that in 2020 Russian emissions would remain 10–15 percent below the 1990 level. It means an increase of emissions by 30–35 percent from the current level. After this target was revised the Russian position was made clear at the Copenhagen Conference in December 2009.

However Russia seems to have softened its position. In February 2010 a plan of emissions reduction was announced. It envisaged 15–25 percent reduction (Safonov 2010: 132). According to Russia such sharp changes in one-sided established targets are not totally driven by any economic calculations but by concerns with climate change induced problems faced by planet earth. Often times these frequent policy changes are driven by local goals and concerns which prevail over global ones. Russian leaders also want to be global good guys, therefore they make important policy announcements prior to international conferences. For example a State Program on energy efficiency was announced only a month before the G8 Summit in Japan where climate change issues were at the center of the agenda and the adoption of the Climate Doctrine took place shortly before the climate conference in Copenhagen (Vatansever 2012: 8).

Many experts are also puzzled by Russia's position on the second commitment period of the Kyoto Protocol. It was stated that Russia would refuse to take on any commitments to reduce emissions during the second period, unless all the leading emitting countries (including China, the United States, and India) would take on such obligations. These states also refused to participate in the second commitment period of the Kyoto Protocol. However, in contrast to Russia, these countries had very high commitments in the first period, but were not able to meet them on their own, and had to incur substantial costs. Russia, by contrast, received additional foreign investments, albeit to a lesser extent than it could have. Russian refusal to participate in the second period of the agreement will cost it to lose these advantages.

It is not surprising that domestic business opposes some of these ad hoc decisions of the President and the Government. The Russian Union of Industrialists and Entrepreneurs, Chamber of Commerce, and the Association of Entrepreneurs "Business Russia" have repeatedly requested the authorities to reconsider their position. They were supported by some ministries (Ministry of Environment and the Ministry of Industry), as well as by green NGOs.

In November 2012 at the climate summit in Doha, the Russian delegation gave its final decision to reject its commitments to the second phase of the Kyoto Protocol. Russia is not satisfied with the current format of the international climate cooperation particularly that major developing countries are exempt from participation. At conferences in Durban (2011) and Doha (2012), Russia even proposed an amendment to change the Annex I of the UNFCCC. The amendment was rejected by developing countries. The Russian authorities believe that international cooperation in the current format is futile. Energy production in China and India is dominated by coal, and their rapid economic growth requires increased demand for coal based energy causing high emissions at a rate that au-

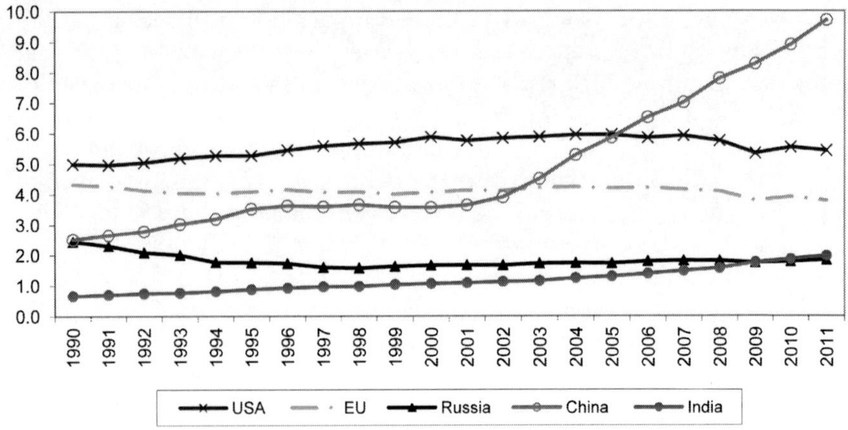

Figure 9.7. CO_2 emissions by countries Gt CO_2
Source: Olivier et al. 2012: 28.

tomatically negates all efforts to reduce emissions (International Energy Agency 2012) (Figure 9.7). In Russia's view China and India should be held accountable for high gas emissions.

To go beyond the goal of maintaining emissions at 2020 below 15–25 percent of the 1990 level, Russia would have to decrease emissions at a rapid pace, quite a difficult undertaking at present. Alexandr Bedritsky, advisor to the President on Climate and head of the Russian negotiating team, realizes that Russia will have to reevaluate the potential benefits occurring from the new system (Verojatnost' smeny pozicii RF 2012). Moreover, after the Doha amendments, "hot air" can be used only to a limited degree: unused quotas can be bought only in the amount of 2 percent of total emissions of an acquiring country in the first commitment period. It seems that for all practical purposes Russia is being cornered by rules established by developed countries which have failed to substantially reduce their own emissions (Figure 9.7).

At the Doha Conference Alexandr Bedritsky observed that Russia plans to create a national emissions trading scheme. The decision is long overdue. At present Russia is the only (except Saudi Arabia) G-20 country which has not developed a domestic emissions trading system. Several people have questioned the effectiveness of this proposal. To be successful it would need further changes on cap levels of 20–25 percent. The 1990 emissions rules actually allow businesses to release more gases.

By international standards Russia's participation in international climate negotiations is hardly constructive. Russia has no long-term negotiating strategy, and its constant reference to its previous achievements (record reducing emissions in the 1990s) to justify its passivity only annoys negotiating partners. Similarly their attempts to downplay

Russian successes in reducing emissions irritate Russian negotiators. As a consiquence, Russia has missed a great opportunity to use the Kyoto Protocol for modernizing its economy. To catch up with other countries moving far ahead in the development of a green economy, Russia needs a set of state regulatory measures based on scientific data and achieved goals. Creation of a domestic emissions trading system, to be gradually integrated with the European carbon market, will be useful. Russia also needs to be more active in the international arena like joining the Climate and Clean Air Coalition to Reduce Short-Lived Climate Pollutants, as well as to cooperate more closely with the BRICS.

CONCLUSION

In conclusion, it could be emphasized that Russia has actually reduced greenhouse gas emissions by one-third, but has refused to take on new commitments. It is largely due to the economic situation of the country which has struggled to reclaim its GDP at 1989 level. Both general population and elites in Russia are still too concerned with economic, social, and political problems facing the country.

Significant amount of coal emissions by Asia and even by the European Union should not be ignored. The IEA correctly notes that chances to reduce emissions by 2035 will be impossible if radical changes in policies and practices around the world are not made by 2017. Russian contribution to solving the climate change problem in the coming decades will fail unless other countries especially the coal dependent economies redirect their energy policies and practices. The official position of Russia is as follows: it is ready to take further steps provided similar actions are taken by other major powers of the world. Russia demands a more coordinated response of the international community to the climate change challenge.

NOTE

1. Abbreviations for Tables 9.6 and 9.7: Central Federal District; NWFD–North-West Federal District; SFD–Southern Federal District; VFD–Volga Federal District; UFO–Ural Federal District; SFO–Siberian Federal District; FEFO–Far Eastern Federal District.

REFERENCES

Brauch H.G., Spring U.O. et al. (editors) 2009. "Facing global environmental change: environmental, human, energy, food, health, and water security concepts." Berlin: Springer.

Buchner B., Dall'Olio S. 2004. Russia: The Long Road to Ratification. Internal Institution and Pressure Groups in the Kyoto Protocol's Adoption Process. "Fondazione Eni Enrico Mattei." Working Paper 151. www.feem.it/userfiles/attach/Publication/NDL2004/NDL2004-151.pdf.

Charap S., Safonov G. 2010. Climate change and role of energy efficiency. in: "Russia after the Global Economic Crisis" (eds. A. Eslund, S. Guriev, and A. Kuchins). Peterson Institute for International Economics, Washington, D.C. 125–150.

McMichael A.J. et al. (editors) 2003. "Climate Change and Human Health: Risks and Responses." WHO. Geneva.

Danilov-Danilyan V.I. 2003. "Kiotskij protokol: kritika kritiki" ESKO, №1.

"DARA and the Climate Vulnerable Forum. Climate Vulnerability Monitor" (2nd edition) 2012. A Guide to the Cold Calculus of a Hot Planet. 119–120. www.daraint.org/wp-content/uploads/2012/10/CVM2-Low.pdf.

DeBardeleben J., Heuckroth K. 2001. Public attitudes and ecological modernization in Russia. In: "The Struggle for Russian Environmental Policy." 49–76.

"Doklad ob osobennostyah klimata na territorii Rossijskoj Federatsii za 2012 god" 2012.

Grigoryev L.M., Kurdin A.A. Energeticheskaja effektivnost'—Kljuchevoj jelement novoj ekonomiki / In: "Doklad o chelovecheckom razvitii v Rossijskoj Federatsii 2013: Ustojchivoe rzvitije: vyzovyRio," UNDP-Russia, M.P. 102–116.

Henry L., Douhovnikoff V. 2008. Environmental Issues in Russia. "The Annual Review of Environment and Resources." Vol. 33. pp. 433–460.

Katsov V.M., Porfiriev B.N. 2011. "Otsenka makrojekonomicheskih posledstvij izmenenij klimata na territorii Rossijskoj Federatsii ne period do 2030 g. i dal'nejshuju perspektivu—Moskva": D'ART: Glavnaja geofizicheskaja observatorija.

Khon V.C., Mokhov I.I., Latif M., Semenov V.A., Park W. 2010. Perspectives of Northern Sea Route and Northwest Passage in the twenty-first century. "Climatic Change," Vol. 100, No. 3. 757–768.

"Klimaticheskaja Doktrina Rossijskoj Federatsii" 2009: www.Kremlin.ru

Korotkov Ju.S. 2005 Postepennaja izmenchivost' parazitarnoj sistemy kleshhevogo jencefalita. "Voprosy virusologii," № 3. 52–56.

Korotkov Ju.S., Akulova L.M., Hazova T.G., Kilina A.I., Kislenko G.S., Chunihin S.P. 1992 Ciklicheskie izmenenija chislennosti taezhnogo kleshha v zapovednike "Stolby." "Med. parazitologija i parazitarnye bolezni." 7–10.

Korppoo A., Vatansever A.A. 2012. "Climate Vision for Russia: From Rhetoric to Action." Carnegie Endowment Policy Outlook, August 2012.

Levada-centr. "Sostojanie okruzhajushhej sredy i aktual'nye jekologicheskie problemy" 2011. www.levada.ru/press/2011060203.html.

Ljutova M. 2012 Rossija mozhet poterjat' kvoty na torgovlju parnikovym gazom. "Vedomosti." 15 oktjabrja 2012 g.

Marland G., Boden T.A., Andres R.J. 2011 "Global, Regional, and National Fossil-Fuel CO_2 Emissions." Carbon Dioxide Information Analysis Center. cdiac.ornl.gov/trends/emis/overview.html.

Mironova V.A. 2006 Tendencii izmenenija klimata i maljarija v Moskovskom regione. "Med. parazitologija i parazitarnye bolezni," No. 4. 20–25.

"Nacional'nyj doklad o kadastre antropogennyh vybrosov iz istochnikov i absorbcii poglotiteljami parnikovyh gazov ne reguliruemyh Monreal'skim protokolom za 1990—2010 gg." 2012 Chast' 1. Moskva.

Oliphant R. 2011 Counting the Cost of Russia's Melting Permafrost. "The Moscow Times," Issue 4739, 6 October 2011.

Olivier J., Janssens-Maenhout G., Peters J. 2012 "Trends in global co2 emissions." 2012 Report. The Hague: PBL/JRC.

Otchet ekspertnoj gruppy № 11 Strategii-2020, 2011 "Zdorov'e i sreda obitanija cheloveka" za fevral'-ijun" 2011. // Internet-ssylka: 2020strategy.ru/g11/documents/32585713.html.

"Otsenochnyj doklad ob izmenenijah klimata i ih posledstvijah na territorii Rossijskoj Federatsii." Obshhee rezjume 2008 Rosgidromet. M.

Pickrel, J. "Ten of the most polluted places on the planet" 2012. www.abc.net.au/environment/articles/2012/07/23/3549975.htm.

Porfiriev B.N., Katsov V.M., Roginko S.A. 2011 "Izmenenija klimata i mezhdunarodnaja bezopasnost."—M.: Otdelenie obshhestvennyh nauk RAN.

Portal "Pogoda i klimat:" www.pogoda.ru.net.

"Proekty sovmestnogo osushhestvlenija v Rossii 2012. CCGS. Analiticheskaja spravka:" ccgs.ru/publications/other/_download/copy_Analyz_25102012.pdf.

Revich B.A. 2010 Izmenenie zdorov'ja naselenija Rossii v uslovijah menjajushhegosja klimata. "Problemy prognozirovanija," №3. 140–150.

Revich B.A. "Izmenenija klimata v Rossii—chem oni grozjat zdorov'ju naselenija?" www.meteoinfo.ru/meteo-med-revithc/.

Revich B.A., Avaliani S.L., Tihonova G.I. 2004 "Jekologicheskaja jepidemiologija." M.: centr "Akademija."

Revich B.A., D.A. Shaposhnikov. 2012 Climate change, heat waves and cold as risk factors for increased mortality in some regions of Russia. "The problems of prediction." No. 2.

Revich B.A., Shaposhnikov D.V., Galkin V.T., Krylov S.A., Chertkova A.B. 2005 Vozdejstvie vysokih temperatur atmosfernogo vozduha na zdorov'e naselenija v Tveri. "Gigiena i sanitarija," No. 2. 20–24.

Sabonis-Helf T. 2006. "Climate Change Policy in Russia, Ukraine and Kazakhstan." Washington D.C.: The National Council on Eurasian and East-European Research.

Stern N. et al. 2006. "The Economics of Climate Change." The Stern Review. Cambridge.

Tynkkynen N. 2010. Russia, a great ecologic Power in global climate policy? Framing climate change as a policy problem in Russian public discussion. "Environmental Politics," Vol. 19, No. 2. 179–195.

VCIOM. "Global'noe poteplenie: mif ili real'nost?" (2007) Press-release No. 665: www.wciom.ru/index.php?id=459&uid=4339.

———. "Grozit li nam global'noe poteplenie?" (2008) Press-release No. 1049: wciom.ru/index.php?id=459&uid=10708.

"Verojatnost' smeny pozicii RF po Kioto-2 blizka k nulju—Bedrickij" 2012 // RIA-Novosti, November 21, 2012.: ria.ru/science/20121121/911644186.html.

"Vlijanie global'nyh klimaticheskih izmenenij na zdorov'e naselenija rossijskoj Arktiki" 2008. OON. M.

WMO. "Provisional Statement on the State of Global Climate in 2012" 2012: www.wmo.int/pages/mediacentre/press_releases/documents/966_WMO-statement.pdf.

"World Energy Outlook 2012" 2012. International Energy Agency. Paris: OECD/IEA.

WWF-Russia. "WWF Russia Strategic Plan 2012–2017 2012" Moscow. www.WWF-Russia.ru.

Zlobin V.I., Danchinova G.A., Suncova O.V., Badueva L.B. 2004 "Izmenenie klimata i zdorov'e Rossii v XXI veke." M.: Adamant. 121–124.

10

+

Climate Change, Its Effects, and the Political Economy of Adaptation and Mitigation

Turkey and the Eastern Mediterranean Region

Paul A. Williams

INTRODUCTION

Climate change, summed up by the popular catchphrase "global warming," has become an increasingly urgent global concern. As the Intergovernmental Panel on Climate Change (IPCC 2007: 30) details, the planet warmed by 0.74°C between 1906 and 2005, with decadal rises in the last half of the twentieth century nearly double that of the full one-hundred years, and 1995–2006 contained eleven of the twelve hottest years since 1850; the ocean absorbed four-fifths of this added heat, causing its mean temperatures to increase as far down as three kilometers below sea level, which has risen 1.8 mm per year over the 1961–2003 period; and precipitation has "declined in the Sahel, the Mediterranean, southern Africa and parts of southern Asia." The frequency and intensity of "extreme weather events," like heat waves, "heavy precipitation events" (more rain from heavy falls), and "extreme high sea level," have been rising (IPCC 2007: 30).

This chapter focuses on the Mediterranean region, specifically Turkey. Ten of the Mediterranean's twelve driest winters since 1902 occurred in the last two decades, pointing to a "hotspot" trend of declining winter rainfall and rising summer temperatures due primarily to anthropogenic greenhouse gas (GHG) emissions and aerosol forcing (IBRD 2012: 15, 16). Climate change will accentuate this region's future climate variability. Annual European temperature increases will likely exceed the global mean, with southern Europe experiencing drops in overall rainfall and number of precipitation days, along with increased risk of summer drought and the highest summer warming (Christensen et al. 2007: 850, 872). There, climate change will "reduce water availability, hydropower potential, summer tourism, and . . . crop productivity" (IPCC 2007: 50). Global mean temperature may increase by 2.8°C (in the likely range of 1.7–4.4°C) from 1980–1999 to 2090–2099 according to the Special Report on Emissions Scenarios (SRES) marker scenario A1B (IPCC 2007: 45), but Southern Europe and the Mediterranean (SEM) could see a median annual increase of 3.5°C (2.2–5.1°C), with a higher summer (June–August) rise of 4.1°C (2.7–6.5°C), as well as a 12 percent (4 to 27 percent) drop in median annual rainfall, including a summer decrease of 24 percent (3 to 53 percent) (Christensen et al. 2007: 854, 873–874). The Potsdam Institute starkly warns that "the warmest July in the Mediterranean region could be 9°C warmer than today's warmest July" (IBRD 2012: xv).

Various Mediterranean countries have sunk into economic and political crises that make them even more vulnerable to the effects of climate change. Syria, for one, experienced a severe drought in 2008 that affected two-fifths of its cultivated land; as a result, 1.3 million people suffered losses and tens of thousands were uprooted (IBRD 2012: 16). The violent civil strife that broke out there in 2011 sent huge flows of refugees to neighboring countries, including Turkey, a comparative beacon of stability and growth. However, climate change could also imperil Turkey's own population and economy. Pressured to meet European Union (EU) conditions in order to become a member despite strong countervailing developmental aspirations, Turkey finally acceded to the 1992 United Nations Framework Convention on Climate Change (UNFCCC) in 2004 and the 1997 Kyoto Protocol in 2009, and has since advanced plans (without targets) to mitigate its GHG emissions as well as adapt to climate change.

This chapter surveys the sources of Turkey's GHG emissions, climate-change trends in the country, and the latter's impact on the country's human security values. The fourth section looks at how official and unofficial stakeholders have approached the issue of "proofing" Turkey's economy against the effects of climate change via mitigation and adaptation measures. The fifth part looks at Turkey's stance towards international agreements on climate change. The last part concludes by examining what Turkey can and should do now.

SOURCES OF TURKEY'S GREENHOUSE GAS EMISSIONS

Economic and Population Growth

Turkey's economy has undergone rapid growth since the end of the Cold War. The July 2011 National Climate Change Action Plan (NCCAP) of the Ministry of Environment and Urbanization (MEU 2011a: 5–6) shows how much the economy, energy use, and GHG emissions have grown. GDP increased 170 percent to reach USD 832 billion in real terms by 2008, raising per capita income to USD 11693, an increase of 114 percent. Population grew at a much lower rate, from 56.20 million in 1990 to 71.08 million in 2008. According to the Turkish Statistical Institute (Turkstat 2012), the country gained 2.64 million more people by 2010, with more than one-third of the total population concentrated in five (out of eighty-one) provinces, and, while mean population density crept up in this two-year interval from ninety-three persons per square kilometer (km^2) to $96/km^2$, it exceeded 2500 people/km^2 in Istanbul, which straddles a 452 km stretch of Turkey's 8300 km coastline. Coastal areas, with over 30 million people, have an average density twice the national average (MEF 2007: 166–167).

This expansion meant more construction, raising direct residential pressure on energy supply and GHG emissions. The number of buildings increased from 4.3 million in 1984 to 8.4 million (on an area of 1.5 billion square meters) in 2009, when total dwellings reached 16.2 million (MEU 2011a: 20). More people have also moved into energy-hungry luxury high-rise apartment complexes or gated community subdivisions. Residential energy use went up 88 percent between 1990 and 2010, increasing from 15 million tonnes (metric tons) of oil equivalent (toe) to 29 million toe; however, as total final energy consumption (TFEC) doubled, residential share (51 percent in 1970) dropped from 37 percent to 35 percent, slipping below industry's share in 1996 (MENR 2012a; MENR 2012b). In 2010, coal, electricity, and gas met 28 percent, one-quarter, and 22 percent, respectively, of residential TFEC, while the latter used 56 percent of coal, half of electricity, and 46 percent of natural gas TFEC (MENR 2012b).

Residential GHG emissions also grew over the two decades ending in 2010. According to Turkstat (2012), they rose to 50 million tonnes of CO_2 equivalent (tCO_2e), more than doubling, with almost all of this sector's GHG emissions consisting of carbon dioxide (18 percent of energy-related CO_2 emissions and 15 percent of all CO_2 emissions) in 2010. Moreover, by 2010, this sector emitted 56 percent of all methane (CH_4) emissions from energy use, albeit only six percent of total CH_4, and one-quarter of nitrous oxide (N_2O) emissions from energy use, but only 3 percent of all N_2O, with all four of these shares down from their 1990 levels.

Energy Use

Turkey's economic growth and rising energy demands exhibit a tight interdependence. According to International Energy Agency (IEA) data compiled in the NCCAP (MEU 2011a: 6), Turkey's total primary energy supply (TPES—roughly equivalent to primary energy demand (PED)) reached 99 million toe in 2008, an increase of 86 percent since 1990, while electricity output rose from 58 billion kilowatt hours (kwh), or 58 terawatt hours (twh), to 198 twh, a growth of 245 percent. In per capita terms, TPES rose by 48 percent to 1.39 (below respective world and OECD averages of 1.83 and 4.56) toe, electricity use increased to 2400 (still below respective world and OECD averages of 2782 and 8486) kwh, and electricity output rose to 2791 kwh. Conversely, per USD energy intensity fell from 0.17 to 0.12 toe and per USD carbon intensity fell from 0.61 to 0.44 tCO_2e.

Turkey's energy uses have propelled the bulk of GHG emissions growth. As the same report (MEU 2011a: 6) details, between 1990 and 2008, total GHG emissions rose by 96 percent, from 187 to 367 million tCO_2e, but carbon dioxide increased by 114 percent, from 127 to 271 million tonnes, and GHG emissions linked to electricity generation ascended by 234 percent, from 30 to 102 million tCO_2e, expanding from 16 to 27 percent of total GHG emissions. Per capita GHG emissions increased by 55 percent, from 3.33 to 5.16 tCO_2e. The carbon intensity of Turkey's energy supply went from 2.39 to 2.75 (above respective OECD and world averages of 2.33 and 2.40) tCO_2e/toe. Given the impact of the global financial crisis in 2008–2009, the NCCAP data understate Turkey's overall trend (MEU 2011a: 7).

Thus, resumption of economic growth put energy use back on its upward trajectory. Turkey's PED climbed to 109 million toe (per capita demand rose to 1.48 toe), an increase that encompassed all key fuels (except nuclear, of which Turkey has none), including hydrocarbons. Coal use nearly doubled, oil demand rose by more than one-third, and gas use saw a tenfold increase (BP 2012). Coal's share of PED fell about two percent (to 28–29 percent) and oil's share plummeted by nearly one-fifth (to 27–28 percent), but natural gas soared from 6 percent to just under one-third (MENR 2012b, 2012c). However, after deducting the 24 percent of PED diverted to power plants, coking plants, oil refining, and losses, the 84 million toe of 2010 TFEC consisted of equal shares (17 percent) of coal and gas and 32 percent for oil (MENR 2012b).

Electricity generation and use also increased. By 2010, 42 percent of coal (including three-fifths of all lignite) and 56 percent of gas were firing power plants (MENR 2012b). Installed capacity trebled to 49.52 gigawatts (GW), from which Turkey generated 211.21 twh of electricity, nearly 2865 kwh per person (TETC 2011). In 1990, two-fifths of electricity came from dams, 35 percent from solid fuels (mainly hard coal and lignite), 18 percent from gas, seven percent from fuel oil, and a minuscule share from geothermal-plus-wind (TETC 2011). Two decades later, hydropower, solid fuels, and fuel oil were supplying smaller fractions of total electricity (one-quarter, 26 percent, and 1 percent, respectively), while biomass-plus-waste, geothermal-plus-wind and gas were generating 0.2, 2, and 47 percent, respectively (TETC 2012: 13). Turkey's First National Communication on Climate Change (MEF 2007: 14) forecasts a decline in the shares of electricity produced by fuel oil, diesel, and natural gas by 2020 (when output may reach 544 twh), but a rise in coal's share.

Expanding energy uses continued to pull up GHG emissions. Turkstat (2012) records that total GHG emissions reached 402 million tCO_2e in 2010, increasing 115 percent since 1990, with per capita GHG emissions edging up to 5.45 tCO_2e. GHG emissions from energy use increased from 132 to 285 million tCO_2e (71 percent of the total), while GHG emissions from electricity generation rose 250 percent to reach 107 million tCO_2e. By 2010, total carbon dioxide reached 326 million tonnes (81 percent of total GHG), with per capita CO_2 emissions rising from 2.26 toe to 4.42 toe. Energy-linked carbon alone reached 277 million tonnes, slipping only slightly from 90 to 85 percent of total CO_2 emissions, and the carbon content of energy supply came to 2.54 tCO_2/toe, a 6 percent intensification, albeit smaller than that of the 1990–2008 period.

Industrialization

Industry still plays a significant role in Turkey's development. Accounting for 20–25 percent of GDP, it includes food, textiles, petroleum products, iron and steel, automobiles, and chemicals, which make up 18.8 percent, 16.3 percent, 8.8 percent, 6.2 percent, 5.8 percent, and 5 percent, respectively, of industrial output (MEU 2011a: 23). It also consumes a lot of energy. Between 1990 and 2010, industrial energy uses doubled from 15 to 31 million toe (edging up from 27 to 28 percent of PED and from 35 to 37 percent of TFEC), accounting for one-fifth of total coal demand (behind power plants and residencies), one-fifth of all natural gas demand (second only to power plants), and 13 percent of total oil demand, behind transportation and agriculture (MENR 2012a; MENR 2012b). In 2010, hydrocarbons continued to dominate industrial fuel supply, with gas, coal, and oil supplying 23 percent, 21 percent, and 13 percent, respectively, of industry's PED, while iron and steel (15 percent based on coal and one tenth on natural gas) made up 22 percent of industrial energy use, and cement (54 percent based on coal), which has a small share of GDP but a large share of industrial carbon emissions, comprised 15 percent of industrial energy use, nearly two-fifths of industrial coal use, and 57 percent of industry's use of hard coal (MENR 2012b).

Manufacturing represents a significant secondary factor in Turkey's GHG emissions. Turkstat (2012) enumerates GHG emissions from "industrial processes" and differentiates them from the emissions due to energy use of "manufacturing industries and construction." GHG emissions from "industrial processes" grew from 15.44 to 29.26 million tCO_2e between 1990 and 2008 (MEU 2011a: 7), but by 2010, they had reached 54 million tCO_2e (a 249 percent increase since 1990)—amounting to 13 percent of Turkey's 2010 GHG emissions and 18 percent of the total 1990–2010 GHG increase. Nine tenths of industrial processes' 2010 GHG emissions consisted of carbon dioxide and these CO_2 emissions expanded from one tenth in 1990 to 15 percent of total CO_2 emissions in 2010. Industry in total spewed an even larger volume of carbon dioxide. While CO_2 emissions from "industrial processes" reached 49 million tonnes in 2010 (more than trebling the 1990 volume), nearly three-fifths of it from cement making, energy-related CO_2 emitted by "manufacturing industries and construction" increased to 57 million tonnes (Turkstat 2012). Thus, all industrial CO_2 reached 106 million tonnes in 2010, just under one-third of total carbon emissions and close to GHG emissions from electricity.

Urbanization

Turkey's industrialization has propelled its urbanization. Most of the world now lives in cities and the urban developing world will drive most of the world's population growth until 2030 (Dupont and Thirlwell 2009: 80), but most people in Turkey have lived in urban areas since the early 1980s (Ozdemir et al. 2012: 80). In 2010, three-quarter of Turkey's population lived in "province and district centers," with only eight out of its eighty-one provinces having more residents of "towns and villages," and other figures show that 85 percent of the total population lives in 2950 municipalities, including the sixteen metropolitan areas (Turkstat 2012). Thus, rising urban residential energy uses, mostly centered in multistory buildings, transportation fuel uses, and municipal waste have increased ambient air pollution as well as GHG emissions.

Transportation

Transportation's energy uses and GHG emissions have grown relatively slowly, with the fastest expansion in motor vehicle (and petroleum) uses. Motor vehicles on Turkey's roads (which actually decreased in number by eliminating most unpaved village roads) quadrupled from 3.75 million (about 15 people per vehicle) to just over 15 million (just under 5 per vehicle) between 1990 and 2010, when privately owned vehicles made up 90 percent (and personal cars comprised 49 percent) of the total (Turkstat 2012). This sector's energy use rose by three-quarters, reaching 15 million toe in 2010, but shrank from 16 to 14 percent of total PED and 21 to 18 percent of TFEC (MENR 2012a; MENR 2012b). Nonetheless, by 2010, transportation consumed half of Turkey's oil demand and oil supplied nearly all transportation energy (MENR 2012b).

GHG emissions from ships, planes, trains, and motor vehicles increased between 1990 and 2010, but mostly diminished as shares of the total. According to Turkstat (2012), this sector's collective emissions of carbon dioxide (nearly all of this sector's GHG), methane, and nitrous oxide increased by 72 percent, from 26 to 45 million tCO_2e, but fell from one-fifth to 16 percent of GHG emissions from energy use and from 14 to 11 percent of total GHG emissions. Transportation's methane emissions increased by four-fifths, but only the growth rate (117 percent) of this sector's nitrous oxide emissions exceeded that of overall GHG emissions. Roads (including motorways) make up most of this sector's GHG emissions, but even this sub-sector, which gained slightly in terms of methane, saw its 93 percent shares of transportation's GHG and CO_2 emissions in 1990 drop four percent by 2010, mostly to air travel and maritime shipping, where CO_2 emissions grew even more rapidly. Per vehicle carbon emissions fell nine percent between 1990 and 2004 via better engine technologies, greater alternative fuel use, and removal of 320,000 old vehicles (MEU 2011a: 30).

Waste

Turkey's municipalities disgorge ever more voluminous amounts of refuse. In 2008, municipal waste collection exceeded 24 million tonnes (MEU 2011a: 34). According to Turkstat (2012), in 2010, the total topped 25 million tonnes, equaling 1.14 kilograms (kg) per capita per day or 416 kg per annum and covering 83 percent of the total population and 99 percent of the total municipal population. However, only 54 percent of this waste went to controlled landfill sites, although this rate averaged 83 percent in the five provinces with the largest total and municipal populations, which collectively make up over two-fifths of the country's municipal population and 44 percent of its combined waste (Turkstat 2012).

Waste emits the most CH_4, which also makes up most of its GHG emissions. Turkstat (2012) indicates that GHG emissions from solid waste disposal and wastewater handling increased by 270 percent, from 9.68 million tCO2e (5 percent of total GHG emissions) in 1990 to 35.83 million tCO2e (9 percent) in 2010. Waste's methane (which expanded from 86 to 95 percent of the GHG it emits, at the expense of nitrous oxide) rose even more rapidly than its entire GHG emissions and grew from one-quarter to nearly three-fifths of Turkey's CH_4 output. Even if controlled waste disposal has improved, its share of methane emissions has risen: While unmanaged waste sites made up three-quarters of the waste sector's CH_4 in 1990, these comprised less than one-third in 2010.

Agriculture and Forestry

Agriculture employs about one-quarter of Turkey's workforce (and a larger fraction of its female workforce), but has less importance in the larger economy and in GHG emissions. Agriculture's share of GDP fell from one-third in 1968 to less than ten percent in 2010 and even its overall land use shrank (Turkstat 2012). That is, "utilized agricultural land" (the sum of "arable land," "land under permanent crops," and "land under permanent meadows and pastures") fell seven percent between 1990 and 2010 to 39 million hectares (half of Turkey's landmass), including a 12 percent drop in arable land to 21 million hectares.

Concomitantly, this sector's non-energy related GHG emissions also fell. Turkstat (2012) reveals a nine percent decrease between 1990 and 2010, from 30 to 27 million tCO_2e, with agriculture's nonenergy share of total GHG emissions falling from 16 to 9 percent and its portion of methane dropping from three-fifths to under one-third. In this period, cattle in Turkey stayed around 12–13 million head per year, but the sheep and goat population fell by over two-fifths and the combined total of five other farm animals fell by three-quarters. However, number of tractors (85 percent of which have thirty-five-plus horsepower engines) rose by 58 percent. Thus, agriculture's total GHG emissions actually increased from 35.6 to 40.3 million tCO_2e, one-third of the latter volume consisting of carbon dioxide from fuel consumption.

In Turkey, forests constitute more of a "carbon sink" than a GHG source. Between 1990 and 2010, forest area grew from 20.2 to 21.5 million hectares (Turkstat 2012). Between 1972 and 2004, tree stock increased by 35 percent (standing at 1368 million m^3 in 2008) and by 29 percent in terms of "current annual volume increment rate" (MEU 2011a: 49). Turkstat (2012) data on forested area, assembled by Turkey's Ministry of Land, Agriculture, and Forestry, includes both "normal" and "spoiled" area, and exceeds that of Faostat (2012a) by a factor of two. But half of Turkey's forest area (10.58 million hectares) is "unproductive," and "one-third of productive forests have a low density" (MEF 2007: 72). Nonetheless, net stock increment rose from 12.02 million tonnes per year (44.08 million tCO_2e of carbon stock) in 1990 to 15.64 million tonnes per year (57.36 million tCO_2e of carbon stock) in 2008 (MEU 2011a: 51).

CLIMATE CHANGE IN TURKEY

While its contiguous southern and southwestern coastlines locate Turkey in the Mediterranean basin, its oft-cited quality as a "bridge" between continents and cultures is climatic as well. The country's 783,000 km^2 territorial land mass, coterminous with latitudes 36–42N and longitudes 26–42E, straddles southeastern Europe and southwestern Asia and forms part of the Alpine-Himalayan ridge. Its Mediterranean macroclimate lies "between the temperate and sub-tropical zones at western parts of the [aforementioned] continents" (MEF 2007: 3, 35). But Turkey consists of distinct regional and subregional microclimates (MEF 2007: 37–38). The southern coastline has classic Mediterranean features—snow-free and frost-free winters and hot and dry summers, with the humid Mediterranean coastline getting about 1000 mm/year in rainfall and the semi-humid Aegean receiving about 600–800 mm/year. The semi-humid Marmara region, north of the Dardanelles Straits and south of the Bosporus Straits that cut through Istanbul, has colder winters and gets 500–700 mm/year. Further in towards the more arid "steppes," Central Anatolia has cold winters and gets most of its 350–500 mm of annual rainfall in the spring, and Southeastern Anatolia receives 350–800 mm of rainfall per year, but experiences hotter, drier, and longer summers. Eastern Anatolia, with the country's coldest winters, gets over 500 mm/year, while the Black Sea region, demarcating the country's northern border, is Turkey's rainiest. As such, climate change will not have a uniform impact on the country's landmass.

Increasing Temperatures

Turkey's climate in general has been getting hotter and drier, a trend that is set to continue. Data from 1940–2008 show Turkey's mean annual temperature, estimated at 13.6°C over the last three decades of the twentieth Century, increasing at a rate of 0.64°C per century (Mengu et al. 2011: 824). Over the course of the century ahead, according to regional climate model simulations based on the A2 emission scenario, average annual temperature will rise 2–3°C for Turkey as a whole, with winter increases higher in the east and summer increases up to 6°C in the west (MEF 2007: 165). This trend may become especially pronounced in the larger municipalities. By expanding the area of impermeable surfaces (viz, asphalt, cement, concrete, etc.) that absorb solar energy during the day and emit it at night, greater concentrations of residents and economic activity in Turkey's municipalities have created an "urban heat island" (UHI) effect, characterized by higher urban temperatures and wider temperature spreads between rural and urban areas, as shown in studies of minimum-degree data over the 1965–2006 period (Ozdemir et al. 2012). Warmer evenings will boost energy use for urban air conditioning (MEF 2009a: 8).

Decreasing Precipitation

Turkey has also been experiencing adverse trends in precipitation. Total rainfall, which averaged 635 mm/year between 1971 and 2000, has been decreasing by 29 mm/century, with Mediterranean precipitation falling one-fifth since the mid-1980s (Mengu et al. 2011: 824). Turkey is projected to encounter regionally divergent, but generally adverse, precipitation outcomes by 2071–2100. Winter and spring rainfall will decrease along the Mediterranean coastline (with the largest drops in the southwest), increase along the Black Sea coastline, and remain relatively constant in Central Anatolia; seasonally, however, while the country as a whole will experience little change in summer precipitation volumes, it may get more in the fall, especially in the Euphrates-Tigris river basin (MEF 2007: 164). "Snow water equivalent" is projected to fall by 200 mm over the Eastern Anatolia steppes and the eastern Black Sea mountain range (MEF 2007: 165).

Rising Sea Levels

Given the length of Turkey's coastline and the large numbers of people living there, the country and its economic values will have no immunity from rising sea levels. Turkey's coasts, along which sea levels have already risen 12 cm or so over the past century, are forecast to face erosion, flooding, inundation, and saltwater intrusion, a problem already exacerbated by population (permanent and seasonal tourist) pressure on groundwater sources and one that could, if Accelerated Sea Level Rise (ASLR) occurs, threaten Istanbul's Buyukcekmece and Kucukcekmece lagoons and the Halic estuary (the so-called "Golden Horn") as well as the city's drinking-water supply (MEF 2007: 167).

IMPACT OF CLIMATE CHANGE ON HUMAN SECURITY IN TURKEY

Population and Health

Greater magnitudes and frequencies of high temperatures will have multiple negative impacts on the health of Turkey's people. Over half of the population lies in the 15–49 age-range (Turkstat 2012), a factor that could lessen the full health-related impact of climate change. Yet, an increase in the number of consecutive extremely hot days (i.e., "heat waves") will increase the incidences of heart attack, diseases of a cardiovascular, metabolic, renal, or respiratory nature, and colorectal as well as gastric cancers (MEU 2011b: 104). Rising temperatures will also exacerbate vector-borne contagions in Turkey, including cutaneous leishmaniasis, malaria, leptospirosis, rat-borne tularemia, West Nile fever, tuberculosis, trachoma, and tick virus-transmitted Crimean Congo Hemorrhagic Fever or CCHF (MEU 2011b: 106). A 1998 prevention program sponsored by the World Health Organization (WHO) and United Nations Development Program (UNDP) curtailed malaria, which peaked at nearly 120,000 cases in 1977 (around Adana in the south) but reached a secondary maximum of over 90,000 cases in 1994 (near Urfa and Mardin in the southeast), with both respective epicenters experienced unusually warm temperatures in those years (MEF 2007: 176–177). In 2002–2004, rural animal husbandry workers in the central provinces of Tokat, Sivas, and Yozgat bore the worst brunt to date of CCHF, which may be correlated with above average April temperatures (MEF 2007: 177–178). Seasonal migrations of agricultural workers, like those from further east who set up tent cities near irrigation canals in the Seyhan River Basin or the energy hub of Yumurtalik (MEU 2011b: 106), could raise the risk of future outbreaks.

Water

Adverse changes in temperature and precipitation have blunt impacts on water supplies. Turkey already faces a deteriorating water-supply situation based on rising population. The FAO's (2012) Aquastat database cites a 2011 figure of 2783 cubic meters (m^3) of actual "total renewable water resources per capita" (based on an overall estimated volume of 212 billion cubic meters (bcm)), but it also cites total "exploitable" water sources of 112 bcm (87.5 percent surface water, 12.5 percent groundwater), thus concurring with Turkish agencies (Williams 2012: 20). This amounts to per capita available volume of 1519 m^3 in 2010, when actual consumption totaled 44 bcm—with irrigation, households, and industry consuming 72 percent, 16 percent, and 12 percent, respectively (DSI 2011: 25), and thus amounted to annual per capita consumption of nearly 600 cubic meters.

Official sources project that available per capita water supply will decline. A population of 100 million would lower per capita "exploitable" quantity to 1,120 m^3, thus drawing the country below what official accounts refer to as "water scarcity" (less than 2,000 m^3 per annum) and just above "water poverty" (DSI 2011: 25). However, even this dire projection may not factor in independent climate change-related constraints on the water budget. Notably, the Aegean region's Gediz and Greater Menderes river basins could lose one-fifth of their surface waters by 2030, 35 percent by 2050, and one half by 2100, while facing larger irrigation water demands due to "increasing potential crop evapotranspiration (up to 10 percent and 54 percent for the years 2030 and 2100, respectively)" (MEF 2007: 19, 170). Urban areas of Ankara, Aydin, Bursa, Istanbul, and Nevsehir face elevated risks of water scarcity (MEU 2011a: 59). Thus, plans to raise irrigation, household, and industrial water uses to 72, 18, and 22 bcm, respectively, by 2023 (Williams 2012: 22–23) could come up short if the water supply were to fall below 112 bcm in line with the projected trends for the analyzed Aegean basins.

Hydropower and Other Renewable Energy Sources

Hydroelectric power (HEP) provides a significant source of electricity for residents of Turkey and non-HEP renewables are also rapidly increasing as sources of power and heat. BP (2012) shows that Turkey's HEP increased by 124 percent between 1990 and 2010 (hovering around 11 percent of Turkey's total energy consumption), while non-HEP renewables, providing about 0.2 percent of Turkey's total energy use in 2010, saw a steeper rise over the same two decades. The government has set ambitious targets for HEP expansion, with official sources citing technically feasible potential of 216 twh, four times more than current average production from 267 HEP plants in operation (DSI 2011: 16).

However, HEP and wind depend on weather-linked cycles that climate change may upset. HEP output growth has not actually kept up with overall electricity generation and even its longer-term overall growth masks salient multiyear drought-induced decreases. Between 1998 and 2001, annual HEP fell from 42.23 to 24.01 twh (with its share of Turkey's total electricity output shrinking from 38 percent to one-fifth), and fell again between 2004 and 2006 from 44.24 to 33.27 twh, whereas wind power rose from 0.08 twh to 3.58 twh per annum between 1990 and 2010 and increased every year between 2003 and 2011 (TETC 2011). But Turkey's burgeoning wind-energy sector could also be vulnerable to the "more likely than not" outcome of "reduced windiness activity" in Mediterranean Europe caused by a northward shift in cyclonic activity (Christensen et al. 2007: 864,878). As one government assessment has put it, "Droughts, extraordinary natural events, and climate system instability may cause expensive malfunctions in Turkey's energy infrastructure" (MEF 2009a: 29).

Agriculture, Food and Forests

Climate change could inflict the most severe and pervasive effects on Turkey's agriculture and thus its food security. For Turkey, "86.5 percent of its total land area and 73 percent of its arable land" are vulnerable to erosion, land degradation and desertification, and the Aegean region's increasing aridity between the 1960s and 1990s (MEF 2007: 178) augurs poorly for ameliorating this factor. Shifts in temperature and rainfall patterns will diminish the quantity and quality of water available for food production, further degrade pasture land, and increase the range and species count of agricultural pests (MEU 2011a: 58), thus leading to "a possible decline of 2–13 percent in [agricultural] productivity throughout Turkey" (MEF 2007: 170). Temperature increases and rainfall decreases, by shortening fruition periods (i.e., vegetative duration and the grain-filling process) and speeding plant water loss, may cut Turkey's winter wheat yields by up to one-fifth (Ozdogan 2011), and severe drought (as in 2007) will exact a toll on the yield of Turkey's olive trees, which produce up to five percent of the world's olive oil (Tunalioglu and Durdu 2012). Turkey's forests also face greater vulnerability to fire. About 12 million hectares (58 percent) of Turkey's forests are located in the fire-prone Aegean and Mediterranean regions, 46 percent consist of calabrian and black pine, and one-fifth have flammable accumulations of brush (MEU 2011b: 94).

Ecosystems and Biodiversity

Turkey's forests serve as hosts to its larger biodiversity heritage. While "forests and natural land," the latter consisting of "macquis, scrubs, weeds, and open ranges," represent 53 percent of Turkey's land mass, wetlands and marshes constitute three percent (MEU 2011a: 49). After Turkey had largely ceased marsh drainage (100,000 hectares between 1955 and 1970), it joined the Ramsar Convention in late 1994, and this accord's criteria apply to 200 of Turkey's wetlands, twelve of them consisting of 200,000 hectares and three of them on the coasts (MEF 2007: 175–176). Turkey's wetlands constitute rich troves of biodiversity. Turkey has "at least 8,650 vascular plant species, 30.9 percent of which are indigenous to Turkey" (MEF 2007: 36), 120 species of mammal, 130 of reptile, and 400 each of birds and fish (with over 1700 inhabiting the surrounding seas), with 3–4 percent of the total being indigenous and six percent of them threatened; thus, 0.7 percent of Turkey's land falls under categories I–V of IUCN protection and the country has been a party to the UN Biodiversity Convention since 1996 (MEF 2007: 174).

Global warming could severely diminish Turkey's biodiversity. Studies suggest that a 3.6°C increase in temperature may halve the count of plant species in the northern Mediterranean and the Mediterranean mountainous region (MEF 2007: 174). Under rising temperature and falling streamflow pressures, the Greater Menderes River Basin, for example, which contains Lake Bafa, an important wetland in the Aegean region, could lose scores of macro-invertebrate taxa that need "low temperature, high dissolved oxygen, [and] high current velocity" (MEF 2007: 174). Similar climate trends will make many of Turkey's 200 lakes, covering over 9000 km^2, richer in nitrogen and phosphorous nutrients, thus fostering algal blooms and lowering diversity of aquatic plants and predator fish and birds via eutrophication (MEF 2007: 175).

Socioeconomic Impacts

Socioeconomic ramifications of climate change in Turkey will be multifaceted but unevenly distributed in terms of geography, sector, and gender. Turkey achieved a Human Development Index (HDI) score of 0.679 in 2010 (below the OECD and even "high human development" group averages), but it registered the second largest negative differential (-26) between GNI per capita rank and HDI rank of the latter group, lowering its "inequality-adjusted" HDI score to 0.518 (UNDP 2010: 143–146,149,153). In Turkey, 47 percent of males aged 25 or older had secondary schooling in 2010 and 70 percent belonged to the workforce in 2009, but females lagged behind at 27 and 24 percent, respectively, dragging it below the "low human development" average for females on both indices and yielding a 2011 Gender Inequality Index (GII) score of 0.443, better than in 2008, but still below the "high human development" average GII score (UNDP 2011: 140–142).

Turkey's population and economy face potential threats from sea-level rise. Over two-thirds (70 percent) of Turkey's industries are found along its coasts (MEF 2007: 57). Nearly three-fifths of Turkey's GNP has come from the heavily industrialized Tekirdag-to-Kocaeli corridor on the Marmara Sea's northern coast, with its epicenter in Istanbul, which displays "high-risk values" in this regard. A "preliminary assessment of vulnerability analysis" suggests that Turkey could suffer capital losses of 6 percent of GNP as well as protection and/or adaptation costs equivalent to 10 percent of GNP—mainly for dike and levy installations (MEF 2007: 167).

Tourism, a major revenue source both for countless businesses and for the government, remains largely seasonal in nature and thus sensitive to climate change. As Turkey's National Climate Change Adaptation Strategy and Action Plan or NCCASAP (MEU 2011b: 12) notes, "It is inevitable that the tourism sector will also be negatively affected from decreasing snow cover in mountainous areas and increasing temperatures in the Mediterranean Region." Some potential loss of touristic value from the economy has already been attributed to climate change, especially via rising sea levels, and this is projected to increase. For example, increased wave and sand-dune activity, in addition to ASLR, could damage or destroy landmark cultural sites like Phaselis and Patara on Turkey's southwestern coastline (MEF 2007: 19). "Flagship" historical sites on Istanbul's shoreline, such as the Dolmabahce Palace and Mosque, the Ortakoy Mosque, the Beylerbeyi Palace and the Kucuksu Kiosk, face similar threats (MEF 2007: 167).

Extremely sensitive to heat and rainfall patterns, agriculture will bear the worst brunt of climate change. As the NCCAP (MEU 2011a: 58) contends, agriculture constitutes a "priority sector for socioeconomic reasons" and a key source of food supply. Soil erosion and drought will require considerable expenditure to fix while lowering farm incomes and exacerbating rural-to-urban migration (MEF 2009a: 10). Stark gender inequalities will further aggravate climate change's socioeconomic effect on agriculture (MEU 2011b: 76). While "unpaid family workers" in 2010 made up one-fifth of the total workforce of 22.59 million and 47 percent of 5.68 million agricultural laborers, they made up three-quarter of females working in agriculture and four-fifths of rural females in that sector (Turkstat 2012).

GOVERNMENT POLICIES AND STAKEHOLDER ACTIONS

On the issue of addressing climate change, Turkish authorities contend that the country faces a more difficult predicament than developed countries and even non-Annex I developing countries (see below). While Turkey had developed a National Environmental Action Plan to integrate environmental issues into the 1996–2000 seventh Five-Year Development Plan (MEF 2007: 6), the EU's 1999 Helsinki decision to nominate Turkey as a candidate member and its 2005 decision to open accession talks bolstered Turkey's environmental efforts. In 2006, Turkey renewed its 1983 Environment Law and passed the EU Integrated Environmental Approximation Strategy, both emphasizing sustainable development (Izci 2012: 191). The ninth Development Plan (2007–2013), National Climate Change Strategy (2010–2020), and in 2011, the NCCAP and NCCASAP, focused more directly on climate change. Although the European Commission (EC 2012: 83) has faulted the NCCAP for lacking targets, the 2011 plans enumerated core sectors (one set for mitigation and another for adaptation) as well as key purposes and objectives for each sector, and involved "consultation" with nongovernmental stakeholders in their drafting and formulation.

Mitigation

Mitigation seeks to cut or limit the sources of GHG emissions. Turkey's NCCAP identifies seven main sectors—energy, buildings, industry, transportation, waste, agriculture, and land use and forestry, as well as "cross-cutting issues in GHG control," such as institutions, technology transfer, financial assistance, data and information collection, capacity building, and mechanisms to monitor and evaluate the plan itself (MEU 2011a: 2–3). Every mitigation sector has delineated key "purposes," each with specified "objectives" and "actions," the latter containing timetables, co-benefits, performance indicators and tasked organizations (MEU 2011a: 66–132).

As discussed above, Turkey's energy sector accounts for 71 percent of its total GHG emissions and 85 percent of carbon dioxide. This area's mitigation foci aim to reduce 2008 energy intensity 10 percent by 2015 (via enhanced energy efficiency); increase the share of "clean" energy sources; limit GHG emissions from coal-fired electricity generation via "clean coal technologies" and higher coal-fired power-plant "cycle efficiencies"; and reduce "losses and illicit use in electricity distribution" to 8 percent by 2023 (MEU 2011a: 66). Annual electricity losses, which exceeded 17 percent of "total network supply" during 1996–2003, rose from 12 to 15 percent of network supply between 1990 and 2010, when they reached 30.22 twh, four-fifths of this from "distribution" (TETC 2011). The EU 2008 progress report on Turkey observed that electricity theft and loss "remains almost twice the EU average" (CEC 2008: 57).

Hydropower, biomass (e.g., firewood and dung), and solar hot water constitute established energy sources in Turkey, but wind, geothermal and advanced solar power have gained greater policy emphasis and legal impetus. The 2001 Electricity Market Law allows renewable power facilities to pay no annual licensing fee in their first eight years in operation, obligates priority grid access to renewable energy, and (its 2008 amendment) waives the production-license requirement for renewable power plants up to 0.5 MW (Tukenmez and Demireli 2012: 7–8). The 2005 Renewable Energy Law requires that 8 percent of retail power licensees' purchases consist of renewable energy source (RES)-certified electricity; guarantees prices for renewables; and authorizes the Council of Ministers to raise those prices by 20 percent annually (Tukenmez and Demireli 2012: 8). By 2010, 853 plants (including 736 HEP, ninety-one wind, eleven geothermal, four waste gas, and seven each of biomass and biogas) had been licensed within this law's scope (MEU 2011a: 19). A 2011 amendment to this law differentiates feed-in tariffs (i.e., USD 0.073 per kwh of hydropower and wind energy, 0.105 for geothermal and 0.133 for biomass and solar power) and incentivizes domestic manufacturing content (Tukenmez and Demireli 2012: 6,8), although the EU has questioned the latter's conformity with WTO rules (EC 2011: 74).

Energy-related GHG mitigation extends to buildings, industry, and transportation as well. Each of these sectors aims to increase energy efficiency, while the residential sector also strives for more use of renewable energy and transportation for greater use of biofuel. A 2009 study by the General Directorate of Electrical Power Resources Survey and Development Administration, tasked with implementing the 2007 Energy Efficiency Law, concluded that Turkey could cut industrial and transportation energy uses each by 15 percent and residential uses by 35 percent, thus surpassing maximum 2020 electrical output from "domestic and clean resources" (MEU 2011a: 17). Turkey has "a lot of unlicensed buildings that are poorly-insulated" (MEF 2009a: 27). Use by 10 million more residences of better insulation materials, including double-paned windows, could save 2.4 twh and 2.3 million toe by 2023; thus, buildings constructed after June 2000 require thermal insulation, and recent energy performance regulations mandate central heating of spaces over 2,000 m^2 and use of renewables and cogeneration for those over 20,000 m^2 (MEU 2011a: 21–22). In transportation, lower GHG-emitting railroad travel in Turkey, "the least railroad-intense country in Europe" (MEU 2011a: 29), has fallen far behind rising demand for greater personal mobility.

Curbing methane emissions requires better waste control. EU laws on waste disposal (especially the 1994 Packaging and Packaging Waste Directive, the 1999 Landfill Directive and the 2008 Waste Framework Directive) led Turkey to pass national legislation and propose, with EU financial help and technical assistance for Environmental Heavy Cost Investment Planning, an Integrated Solid Waste Management Plan that would close unmanaged waste sites (half by 2015 and all by 2023) and regulate recyclable packaging waste (MEU 2011a: 35–36). In terms of the European Union's 1999 Landfill Directive mandating member states to dispose of declining fractions of 1995 biodegradable waste in landfill sites in 2006, 2009, and 2016, Turkey uses 2006 as the baseline year and lags about nine years behind EU targets (MEU 2011a: 36). Turkey lacks sufficient revenue from the Environmental Cleaning Tax (ECT), paid as part of (often uncollectible) local water bills, and adequate data on unmanaged waste sites (MEU 2011a: 37–38). Conversely, seven municipal sites, including Ankara's mammoth Mamak landfill, now generate biogas power (MEU 2011a: 38). The European Commission has found Turkey to have made "good progress" in this sector (EC 2011: 100).

Turkey's GHG abatement also requires further agricultural reform. Unsustainable practices like over-fertilization, stubble incineration and excess tillage, inefficient irrigation methods, and scattered plot distribution consume 50 percent of agricultural energy (MEF 2007: 6). Improved land consolidation could save about one-quarter of agricultural fuel use and better cultivation techniques like "zero-tillage" could reduce energy use and enrich soil organic content (MEU 2011a: 44–46). Better management could enrich the carbon-sink capacity of Turkey's pastures, natural meadows, and "vegetation change areas" and boost Turkey's competitiveness in carbon trading (MEF 2007: 12; MEU 2011a: 42–43). Turkey's forests need resuscitation as well. The National Forestation Action Plan envisages rehabilitation of about 1.7 million hectares of "degraded" forest, thus increasing annual sink capacities from 41 to 222 million tonnes of carbon, although this remains a minority fraction of degraded forest and tourism and mining threaten forest biomass (MEU 2011a: 52–53).

Identifying tangible benefits play a crucial role in selling mitigation. As costs are loaded up front, GHG cuts require "framing through local issue bundling, or tying together the co-benefits of previously distinct public concerns" (Koehn 2008: 54). Turkey's 2011 NCCAP contains pervasive co-benefits on energy (energy security, lower energy costs, higher energy savings, and lower import dependency); EU alignment; capacity building; competitiveness; "green employment"; lower air pollution; and human health (MEU 2011a: 76–116). Agriculture and forestry focus on augmenting carbon sequestration, with cited co-benefits implicitly integrated with adaptation goals: improved natural resources management; sustainable use of soil and water resources; erosion control; healthy food production and food security; biodiversity conservation; and ecosystem services (MEU 2011a: 118–132).

Adaptation

While the NCCAP includes adaptation, the ensuing NCCASAP covers it more fully. Citing the long lag time for the elimination of existing atmospheric GHG concentrations, the latter plan states that, "adapting to the impacts of climate change is necessary regardless" of success in reducing global emissions (MEU 2011b: 12). The interconnected vulnerability areas consist of water resources management; agriculture sector and food security; ecosystem services, biodiversity, and forestry; natural disaster risk management; and public health; as well as "cross-cutting issues in adaptation" (MEU 2011a: 3; MEU 2011b: 15–16).

The first area emphasizes Integrated Water Resources Management (IWRM). This principle calls for holistic basin-centric water resources development capable of flexibly responding to shifting needs and covers effective agricultural water management, a key goal of the second vulnerability area (MEU 2011a: 71–72; MEU 2011b: 79–81). Turkey's early experiences in IWRM came via two World Bank-funded projects, one in 1993 to "restore sustainable range, forest, and farming activities" in the upper Euphrates catchment and the other in 2004 to improve "sustainable natural resource management" in 28 Black Sea-region micro-catchments (Dargouth et al. 2008: 112). Research continues into climatic effects on four crucial river basins—Gediz, Greater Menderes, Konya, and Seyhan (MEU 2011a: 60), the latter being the focus of a 2009–2010 systems approach-based program that utilized "stakeholder-livelihood-ecosystem" analysis and produced basin-level pilot eco-efficiency and "clean production" schemes (MEU 2011b: 43–36). IWRM also encompasses curbing illegal uses of underground water resources and urban water losses and theft (MEU 2011b: 66).

Basin water resource management has obvious ramifications for agriculture and food security, both of which intersect with ecosystems, biodiversity, and forestry as well as with natural disaster risk management and public health. While Turkey has been indirectly addressing adaptation needs, for example, by supplanting traditional flood techniques with sprinkler and drip mechanisms, its "[e]xisting strategies, policies, plans, and programs for the management of water resources . . . do not directly cover climate change adaptation targets" and it lags on modern water-use techniques, irrigation efficiency management, and developing low-water-use and drought-resistant crops (MEU 2011b: 25, 33). Poor cultivation practices (e.g., plowing and over-grazing pastures, marsh drainage, dam inundation of vulnerable habitats, over-use of aquifer water, and polluted runoff) have damaged natural areas and indigenous species varieties (MEU 2011b: 82). These areas thus face greater drought risk, elevating the urgency of developing climate-relevant "soil and land database and land information systems" and integrating agricultural drought into disaster planning and management activities (MEU 2011a: 71). Soil erosion jeopardizes cropped and forested land as well as biodiversity content and rural communities by aggravating the risk of natural disasters like fire, floods, and drought, which in turn can lead to desertification (MEF 2009a: 10). The NCCASAP thus aims to improve female farmer training, diversify forest villagers' livelihoods, and bolster insurance mechanisms (MEU 2011b: 126, 132, 135).

Stakeholder Actions

As elsewhere, Turkey's stakeholders in climate change encompass bureaucrats, corporations, unions, universities, community groups, and nongovernmental organizations (NGOs). NGOs, along with various international organizations (IOs) and international financial institutions (IFIs), have played key roles in placing environmental issues on Turkey's policy agenda, but governments, influenced by development imperatives, have had tenuous relationships with these NGOs, including, *inter alia*, Greenpeace Turkey, TEMA (a member of the European Environmental Bureau), and the "climate platform" (Izci 2012: 195). The Ministry of Environment and Forestry designated the Turkish branch of an IO known as the Regional Environmental Center (REC) as focal point for UNFCCC's Article 6 (Education, Training, and Public Awareness) during 2005–2008, but took over this duty in 2008 (MEU 2011b: 42).

Turkey's government has advanced mixed assessments of its own and NGO roles in increasing environmental consciousness. It has stressed the need for technocratic "soft adaptation measures," praising the establishment of the country's first certified education and training program on climate change and adaptation at Ankara's Middle East Technical University (MEU 2011b: 31, 43). But it has also critiqued the larger number of Turkish NGOs focused on mitigation instead of adaptation (MEU 2011b: 86, 114) and pointed to the "leading role" of the Education Ministry in raising "public awareness" on climate change. However, it concedes that, "Turkey still needs to strengthen participatory processes for adaptation to climate change" (MEU 2011a: 64), concurring with European Commission opinion that there exists at all levels "a significant need for awareness-raising on opportunities and challenges of climate change" (EC 2011: 101).

The government has engaged in public consultation, albeit generally on its own terms. The NCCASAP's own drafting and formulation involved only four months of "spirited stakeholder discussion," resulting in a "Participant Vulnerability Analysis" (a means of identifying local vulnerability areas "with the help of a participatory approach") that also highlighted a lack of urgency by province-level administrators, poor inter-agency coordination, bureaucratic obstacles to "implementation of effective action," and weak engagement by local universities and research institutes (MEU 2011b: 47). In a larger sense, though, civil society groups' access to information tends to be limited and their participation and consultation sought out merely to satisfy "procedural requirements" (Izci 2012: 196).

TURKEY'S STANCE ON CLIMATE-CHANGE TREATIES

Turkey has signed key climate change-related accords, although belatedly and only after securing specific exemptions from their most stringent obligations. Despite its "relatively well developed environmental policy and administration" and membership in numerous international environmental regimes since the 1970s, Turkey salient identity as a developing country has made it highly ambivalent about incurring the obligations of these regimes, including those on climate change (Izci 2012: 182–183).

Turkey initially balked at signing the 1992 UNFCCC. Fearful that its Cold War-influenced OECD membership would burden it with Annex-I GHG emissions-reduction pledges (applying to 15 Economies in Transition (EIT) and all OECD states except New Zealand) and Annex-II aid obligations (for all OECD states, including New Zealand), it did not join until 2004, over two years after the Seventh Conference of the Parties (COP7) in 2001 agreed to remove it from Annex II and recognize that

Turkey's "special circumstances" place it "in a situation different from that of other Parties included in Annex I to the Convention" (Erdogdu 2010: 1114). Conversely, as Turkish governments have also stressed, major non-Annex I (developing-country) parties lack emissions targets, but account for larger shares of 1850–2002 "cumulative CO_2 emissions" and emit higher per capita GHG volumes (MEF 2009a: 21–24, 32).

The European Union accession process has had a dualistic influence on Turkey, motivating its authorities both to approve the Kyoto Protocol and to delay that approval for as long as possible. Even after its parliament ratified it in February 2009, Turkey stayed out of Annex B and thus incurred no "quantified emissions reduction and limitation commitments" (MEF 2009a: 31–32) or QERLCs. While Turkish officials worried that the €60-billion costs of adopting the environmental chapter of the EU *acquis* (body of law) and accepting Annex-B targets would harm its competitiveness, improved prospects of eventual EU membership conveyed by the official launch of accession talks in 2005 and the opening of environmental chapter negotiations in December 2009 allayed these concerns somewhat (Izci 2012: 184–185, 190–191). Turkey also wanted to avoid losing a say in any post-2012 climate regime that would be influenced by the EU's 2006 "20–20–20 Plan" to cut 1990 GHG emissions by one-fifth and up to 30 percent, conditional on others' commitments, by 2020 (Erdogdu 2010: 1115, 1117). The EU also aims for 50 percent cuts by 2050 (MEF 2009b: 48).

Nonetheless, Turkish officials have invariably stressed the country's inability to take on obligations commensurate with Annex B or Kyoto's post-2012 successor. In 2009, after Turkey had joined the protocol and the Bonn-I Climate Change Talks had ended, the government issued an official opinion that "each country should decide on the suitable commitment for itself" after 2012 and that "there should not be any imposition upon the Parties" (MEF 2009b: 57). Prime Minister Erdogan, in a video statement to the September 2009 New York Summit on Climate Change, underscored Turkey's willingness to pursue "nationally appropriate mitigation actions" (NAMAs)—not QERLCs—according to the principle of "common but differentiated responsibilities" and its own "national circumstances" (MEF 2009b: 43). In the run-up to COP15 (Copenhagen), another government report bluntly stated that "every possible effort shall be made not to have Turkey's name under Annex B of the Protocol" (MEF 2009a: 35). COP16 in 2010 (Cancun) and COP 17 in 2011 (Durban) endorsed (Marrakech) COP7's original differentiation of Turkey from its Annex I partners (EC 2011: 101; EC 2012: 83). Yet, even if Turkey cannot be coerced by the post-2012 regime forged at Doha's COP18 to forgo adherence to NAMA-type flexibility, it knows that the European Union wants candidate members to adopt "ambitious binding" targets (MEF 2009b: 48).

CONCLUSION: PRESCRIPTIONS FOR TURKEY

Since 1983, when democratic government was restored to Turkey following the post-1980 junta rule and parliament passed the original Environment Law, affluence and effluents have grown apace, along with environmental concerns. Turkey's authorities, compelled to address the country's rising GHG contributions to climate change, which has exacerbated Turkey's Mediterranean-type weather and rainfall patterns and could threaten its rising standard of living, have elevated climate change and its impacts on human security higher on a policy agenda that classic security issues once monopolized and still dominate.

Turkish policy officials and civil society actors have grown more conscious of the ramifications of climate change, but the European Union has demanded even more of Turkey as a candidate member. The European Commission has decried Turkey's weak institutional capacity. Turkey has many laws that apply directly or indirectly to climate change as well as agency coordination mechanisms, like the Climate Change Coordination Board established in 2001 under the Ministry of Environment and Urbanization's aegis, but has no environmental protection agency with the central mandate to subject large infrastructure projects to impact assessments and public consultation (EC 2011: 101, 102). Given the place of wetlands and biodiversity in mitigation, the European Union has also expressed concerns, *inter alia*, over the law on the privatization of degraded forest land, the weakening of wetlands protection, and absences of such items as nature-protection framework legislation, delineation of agency responsibilities for nature protection, a national biodiversity strategy and action plan, and identification of Natura 2000 sites facing harm from planned HEP installations (EC 2012: 82).

Turkey has adamantly pushed for recognition of its "special circumstances." The European Commission, however, questions this stance, critiquing Turkey's failure to "put forward a greenhouse gas emissions reduction target for 2020" (EC 2012: 83). Without binding Kyoto targets or non-Annex I status, Turkish economic entities cannot avail themselves of Kyoto's flexibility mechanisms for trading in emissions credits or attracting Clean Development Mechanism investment monies. State planners have estimated that bringing Turkey's energy-efficiency, renewables, and solid waste projects into the EU's Emissions Trading System (ETS) could earn Turkey up to USD 166 billion by 2020, or 49 percent of project capital costs (Ulgen 2012: 23–24). Indeed, the EU has praised Turkey's "steps to enhance cooperation on emissions trading, including to raise awareness of the EU Emissions Trading System amongst the different stakeholders in the country" (EC 2011: 101). Whether this activity whets stakeholders' willingness to push for stricter GHG emissions-reduction commitments remains to be seen.

REFERENCES

BP. 2012. *Statistical Review of World Energy June 2012*, available at www.bp.com/statistical review.

[CEC] Commission of the European Communities. 2008. *Turkey 2008 Progress Report*, available at ec.europa.eu/enlargement/pdf/press_corner/key-documents/reports_nov_2008/turkey_progress_report_en.pdf.

Christensen, J.H., et al. 2007. "Regional Climate Projections." In *Climate Change 2007: The Physical Science* Basis, edited by S. Solomon et al., pp. 847–940. Cambridge, UK and New York, NY: Cambridge University Press, available at www.ipcc.ch/pdf/assessment-report/ar4/wg1/ar4-wg1-chapter11.pdf.

Dargouth, Salah, et al. 2008. "Watershed Management Approaches, Policies, and Operations: Lessons for Scaling Up." Water Sector Board Discussion Paper Series: Paper No. 11. Washington, DC: The International Bank for Reconstruction and Development, available at siteresources.worldbank.org/TURKEYEXTN/Resources/361711-1216301653427/5218036-1267432900822/WatershedExperience-en.pdf.

[DSI] State Water Works General Directorate. 2011. *2010 Activity Report* [Turkish]. Ankara: Ministry of Forest and Water Works, available at www.dsi.gov.tr/docs/stratejik-plan/ds%C4%B1-2010-faal%C4%B1yet-raporu.pdf?sfvrsn=2.

Dupont, Alan, and Mark Thirlwell. 2009. "A New Era of Food Insecurity?" *Survival* 51: 71–98.

[EC] European Commission. 2011. *Turkey 2011 Progress Report*, available at ec.europa.eu/enlargement/pdf/key_documents/2011/package/tr_rapport_2011_en.pdf.

[EC] European Commission. 2012. *Turkey 2012 Progress Report*, available at ec.europa.eu/enlargement/pdf/key_documents/2012/package/tr_rapport_2012_en.pdf.

Erdogdu, Erkan. 2010. "Turkish support to the Kyoto Protocol: A reality or just an illusion." *Renewable and Sustainable Energy Reviews* 14: 1111–1117.

[FAO] Food and Agriculture Organization. 2012. *Aquastat*, available at www.fao.org/nr/water/aquastat/data/query/results.html.

[Faostat] Food and Agriculture Organization. 2012a. "Land," available at faostat.fao.org/site/377/DesktopDefault.aspx?PageID=377#ancor.

[IBRD] International Bank of Reconstruction and Redevelopment. 2012. "Turn Down the Heat: Why a 4°C Warmer World Must be Avoided." A Report for the World Bank by the Potsdam Institute for Climate Impact Research and Climate Analytics. Washington, DC: International Bank for Reconstruction and Redevelopment, available at climatechange.worldbank.org/sites/default/files/Turn_Down_the_heat_Why_a_4_degree_centrigrade_warmer_world_must_be_avoided.pdf.

[IPCC] Intergovernmental Panel on Climate Change. 2007. *Climate Change 2007: Synthesis* Report, available at www.ipcc.ch/pdf/assessment-report/ar4/syr/ar4_syr.pdf.

Izci, Rana. 2012. "Europeanisation of Turkish Environmental Policy with Special References to Sustainability Discourse." In *Turkey and the European Union: Processes of Europeanisation*, edited by Cigdem Nas and Yonca Ozer, 181–203. Houndmills, UK, and New York, NY: Palgrave Macmillan.

Koehn, Peter H. 2008. "Underneath Kyoto: Emerging Subnational Government Initiatives and Incipient Issue-Bundling Opportunities in China and the United States." *Global Environmental Politics* 8: 53–77.

[MEF] Ministry of Environment and Forestry. 2007. *First National Communication of Turkey on Climate Change*. Ankara: Ministry of Environment and Forestry, available at www.dsi.gov.tr/docs/iklim-degisikligi/iklim_degisikligi_%C4%B1_ulusal_bildirim_eng.pdf?sfvrsn=2.

———. 2009a. Assessment Report on Turkey Within the Scope of United Nations Framework Convention on Climate Change (UNFCCC). Ankara: Ministry of Environment and Forestry, available at iklim.cob.gov.tr/iklim/Files/Raporlar/special_stitation.pdf.

———. 2009b. *Post-2012 Climate Change Negotiations Guidebook Turkey*. Ankara: Ministry of Environment and Forestry, available at www.dsi.gov.tr/docs/iklim-degisikligi/post_2012_climate_change_negotiations_guidebook_turkey.pdf?sfvrsn=2.

Mengu, Gulay P., et al. 2011. "Impact of Climate Change on Irrigated Agriculture." *Fresenius Environmental Bulletin* 20 (3a): 823–830.

[MENR] Ministry of Energy and Natural Resources. 2012a. Sectoral Energy Balance [Turkish], available at www.enerji.gov.tr/EKLENTI_VIEW/index.php/raporlar/raporVeriGir/4043/2.

———. 2012b. 2010 Primary Energy Resources Consumption [Turkish], available at www.enerji.gov.tr/EKLENTI_VIEW/index.php/raporlar/raporVeriGir/62173/2.

———. 2012c. 1990 Primary Energy Resources Consumption [Turkish], available at www.enerji.gov.tr/EKLENTI_VIEW/index.php/raporlar/detayGoster/4314.

[MEU] Ministry of Environment and Urbanization. 2011a. *National Climate Change Action Plan 2011–2023*. Ankara: The NCCAP Project Team, available at iklim.cob.gov.tr/iklim/Files/IDEP/%C4%B0DEPENG.pdf.

———. 2011b. *Turkey's National Climate Change Adaptation Strategy and Action Plan (Draft)*. Ankara: The Ministry of Environment and Urbanization, available at www.undp.org.tr/energEnvirDocs/00058944_Draft%20Climate%20Change%20Adaptation%20Strategy.pdf.

Ozdemir, Huseyin, et al. 2012. "Quantification of the urban heat island under a changing climate over Anatolian Peninsula." *Theoretical and Applied Climatology* 108: 31–38.

Ozdogan, Mutlu. 2011. "Modeling the impacts of climate change on wheat yields in Northwestern Turkey." *Agriculture, Ecosystems and Environment* 141: 1–12.

[TETC] Turkish Electricity Transmission Corporation. 2011. Electricity Generation and Transmission Statistics of Turkey, available at www.teias.gov.tr/TürkiyeElektrikİstatistikleri/istatistik2011/istatistik%202011.htm.

———. 2012. *Annual Report 2011*, available at 212.175.131.171/Faaliyet2011/ING_Teias.pdf.

Tukenmez, Mine, and Erhan Demireli. 2012. "Renewable energy policy in Turkey with the new legal regulations." *Renewable Energy* 39: 1–9.

Tunalioglu, Renan, and Omer Faruk Durdu. 2012. "Assessment of future olive crop yield by a comparative evaluation of drought indices: A case study in Western Turkey." *Theoretical and Applied Climatology* 108: 397–410.

[Turkstat]. Turkish Statistical Institute. 2012. *TurkStat* (Database) available at www.turkstat.gov.tr/Start.do.

Ulgen, Sinan. 2012. "Avoiding a Divorce: A Virtual EU Membership for Turkey." The Carnegie Papers. Washington, DC: Carnegie Endowment for International Peace, available at carnegieendowment.org/files/avoiding_divorce_paper.pdf.

[UNDP] United Nations Development Programme. 2010. *Human Development Report 2010*. Houndmills, UK and New York, NY: Palgrave Macmillan, available at hdr.undp.org/en/media/HDR_2010_EN_Complete_reprint.pdf.

———. 2011. *Human Development Report 2011*. Houndmills, UK and New York, NY: Palgrave Macmillan, available at hdr.undp.org/en/media/HDR_2011_EN_Complete.pdf.

Williams, Paul. 2012. "Euphrates and Tigris Waters: Turkish-Syrian and Iraqi Relations." In *Water Resource Conflicts and International Security: A Global Perspective*, edited by Dhirendra. K. Vajpeyi, 19–49. Lanham: Lexington Books.

11

The Ecological Paradox in Russia

Political, Social, and Economic Issues

Natalia Eremina and Igor N. Barygin

Climate Change issues are exerting considerable influence on all spheres of human existence. But debating the problem as well as finding viable solutions to it depend in a great measure on how this problem is conceived by the population at large, by relevant policy makers, members of the academic community, and business elites. All the involved parties in the climate debate strive to take cognizance of the new challenges, opportunities, and dangers posed by it. The following discussion will present the trends in climate change and factors responsible for it in the Russian Federation. It analyzes the linkages between climate change and the current state of sociopolitical dialogue in Russia.

As a geographic entity the Russian Federation encompasses one-sixth of the earth's dry land spreading over eight time zones, a variety of natural resources, enormous forests—particularly of the taiga variety—huge water reserves, and free access to the Arctic as well as the Southern Seas. These factors make Russia a major contributor to the maintenance and sustenance of the global ecological system, and a geopolitical actor whose participation in the grand debate on climate change is absolutely indispensable. Developments in Russia will not only affect Russian security and well being of its citizens, but also impact climate conditions elsewhere.

Due to Russia's large size and multiple climate zones the issue of climate change presents particularly acute and many faceted problems.

They are further compounded by enormous complexities of socioeconomic situations and difficulties as well as various internal perturbations which politicize the issue. Information on climate change is sparse, and often is presented as not a serious problem which potentially could affect the national economy and the day-to-day lives of the people. It presents an interesting situation. On the one hand the Russian Federation is an active participant in the world climate change policy making and implementation while on the other hand the absence of insufficient concern in Russian public opinion leaves much needed room for domestic policy makers and climate change 'deniers' to not actively pursue policies in Russia itself. However, despite public opinion apathy the issue is garnering greater attention and interest among experts on various aspects of climate change. As a result policy level measures are gaining momentum. Russia regularly carries out national estimation of its geographical and socioeconomic features to prepare reports—sort of White Papers. These reports are prepared by the National Hydrometeorological Service (Roshydromet) mapping strategies and their potential impact on the country's economy. It also monitors the World Meteorological Organization's convention on current programs under the auspices of the World Weather Watch, Global Atmospheric Watch, Intergovernmental Oceanographic Comission, and the Global Terrestrial Observing System. The mainland meteorological network of Roshydromet has 1627 stations, but in a number of Russian regions the number of these meteorological networks is insufficient (Doklad ob izmeneniah klimata i ih posledstviah na territorii Rossiyskoy Federatsii 2008: 3–5). Roshydromet also collects valuable data and research in several climate zones. It constantly analyzes data relating to several regions from the Arctic territories to the climatic conditions in the Southern Seas (the Black sea, the Azov sea, and the Caspian sea). Analysis of all the data is compiled into an integrated data bank which allows for a better understanding of climate trends and changes. The data collected by these meteorological stations allows us to take into account crucial parameters on the state of the permafrost, and the rate of thawing of the Arctic Sea ice fields, which most vividly illustrate the global climate change.

It is worth noting however the passing definite judgments on the current trends with respect to the climate change in the Russian Federation is possible only when the analysis of various climate estimates are longitudinal, at the very least thirty years. It has been observed by the experts that it is difficult to construct one clear picture of Russia's climate change because of external influences, and uncertainties of the forecasts linked to the technological developments and energy complex enlargement. It is estimated that Russia's temperature is expected to be higher than in other parts of the world. By the middle of the twenty-first century the temperature will rise to 2.6 +/- 0.7° C, and in the winter 3.4 +/- 0.8° C

(Doklad ob izmeneniah klimata i ih posledstviah na territorii Rossiyskoy Federatsii 2008: 14–15). It is also forecasted that the annual difference between the highest and the lowest temperatures throughout entire Russia will amount to an overall increase in the rainfall during winter period, contraction of the permafrost area that now comprises in excess of 60 percent of Russia, reduction in ice-covered area in the Arctic by 2 percent over a ten year period, successive changes of ecosystems in various regions, growth of such dangerous hydrometeorological phenomena such as floods and droughts in various regions, hence possible decline in agriculture production (Doklad ob izmeneniah klimata i ih posledstviah na territorii Rossiyskoy Federatsii 2008: 15–29).

The temperature rise trends in Russia have been monitored since 1975 but it was not until 2004 that the rise of the surface air temperature began to affect various economic sectors such as crops productivity in the majority of Russia's regions, and making it easier to navigate in the Arctic and Bering seas. However melting ice in these two Northern Waterways has also contributed to floating icebergs making navigation more hazardous.

Changes of the temperature regime have caused contraction of Siberian forests area, which in turn have led to even more significant negative impact on human livelihoods and health. It is also possible that these negative consequences will affect Northern Caucuses, Eastern Siberia, in the South of the Far East, Astrakhan Oblast, Calmyk, Republic and several other regions. Different zones create different climate related challenges—a nonuniform situation. It is exactly for this particular reason that solutions to Russia's ecological problems must therefore take into account the regional facet of the problem. Russia needs close cooperation and active participation of international organizations and its own citizens whose well being depends on a healthy environment.

CLIMATE CHANGE DEBATE IN THE RUSSIAN FEDERATION

The Government of the Russian Federation conducts periodic meetings devoted to environmental problems including the Climate Change. There is a realization that the issue is crucial to the country's sociopolitical stability and economic well-being. Unfortunately it has also become a political slogan used as a battle-cry by opposition movements against the government. The opposition argues that it is government inattention to the problem of climate change that adversely affects economic and social spheres. In response to these charges the government uses various social, political forums and personalities to counterattack. For example a meeting was held under the chairmanship of Medvedev on June 27, 2012 to discuss the climate situations in different regions which were adversely affected from high temperatures. Several preventive measures were suggested to preempt damages in the future by providing adequate insurance schemes (D. Medvedev provyol soveschanie po povodu situatsii v regionah, podvergshihsya vozdeystviyu anomalno vysokih temperature 2012). Under pressure from various political and social forces and public at large the government abandoned several huge construction projects that could have damaged the environment. One of the most vivid examples of this is the notorious scandal over plans to erect the so-called "Gazprom Tower" (Okhta-Center) in St. Petersburg.

While discussing the ongoing political debates it is meaningful to group climate change participants into three categories: (1) Members of the political establishment (2) Members of the scientific (academic) community, and (3) Representatives of the public at large. It is worth noting that these groups are not homogenous. Within every group there are both deniers and supporters of the climate change.

Representatives of the Russian political establishment are well aware of the necessity to strictly adhere to Russian's already existing international obligations by sharing information relating to climate changes. Other political institutions are also interested in mutually beneficial information exchanges (Intervyu glavy Rosgidrometa A. Frolova 2012). But while recognizing the reality of climate change the Russian political elites do not consider it as a menacing problem and do not engage in serious dialogues with the populace at large (Fedenko 2009). It seems that only those political institutions and leaders directly involved in climate-related issues, and dangerous hydro meteorological phenomena seriously attempt to stimulate the Russian government's interest in climate matters by prompting it to take concrete measures in solving outstanding multifarious climate change issues. The major inaction in implementing climate change measures come from bureaucracy. The academic community on the other hand recognizes the reality of the Climate Change and participates in several international efforts studying its impact (Intervyu glavy Rosgidrometa A. Frolova 2012) however, like many other countries, Russia is divided on the issue. Some of them believe that climate change debate has become a matter for unseemly bargaining and political speculations and political football (Lukyanov 2012).

By and large the Russian population does not have a uniform opinion on the issue. Nevertheless more and more Russians have started to express their concern on these environmental issues, particularly on climate change. A very rarely conducted public opinion poll in 2008 found that 67 percent of the respondents felt that Climate Change was a reality, but 83 percent of those polled confessed that they had no idea how the situation could be remedied or even tackled in any meaningful way (Fedenko 2009). The general public is too involved and 'overburdened' with day to day life issues. The role of the Russian NGOs has been quite prominent in educating public opinion, and actively engaged in activities related to climate change. More than twenty NGOs set up a "Coalition of Ecological Movements." Among this Coalition the most active ones have been "EKO-SOGLASIE" ("Eco-Agreement"), "Russian Regions' Ecological Center," "Centre for Ecological Policies," and the World Wildlife Fund. (Deyatelnost rossiiskih NPO po probleme izmenenia klimata 2000). The climate problem does not get much coverage by the Russian media. Like media in many other countries information and motivation to face climate issue challenges are often not sensational and therefore sparse and weak coverage.

FACTORS RESPONSIBLE FOR CLIMATE CHANGE

Both natural and anthropogenic factors exert tangible influence on climate in Russia. Among these it is necessary to single out the following:

Anthropogenic growth of the greenhouse gas concentration in the atmosphere
Enhancement of the radiation heating of the atmosphere (the greenhouse effect)
Growth of the mean temperature
Increase in the number of negative climatic phenomena
Changes in carbon dioxide balance

Human activities also directly affect global environment through emission to atmosphere of various greenhouse gases. Russia in accordance with the United Nations Framework Convention on Climate Change and the Kyoto protocol regularly prepares National Reports on the audit and registration of the anthropogenic leaks into environment. These periodic reports provide much needed data on the climate change.

In addition to governmental efforts to monitor the greenhouse gas emissions the Russian scientific community analyzes so called short-lived factors such as methane, soot, and hydroflorocarbonats leaked into the atmosphere. As a member of the Global Initiative, Russia is committed to take effective measures to control these factors. At present Russia occupies sixth place in the world after China, India, Brasilia, Indonesia and the United States in this context (Rossiya vozmyotsya za korotkozhivuschie faktory izmenenia klimata 2012). Other factors such as population growth, deforestation, urbanization are not perceived as important. Russia views that climate change problems are predominantly local. Russia as blessed with vast areas of forests which absorb a large amount of carbon dioxide.

Out of 500 billion global tons of carbon contained in live vegetation, Russia's share is thirty-four billion tons, 13 percent out of that total percentage is represented by pine-needle lands. If one adds to this data on the soil belonging to the lands interspersed with the forests, data on the soils covered with woods, on the dead substances of various origin that bestrew the woods' surface one cannot help but to conclude that total carbon reserves of Russia amount to 427 billion tons, about 8.5 percent of the world's carbon reserves. Due to this, Russia's forests absorb 8 percent of the anthropogenic polluting emissions (6.5 billion tons). Correspondingly, Russia's forests, similar to Brasilia's forests are the two dominant donors of oxygen for the planet (Prirodnye i antropogennye factory, vliyayuschie na izmenenie klimata 2010), but as the average global temperature continues to rise, in the long run, the area covered by Siberian pine-needle forests (taiga) might decrease. Thus, the main task of society at large, scientific community and of the Russian government is to find ways to effectively change the pattern of anthropogenic factors' influences on the climatic system—namely, to reduce the amount of greenhouse gas emissions into the atmosphere.

It is to be noted that between 1991 and 1998 the volume of polluting greenhouse gases dropped mainly due to the collapse of industrial production. The emission increased during 1999–2010 when the industrial activities picked up, however this increase did not exceed the volume prior to 1991 (Natsionalnay doklad Rossii o kadastre antropogennyh vybrosov b absorbtsii poglotitelyami parnikovyh gazov, ne reguliruemyh Monrealskim protokolom 2012). The low population growth and efficient management of vast energy sources are supposed to not increase the emission of greenhouse gases in Russia's atmosphere. Experts predict that by 2050 the emission volume will amount to about 10 percent almost equal to 2003 levels (Gricsevitch, Kokorin, Lugovoy, Safonov 2006). The biggest share of the emissions is caused by the energy sector—more than one-third of the overall volume of the polluting leaks. It is for this reason that development and implementation of energy efficiency programs will drastically reduce the volume of released greenhouse gases' into the atmosphere. At present the energy consumption rate of the Russian economy exceeds the world average by 2.3 times (Snizhenie vybrosov 2009); ("Main Directions of the Russian Federation Government Activities for the Period up to the Year 2012;" "Long Term Concept for Social-Economic Development of the Russian Federation"; "Directive as of 08, January 2009 N 1, on the Main Directions of Government Policies on the Bases of Using Renewable Energy Sources for a period up to 2020").

Besides the government, several measures and programs aimed at compiling an inventory of greenhouse gas emissions are carried out by large energy producers, such as Gazprom. As a result several positive steps in reducing the volume of greenhouse gas emissions in the energy sector have been undertaken. In 2011 Russia for the first time traded its carbon quota to Japanese Companies Mitsubishis and Nippon Oil. It was made possible due to the Kyoto Protocol implementation, government Directives, and measures undertaken by Gazprom. This constitutes only the beginning of a long and arduous road, for insuring the continued movement to maintain the achieved momentum of close cooperation between all the levels of authorities with public in general and the large-scale energy producers.

EVIDENCE OF CLIMATE CHANGE IN THE RUSSIAN FEDERATION

Any estimation of the scope of Climate Change in Russia over a long period of time should take the following factors into consideration: 1) Volume of precipitation, 2) Snow cover, 3) Sea ice area in the Arctic basin, 4) Changes in the permafrost zone, 5) and the sea level.

Surface air temperature: Between 1976 and 2005 the average temperature throughout the Russian Federation rose by 1.33° C. Surface air temperature continued to rise until 2012. Despite a very cold winter in 2005 the years 2001 and 2005 turned out to be warm. The year 2007 was the warmest year, while 2010 witnessed an extremely cold winter, and summer with record-high temperatures, as well as extremely warm Autumn (Doklad ob osobennostyah klimata na territorii Rossiyskoy Federatsii za 2010. 2011: 5–6). For a while it seemed that an apparently well-set pattern had emerged: hot summer—cold winter—but this trend did not last for too long. The year 2012 was conspicuously warm in winter (higher than the norm by 0.86° C), very warm in Spring (higher than the norm by 1.63° C), and exceptionally hot in summer all over Russia comparable only with the hot summer of 2010 (Izmenenia klimata 2012. Zima 2011–2012. 2012; Izmenenia klimata 2012. Vesna 2012. Mart-May. 2012; Izmenenia klimata 2012. Leto 2012. Iyun-Avgust. 2012).

Precipitations: Any discussion of precipitation patterns in Russia is complicated and poses a variety of problems. While the annual net volume of precipitations for 1976–2006 period had increased (at the rate of 7.2 mm/10 years) the data does not include considerable variations from region to region. The periodic variation most pronounced in precipitations volume was in the Western Siberia, in the Far East, and in the European part of the Russian Federation, then the years 2010 and 2011 were rather close to the norm (Doklad ob osobennostyah klimata na territorii Rossiyskoy Federatsii za 2010. 2011: 23; Doklad ob osobennostyah klimata na territorii Rossiyskoy Federatsii za 2011. 2012: 29), and in 2012 there was more precipitation than the average (Izmenenia klimata 2012. Zima 2011–2012. 2012; Izmenenia klimata 2012. Vesna 2012. Mart-May. 2012; Izmenenia klimata 2012. Leto 2012. Iyun-Avgust. 2012).

Snow Cover: In the last thirty years, a noticeable reduction in the snow cover in the northern hemisphere was noticed but in several other regions winter temperatures were mild yet there was an increase in the snow cover particularly along the entire Russian Arctic coast. Since 2005 the number of days when the snow cover exceeded 0.2 m in large parts of Russia (Doklad ob osobennostyah klimata na territorii Rossiyskoy Federatsii za 2005. 2006: 5). Again in 2010 regional variations have been observed. In the European part of Russia, in Western Siberia the snow cover decreased, but it increased in the Ural mountains in Southern Siberia, and the Far East (Doklad ob osobennostyah klimata na territorii Rossiyskoy Federatsii za 2010. 2011: 30). Similar variations were reported in 2011 (Doklad ob osobennostyah klimata na territorii Rossiyskoy Federatsii za 2011. 2012: 30). The winter of 2012 had less snow than the preceding years (Izmenenia klimata 2012. Zima 2011–2012. 2012: 24).

Sea Ice are in the Arctic basin: Over the last twenty-eight years a significant reduction in the sea ice area has been observed. In 2007 it was reduced to 4.3 million sq. km. despite the fact that there was a prolonged rise in temperature of the Arctic Seas. It should be noted that between 2009 and 2010 the decrease in the Atlantic water temperature played a major role on the Arctic temperature, hence the Arctic Sea ice did not contract (Доклад об особенностях климата на территории Российской Федерации за 2010 г. 2011: 37). However the area contracted due to the increase in temperature in the regions close to the north part of the Atlantic Ocean and the Pacific Ocean from 5.90 million sq. km. in 2010 to 4.9 million sq. km in 2011 (Doklad ob osobennostyah klimata na territorii Rossiyskoy Federatsii za 2010. 2011: 37). In 2012 melting of the ice sharply increased due to the temperature rise in the winter, spring, and the summer. It is estimated that this sea area could be reduced to 4 million square km in the coming years. On September 16, 2012 the Arctic Sea ice covered area reached its record minimum since 1979. But during the month of October 2012 it showed an increase at a record rate, and doubled in size. By November 2012 the ice-covered area comprised of 8.22 million square meters, however it was lower than in 2007 (Ploschad lyda udvoilas za oktyabr 2012)

Changes in the permafrost zone: A marked increase in temperature of the upper crust of the permafrost soil has been registered over a long period of time. On the average the temperature of the upper layer of the permafrost in the north of the Western Siberia between 2001 and 2012 rose (Doklad ob izmeneniah klimata i ih posledstviah na territorii Rossiyskoy Federatsii 2008: 10).

Sea Level: The data on the Black Sea, the Baltic and Caspian Seas provide somewhat different trends. Normally the Black Sea water rises from several millimeters to 3 centimeters a year (Ginzburg, Kostyanoy, Sheremet, Lebedev 2009: 355). These changes might in the near future create difficulties for the inhabitants of the Southern sea shores (Doklad ob izmeneniah klimata i ih posledstviah na territorii Rossiyskoy Federatsii 2008: 21).

Climate Change and Human Security

As discussed above that due to Russia's tremendous climate zones diversely specific impacts on human population will be different, however there is no doubt that irrespective of these variations climate change is taking place and is bound to impact the quality of Russia's environment, its population, and its socioeconomic and political development. Climate changes have impacted several aspects of human health such as proliferation of insects and ticks which carry contagious illnesses, changes in seasonal distributions of allergenic species of plant pollen, mortality rates and increased trauma.

As discussed in Chapter One the impact of climate change will not be uniform. Some regions and countries will have negative effects while others may even benefit from these changes. In Russia certain regions will suffer from its negative impact while the country as a whole might economically benefit from "possible positive consequences" such as thawing of the Arctic sea (Zgurovsky 2008). It will give increased and strategic advantage to Russian economic and strategic activities by revitalizing the 'Great Arctic' northern route.

On the other hand, warming of the Arctic sea will seriously affect the fragile ecosystems. Russia's Arctic coastal area—where about 5 percent (6.4 million) of the people live includes minority indigenous people. (Regiony Rossii 2011: 54–56). As a result of the constant change of climate's cooling and warming periods of the Arctic ecosystem will make it difficult for these populations to adapt and adjust to new conditions. These seasonal functions of climate will also affect the size of fish populations and their migration routes. For example, a decrease in warmth of the Barents Sea has already led to the decline in the number and size of cod shoals which is one of the most popular species of fish among Russia's minority people (Makarevitch, Ishkulov 2010: 633–639). It also seriously affects the traditional way of life for several of these indigenous minority people. It has been reported that cases of severe cardiovascular ailments, traumatism, and high death rates have increased in the northern regions of Russia. In southern Russia the rise in temperatures has resulted in droughts.

Scientists at the Russian Academy of Medical Sciences (RAMS), and the Russian Academy of Sciences (RAN) together with Ministry of Health and Social Development (Minzdravsotsrazvitie) carry out quantitative scientific research to find trends/regular dependencies (laws) on the impact of climate change on human health and security. It has been also reported that about 1000 people perish annually due to unfavorable climatic conditions, and from negative climate related phenomena specially in Yakutia, Stavropolie, Krasnodar regions of the Far East, and the northern Caucasus. It is the elderly people over the age of seventy-five years, and children who suffer from chronic ailments. Physicians have reported an increasing number of fourteen and younger children born with asphyxia, poor health, and diseases usually associated with people twenty years and older. Also people living in the northern territories where the air temperature is +29° C find themselves particularly at risk. Even minor increases in temperature cause frequency of stomach infections, and higher mortality rates. Several experts believe that at present a temperature increase of 10° C will increase mortality rates by 10 percent from four thousand to 28 thousand (Otchet o provedenii seminara po izmeneniyu klimata i zdoroviu naselenia v Rossii v XXI. 2012). Several measures and preventive steps have been taken by regional scientific organizations to help these groups (Analitichesky vestnik Soveta Federatsii Federalnogo Sobrania Rossiiskoy Federatsii 2008).

Agriculture: Another area of major concern due to climate change in Russia is its agricultural sector, crucial to the health of its national economy and well-being. As far as agriculture is concerned, the Climate Change carries with it certain paradoxically interconnected positive and negative consequences. For example, recently under more favorable temperature conditions the area for winter crops in the Northern Caucuses, in the steppe regions of the Volga river basin, in the Southern Ural Mountains region, and several regions of Western Siberia was enlarged resulting in bigger agricultural yields. In the Stavropolie region over the last twenty years the yields have grown by 30 percent. In the Asian part of Russia the increase in temperature does not entail automatic growth in crop yields. For example, in Zabaykalie ("The Beyond Lake Baykal Region") as well as in areas near Lake Baykal ("Pribaykalie")—agricultural production has dropped. However, despite these regional variations it is predicted that in the long range Russia will have positive stable agricultural growth (Strategichesky prognoz izmeneniy klimata Rossiyskoy Federatsii do 2010–2015 i ih vlianie na otrasli ekonomiki Rossii 2005: 18–19).

Other factors related to Russia's agriculture and climate change are aridity and droughts. Increase in temperatures may contribute to periodic droughts in the northern Caucuses, the Povolzhie, Rostov, Volgorad oblasts, and the steppe of the Ural. Concrete measures such as construction of irrigational facilities, drought-resistant varieties of crops like sunflowers must be introduced. It is also important to consider in water short regions of Russia that further water scarcity and crop failures might take place. Serious efforts are being made to face these issues. By 2015 an increase of about 8 percent to 15 percent of renewable water resources from lakes, rivers, ponds, and artificial basins is expected. The per capita volume of the available water is predicted to increase by 12–14 percent (Strategichesky prognoz izmeneniy klimata Rossiyskoy Federatsii do 2010–2015 i ih vlianie na otrasli ekonomiki Rossii 2005: 20–21).

Energy

Energy sector is Russia's most important source of economic well-being. Adverse climate changes have the potential to negatively affect its socio-economic and political development. The decrease in the number of cold days, particularly in regions such as Kamchatka, Sakhalin, in the Far East (Primorie) will help the exploration and management of energy resources (Strategichesky prognoz izmeneniy klimata Rossiyskoy Federatsii do 2010–2015 i ih vlianie na otrasli ekonomiki Rossii 2005: 13). Russia possesses great potential in enhancing the effectiveness of energy use and in the development and implementation of renewable energy technologies. According to the "Strategic Program of Russia in the field of Energy" adopted in 2000 and is valid and effective through 2012. Russia plans to increase gas extraction and production because gas is a cleaner fuel and energy source and the global demand for gas is increasing. The rise in temperatures will influence the patterns of outflows and inflows of water into various basins, changes in rivers' water flow regimens, ice regimen—the phenomena that are sure to affect the burden on all kinds of ducts and pipe-lines (Strategichesky prognoz izmeneniy klimata Rossiyskoy Federatsii do 2010–2015 i ih vlianie na otrasli ekonomiki Rossii 2005: 14).

The impact of climate change on Russia's energy sector is a combination of both positives and negatives. The rise in temperature will melt the permafrost disrupting the stability of infrastructure, specially the transport infrastructure essential to the efficient functioning of the gas pipelines in western Siberia where oil and gas processing plants and complexes are situated. The reduction in permafrost presents a particularly grave danger for Novaya Zemlya Island and the Yamal Peninsula. It is on these two locations that depots with nuclear waste are situated. These facts require great unremitting attention to stabilization of the bases/foundations of buildings (Strategichesky prognoz izmeneniy klimata Rossiyskoy Federatsii do 2010–2015 i ih vlianie na otrasli ekonomiki Rossii 2005: 16–17).

The issue of effective energy usage is closely linked with the ozone layer problem (chapter 2). Over the last decade a marked decrease in the ozone layer over the Arctic region as well as over the adjacent areas of Russia has been observed. In Russia monitoring of the ozone layer is carried out by twenty Roshydromet centers (Sostoyanie ozonovogo sloya 2012). Observed data leaves no doubt at all that there is a conspicuous connection between the ozone layer parameters and climatic factors, with anthropogenic influences becoming an additional cause of ozone area parameters to rapid change.

WHAT IS BEING DONE?

The main documents reflecting the evolution of the Russian Federation's Climate policies are the National Climate Doctrine, fiats (decrees) issued by the President, his directives as well as Russia's agreements with its partners in the region and abroad. On August 11, 2012 Russia's President Putin signed a Decree "On Conducting in the Russian Federation of the Protection of the Environment Year." Prior to 2012 another very important document "the Fundamental Climate Doctrine" was signed into law on December 17, 2009, by the then President of Russia D. Medvedev. This Doctrine defined and specified priorities as well as peculiarities of the Russian Federation's policy with respect to the problem of Climate Change and measures to meet the Challenge.

The Medvedev Climate Doctrine emphasized the need for cooperation with other States and organizations relevant to issues of climate change in Russia (Ukaz o provedenii v Rossii goda ohrany okruzhayuschey sredy 2012). Since then there has been steady policy cooperation between Russia, and other countries of the Commonwealth of Independent States' (CIS), and the Chief of the CIS executive Body. On November 12, 2012 Sergey Lebedev observed that year—2013—"will become the Year of Ecological Culture and Environment Protection throughout the CIS" and will have first priority as far as cooperation activities among the Commonwealth States are concerned for 2013 (Deyatelnost Partii zelyenyh 2012).

ROLE OF THE RUSSIAN FEDERATION IN GLOBAL CLIMATE INITIATIVES

Russia's role in global and regional programs and initiatives, on the whole, is determined by the fact that the country is defined as a developing country with transitional national economy—part of the BRIC's. Russia is supposed to receive benefits from global ecological projects but at the same time provide economic and technological help to other developing economies. While Russia was not among the main authors of the UN-sponsored Framework Convention on Climate Change

(UNFCCC) adopted at "The Earth Summit" in Rio de Janeiro in 1992, it did ratify the Convention in 1994.

As for the Kyoto Protocol, Russia at first greeted this global initiative with considerable enthusiasm. A Federal Law "On Ratification of the Kyoto Protocol to the United Nations Framework Convention on Climate Change" was enacted by the State Duma on October 21, 2004 and ratified by the Federation Council on October 24, 2004. President Putin signed the document on November 4th 2004. The Protocol became effective on February 16, 2005, ninety days after the official transfer of Russia's Protocol Ratification credential to the FCCC by Secretariat on November 18, 2004. However, several difficulties and problems of practical nature have made full implementation of the Kyoto Protocol impractical. National exchange for trading Quotas was suspended for an unspecified period. It was found that Russia's industrial plants and enterprises did not meet the ecologically clean and productively more efficient standards. In view of this situation, in March 2006 the Russian government issued a directive to the Ministry of Social Development and Trade and other Federal authorities to prepare blueprints for laws aimed at finding a solution to the problem and prepare a Concept of a blueprint for the Law that would regulate the application of Article 6 of the Kyoto Protocol under which Russia would be allowed to attract external investments for projects of joint implementation. The progress to enact laws and relevant decrees was slow at best. It was on May 28, 2007 that the government of the Russian Federation gave final approval to the Decree No. 332 "On the Order of Passing Approval Fiats for Implementation Verification Procedures Relating to the Projects Being Put into Practice in Accordance with the UN Framework Convention on Climate Change." Same year (2007) the Open Liabilities Society (OAO) "Uralskaya Stal" ("Ural Steel Works") Metallurgical branch of Russian industry, successfully implemented the Kyoto Protocol mechanism and procedures to curtail the greenhouse gases into the atmosphere. By the beginning of 2008 about fifty Joint Venture type projects for Russia were undertaken. Several international corporations such as Consulting Firms SAMCO and Global-Carbon, a Body for carrying out independent auditing of projects aimed at reducing leaks—SGS, as well as one of the largest quota buying concerns—Tricoroina Ab (Trikorna OAO) started their operations.

In 2009, the Minekonomrasvitie (Ministry of Economic and Social Development) received about 125 applications from various Russian corporations but not a single one of them was confirmed. The "Novaya Gazeta" newspaper reported, quoting Russian government sources that the real cause for the delay in implementing the realization of the Trade in Quotas Agreement was political. Putin took exception that both Germany and France had failed to help Russia's bid to enter WTO which they had promised to do in exchange for Russia's signing of the Kyoto Protocol (Kiotskiy protocol 2009). In October 2009, the Russian government issued a Decree (No. 843) so that the power to take part in any proceedings and actions aiming at obtaining, transfer carbon quota to third parties, and purchasing of the Units of greenhouse gases leaks was vested with the Sberbank of Russia. Procedures to conduct relevant Application Competitions (Auctions) and the further application of expert assessment procedures were Sberbank's responsibility. It was also stipulated in the Decree that the ultimate authority to accept or reject the results of the expert application assessments was with the Minekonomrazvitie (Postanovlenie pravitelstva Rossiiskoy Federatsii № 843. 2009). At the end of July 2010, Minekonomrazvite confirmed the first fifteen Joint Venture projects which resulted in reducing the emissions resulting from the realization of the projects—the volume of 30 million tons in CO_2-equivalent. In December 2010, the first ever sale of a carbon quota took place when Russian Gasprom company sold its quota to Japanese partners (Громова 2011).

At Doha Conference in December 2012 delegates to the eighteenth Conference of the Parties to the Kyoto-2 UN Framework Protocol on Climate Change bickered on the issue of global warming. Russia, the Ukraine, and Belorussia expressed their disappointment with the voting procedure that led to the adoption of the final—"the summing up"—documents of the Conference. While Russia did not opt out of the Kyoto-2 Protocol it made it very clear that Russia could not participate in the Joint Projects Implementing the Protocol. As a result Russia has forfeited opportunities in participating in 150 projects envisaged at the Conference as Kyoto-2 Protocol target projects.

The policy of "Discrete distancing from the Kyoto-2 Protocol" is nowadays the leading trend of Russia's policy in this direction. Russia has no pressing reason for abandoning its participation in the Kyoto-2 Protocol. Economic loss from the Protocol participation is more then compensated by the absence of burdensome obligations. But it is very important to emphasize at this point that it is from the political point of view that Russia's continued participation in the Kyoto-2 Protocol is important as a symbol of Russia's commitment to cooperative attitude to various Global projects. Hopefully a new UN Framework Coalition "Climate and Clean Air" will allow Russia to participate in the Soot emission problem. Russian perception and participation in the Kyoto Protocol implementation ran into several difficulties. Russian Deputy Economics Minister Mr. A. Klepach went on record that the Kyoto Protocol implementation would surely become the beginning of the end for the Russian national economy (Rossii ne nuzhen Kiotsky protocol 2012).

Russian actions were criticized both by international and domestic actors as inconsistent and illogical. There were numerous calls "to kill" the Kyoto Protocol, but, at the same time preserving all the advantageous elements for Russia such as its quotas and possibilities of participating in Joint Venture Projects. There were several proposals to just postpone Russia's participation in various buy-and-sell carbon CO_2 activities until 2012. This contradictory and illogical position caused "stupor" among many parties concerned with climate change conditions under the Kyoto Protocol. Russians want to keep their cake and eat it too. The Russian Prime minister Medvedev and other policy makers are opposed to Russia's continued participation in Kyoto-02 Protocol. In Medvedev's opinion the scheme is not commercially profitable for Russia. According to Vice president A. Dvorkovic the economic hurdles compel Russia from joining the Kyoto-2 Protocol (Medvedev D.: Pomashem Kiotskomu protokolu ruchkoy 2012).

THE ROLE OF NON-GOVERNMENTAL PUBLIC ORGANIZATIONS

Ecological Green parties and nongovernmental public organizations in Climate Change represent two different types of organizations in Russia. The first group could be called "the controlled ones" always ready and eager to the so-called "constructive" cooperation with all levels of authority. The "Green Commonwealth"—a community of Russia's more than a hundred of nongovernmental ecological organizations is a good example.

The main aim of the "Green Commonwealth" is to coordinate actions of the so-called "green" section of the public by carrying out policies to make healthier society for the Russian people and by meeting the requirements of the real ecological situation in the country. Solutions to the ecological problems are sought at the regional, federal and international levels in accordance with the laws of the Russian Federation, the relevant international agreements, and the main values of Russian civil society. The most active registered group at the national level is self-styled "Green Alliance"—the People's Party. It is narrowly focused on Russian ecological issues. There is no reference to the "Kyoto Protocol" in the party's Program (Rossiyskaya ekologicheskaya partiya zelenyh 2012), though there is a paragraph about it on the official website.

The second type of organizations are represented by local chapters of the "Green Peace" of "Bellona" etcetera. They operate and follow the Green Peace's global agenda. The first project of the World Wild Life Fund (WWF) began as early as 1988, and it was in 1994 that the Russian Mission of the

Fund was opened. Since that time, the WWF has successfully implemented more than 150 field projects in forty regions of Russia, and has invested in excess of 30 billion dollars to preserve and enhance Russia's natural wealth. In 2004, the WWF Fund became the Russian National Organization.

The ecological organization of the first type (the 'controlled ones') basically support Russian State policies on environmental issues and keep this distance from the Kyoto-2 Protocol. The second type of activity (the Green Peace) supports the Kyoto-2 Protocol agenda. They concentrate their national efforts on implementation of Ecology 2020 Program worked out by the Government of the Russian Federation in accordance with Russia's Ecological Program. Prime Minister Medvedev clarified the program on October the 18, 2012 by presenting an ominously grim picture of the current ecological situation in the country: "15 percent of our country's territory exists under ecologically critical conditions; the level of air pollution in 155 cities is either "high" or even "very high," all kinds of industrial waste and garbage of all sorts are kept in dumps and squalid cesspits under unsatisfactory conditions for decades!" Russia's Minister for Natural resources S. Donskov drew even a more gloomy and darker picture of national environment by pointing out that since the Soviet period there has accumulated heaps of refuse to the tune of 30 billion tons in weight, whereas the level of the refuse substances recycling and reuse is no more than 10 percent. And that is the situation in Russia compared to other more "cognizant with respect to matters ecological" countries. In Russia it is at the very least four times greater (Medvedev D.: Pomashem Kiotskomu protokolu ruchkoy 2012).

The situation could be remedied through monetary infusions of the State funds: overall investments in the Governments programs will add up to 336 billion rubles through 2020. The money will be spent, among other things, in implementing recourses-saving technologies—technologies with the prefix "bio" among them. Among other measures proposed included liquidation of unauthorized dumps of industrial waste, rehabilitation of lands, creation of thirty new national reserves (Medvedev D.: Pomashem Kiotskomu protokolu ruchkoy 2012).

As discussed above Russia faces enormous socioeconomic and political challenges, and the role of civic society is in a transitional phase. Active participation in issues such as climate change will require patience, time, and improved political environment. Hopefully it will be soon.

REFERENCES

Alekseev, G.V., 2012, "Letnee sokraschenie morskih lydov v Arktike v 2012," *Izmeneniya klimata*, Ezhemesyachniy informacionniy byulleten Rosgidrometa No 37 (Moskva: Rosgidromet).

"Analitichesky vestnik Soveta Federatsii Federalnogo Sobrania Rossiiskoy Federatsii," 2008, *Dokumenty Federalnogo sobrania*, Dokument Federalnogo sobrania Rossiiskoy Federatsii No 4(349) (Moskva: Pravitelstvo Rossiiskoy Federatsii).

"Deyatelnost Partii zelyenyh," 2012, *Ofitsialniy sait partii Kedr*, at www.dkedr.ru/news_id.php?id=2012111812522210 (10.12.2012).

"Deyatelnost rossiiskih NPO po probleme izmenenia klimata," 2000, *Eco Soglasie*, at www.ecoaccord.org/climate/2000/next_step.htm (21.09.2012).

D. Medvedev provyol soveschanie po povodu situatsii v regionah, podvergshihsya vozdeystviyu anomalno vysokih temperature," 2012, *Izmenenia Klimata*, 27 (35).

"Doklad ob osobennostyah klimata na territorii Rossiyskoy Federatsii za 2005," 2006, *Doklady Rosgidrometa* (Moskva: Rosgidromet).

"Doklad ob izmeneniah klimata i ih posledstviah na territorii Rossiyskoy Federatsii. Rezume," 2008, *Doklady Rosgidrometa* (Moskva: Rosgidromet).

"Doklad ob osobennostyah klimata na territorii Rossiyskoy Federatsii za 2010," 2011, *Doklady Rosgidrometa* (Moskva: Rosgidromet).

"Doklad ob osobennostyah klimata na territorii Rossiyskoy Federatsii za 2011," 2012, (Moskva: Rosgidromet).

Fedenko, P. 10.12.2009: "Rossiya poteplela k izmeneniyu klimata, *Russkaya sluzhba BBC*, at www.bbc.co.uk/russian/russia/2009/12/091204_russia_climate_fedenko.shtml (21.09.2012).

Ginzburg, A.I. Kostyanoy, A.G. Sheremet, N.A. Lebedev, S.A., 2009, "Izmenchivost temperatury poverhnosti i urovnya Chernogo, Mramornogo i Egeyskogo morey po sputnikovym izmereniam," *Raboty Instituta Okeanologii*, at d33.infospace.ru/d33_conf/2009,1/349-358.pdf (21.09.2012).

Gricsevitch, I.G. Kokorin, A.O. Lugovoy, O.B. Safonov, G.B., 2006, *Razvitie energetiki i snizhenie vybrosov parnikovyh gazov*, Moskva, WWF Rossii.

Gromova, A. "Poteryanniy vozduh Kioto," 2011, *Himichesky zhurnal*, 1–2.

"Intervyu glavy Rosgidrometa A. Frolova," 13.11.2012, *Ekologia i zhizn*, Интервью главы Росгидромета А.Фролова. 2012. *Экология и жизнь*, at www.ecolife.ru/intervju/9886/ (21.09.2012).

Izmenenia klimata 2012. Zima 2011–2012," 2012, Doklady Rosgidrometa (Moskva: Rosgidromet)

"Izmenenia klimata 2012. Vesna 2012. Mart-May," 2012. Doklady Rosgidrometa (Moskva: Rosgidromet)

"Izmenenia klimata 2012. Leto 2012. Iyun-Avgust," 2012. Doklady Rosgidrometa (Moskva: Rosgidromet)

Kikilo, B. 21.07.2011: "Klimaticheskie debaty," *Agentstvo Itar Tass*, at www.itartass.ur.ru/details/adekvatnaya_reaktsiya.html (21.09.2012).

"Kiotskiy protocol," 2009, 09.06.2009, *Novaya gazeta*, at www.novayagazeta.ru/politics/44425.html (12.12.2012).

"Klimaticheskaya doktrina Rossiiskoy Federatsii," 2009, *Ofitsialniy sait Prezidenta Rossiiskoy Federatsii*, at www.kremlin.ru/acts/6365 (01.12.2012).

Lukyanov F., 10.08.2012: Klimaticheskiy klinch," *Rossiia v globalnoy politike*, at www.globalaffairs.ru/redcol/Klimaticheskii-klinch-14973 (21.12.2012).

Makarevitch, P.R. Ishkulov, D.G., 2010, "Struktura b vidovoye raznoobrazie pelagicheskih I donnyh biocenozov v Barentsovom more," *Vestnik MGTU*, 13(4/1): 633–640.

"Medvedev D.:Pomashem Kiotskomu protokolu ruchkoy," 18.10.2012, *Moscow Komsomolets*, at www.mk.ru/politics/article/2012/10/18/763124-medvedev-pomashem-kiotskomu-protokolu-ruchkoy.html (12.12 2012).

"Natsionalnay doklad Rossii o kadastre antropogennyh vybrosov b absorbtsii poglotitelyami parnikovyh gazov, ne reguliruemyh Monrealskim protokolom 1990–2010," 2012, *Izmenenia klimata*, at www.global-climate-change.ru/index.php/ru/official-documents/kadastr-report (21.12.2012).

"Otchet o provedenii seminara po izmeneniyu klimata i zdoroviu naselenia v Rossii v XXI," 21.05.2012, *Seminar po klimaticheskim izmeneniam*, at erh.ru/protokol/st09.php (21.09.2012).

"Ploschad lyda udvoilas za oktyabr," 2012, *RIA News*, 6 November.

"Postanovlenie pravitelstva Rossiiskoy Federatsii № 843," 2009, *Pravitelstvo Rossiiskoy Federatsii*, at government.ru/special/gov/results/20244 (10.12.2012).

"Prirodnye i antropogennye factory, vliyayuschie na izmenenie klimata," 08.02.2010, *Izmenenie klimata i glibalnoye poteplenie, глобальное потепление*, at climaty.ru/node/19 (21.12.2012).

"Problema sostoyania ozonovogo sloya v Arktoke obostryaetsya," 2012, *RIA Novosti*, 23 March.

"Regiony Rossii 2011. Sotsialno-ekonomicheskie pokazateli," 2011, *Federalnaya sluzhba gosudarstvennoy statistiki*, at www.gks.ru/bgd/regl/b11_14p/Main.htm (21.09.2012).

"Rossii ne nuzhen Kiotsky protocol," 2012, *Obschestvennoye mnenie*, at mnenia.ru/rubric/finance/rossii-ne-nujen-kiotskiy-protokol (01.12.2012).

"Rossiya vozmyotsya za korotkozhivuschie faktory izmenenia klimata," 25.06.2012, *Rossiiskiy sotsialno economichesly soyuz*, at rusecounion.ru/klimat_25612 (21.12.2012).

"Rossiyskaya ekologicheskaya partiya zelenyh," 2012, *Ofitsialnyi sait partii Zelenye*, at www.greenparty.ru/news.php?anonse=news (11.12.2012).

"Snizhenie vybrosov," 2009, *ClimateChange.Ru*, at climatechange.ru/node/55 (21.09.2012).

"Sostoyanie ozonovogo sloya," 2012, *Severnoye upravlenie po gidrometeorologii i okruzhayuschey srede*, at www.sevmeteo.ru/articles/7/19.shtml (21.09.2012).

"Strategichesky prognoz izmeneniy klimata Rossiyskoy Federatsii do 2010–2015 i ih vlianie na otrasli ekonomiki Rossii," 2005, *Doklady Rosgidrometa* (Moskva: Rosgidromet).

"Sut problemy izmenenia klimata v poslednee desyatiletie," 2012, *Vsya Evropa*, at www.alleuropa.ru/sutj-problemi-izmeneniya-klimata-v-poslednie-desyatiletiya (29.12.2012).

"Ukaz o provedenii v Rossii goda ohrany okruzhayuschey sredy," 2012, *Ofitsialny sait Prezidenta Rossiiskoy Federatsii*, at www.kremlin.ru/acts/16219 (02.12.2012).

Zgurovsky, K., 2008, "Rybolovstvo," *Vozdeystvie izmenenia klimata na rossiiskuyu Arktiku*, Fond dikoy prirody Rossiiskoy Federatsii (Moskva: WWF Rossii).

12

Climate Change in Bangladesh and Nepal

Issues, Challenges, and Strategies

R. K. Mishra and P. S. Janaki Krishna

The earth's climate has been evolving continually over millennia. However, alarmingly the last two centuries have witnessed the development of the greenhouse problem, which threatens to change climate in an unparalleled manner. As a consequence, "climate change" is recognized far and wide as a major environmental challenge facing the world. The impacts of climate change, including increases in extreme weather events, elevated temperatures, sea level rise, biodiversity loss, and declines in water availability and supply impacting agriculture, pose threats to people and their livelihoods. With their scarce resources, infrastructure, and dependence on subsistence agriculture, developing countries are acutely vulnerable to the possible impacts of climate change would increase energy demand, adversely affect transport, and decrease tourism in some parts of Asia. It would also intensify threats to biodiversity due to land use and land cover changes and population pressure. Poleward movement of the southern boundary of the permafrost zones of Asia would result in a change of thermal erosion, bringing negative impacts on social infrastructure and industries.

These threats to humanity and ecosystem are duly recognized by the global community with the adoption of UN Framework Convention on Climate Change (UNFCCC). As Parties to UNFCCC, countries like Bangladesh and Nepal have become conscious of the need to meet the

Table 12.1. Summary of Climate Risks in South Asia

	Afghanistan	Bangladesh	Bhutan	India	Nepal	Pakistan	Maldives	Sri Lanka
Sea Level Rise	-	High	-	Modest	-	Modest	High	High
Glacier Retreat	High	High	High	High	High	High	-	-
Temperature Increase	-	High	High	High	High	High	Modest	High
Floods more Frequent		-	Likely	High	High	Likely	High	
Drought more Frequent	Likely	High some areas		High		Likely	-	

Source: Zia Mustafa 2011.

obligations towards reducing Greenhouse Gas (GHG) emissions. Both Bangladesh and Nepal have refined their environmental policies and are supporting the legal regime to implement environmental management activities and programs. While these countries are geographically situated in south Asia, they have quite different ecological systems, and challenges they face in environmental spheres. The following discussion analyzes these challenges and policies being undertaken by their governments to meet specific issues related to climate change (Table 12.1).

CHANGE IN BANGLADESH AND NEPAL

The impacts of 'climate change' are evinced in several ways. Generally, climate change is likely to affect the movement of people in at least four ways: 1) the intensification of natural disasters—both sudden and slow-onset—leading to increased displacement and migration; 2) the adverse consequences of increased warming, climate variability and of other effects of climate change for livelihoods, public health, food security, and water availability; 3) rising sea levels that make coastal areas uninhabitable; and 4) competition over scarce natural resources potentially leading to growing tensions and even conflict and, in turn, more displacement. (IOM, 2010).

Bangladesh

Bangladesh is one of the most vulnerable countries to the impacts of climate change in the world. In Bangladesh vulnerability to climate change is due to complex relations between biophysical, social, economic and topographical population density, and dire poverty. Bangladesh is already experiencing migration from rural to urban areas thereby leading to unplanned urbanization in city centers such as Dhaka impacting human security and producing tensions due to overcrowding and lack of basic necessities. Often migration is used as an adaptation strategy, however in the long term this could prove to be dysfunctional. Climate change in Bangladesh in already witnessing increased frequency and intensity of natural disasters such as Cyclone Aila. A warmer and wetter future climate that goes further than the typical historical variations will worsen existing climatic risks and increase vulnerability by expanding the scope and intensity of inundation from flooding, storm surges, and by reducing arable land due to sea level rise and excessive salinity. Several anticipated adverse impacts of climate change including sea level rise, higher temperatures, increased monsoon rain and runoff, reduced dry season precipitation, and an increase in cyclone intensity would in fact aggravate many of the existing stresses, which already pose a serious impediment to economic development in Bangladesh. The climate change policy, particularly adaptation thus becomes a part and parcel of the development policies of the country (NIDOS 2009). The general circulation models (GCMs) predict that by 2050 Bangladesh will be 1.5°C warmer and 4 percent wetter. Harsh cyclones originating in the Bay of Bengal are also expected to be more frequent as a result of warmer ocean surface temperatures. Cyclone-induced storm surges are further exacerbated by a potential rise in sea level of over 27 cm by 2050. (World Bank, 2010). These developments will have severe negative consequences on the country's agricultural sector. According to the World Bank Group 2010 EACC report the median temperatures by 2030 in July, August, and September will surpass the 90th percentile of the historical temperature variability. The median warming predictions for Bangladesh across the models by the 2050s is 1.55°C. Most climate models indicate an increasing trend of monsoon rainfall and greater inflows into Bangladesh, the extent of flooding is likely to increase. In flood-prone regions of Bangladesh monthly water levels are estimated to increase for each grid point for the years 2030s and 2050s. Also, increasing sea level rises are one of the most critical climate change issues for coastal areas. The IPCC (2007a) projected that an average temperature rise of 9 to 88 cm could be expected by the end of the century. Recent projections suggest even more substantial rises. Allison et al. (2009) further observe that increasing temperatures will result in sea level rise by the thermal expansion of water and through the addition of water to the oceans from the melting of continental ice sheets. A one meter sea level rise is estimated to impact 13 million people in Bangladesh. As mentioned above Bangladesh is exposed to a wide array of climate change impacts and induced hazards including river floods, flash floods, cyclones and tidal surges, drought, salinity intrusion, water-logging and drainage congestion. Current climate variability in Bangladesh already represents a huge threat to the livelihoods of the poor as they often reside in areas of highest risk. To estimate the consequences of climate variability and change on economic growth and household welfare, dynamic recursive computable general equilibrium (CGE) models have been developed. These models infer that climate change also has broader implications on several sectors of national economy. Natural calamities will cost Bangladesh $129 billion in total GDP over a forty-five-year period between 2005–2050, equivalent to $2.9 billion overall loss each year due to climate change, or alternatively an average annual 1.6 percent reduction in total GDP. These climate risks will have severe implications for household welfare. Around 80 percent of total losses will be felt by consumers. About 80 percent of the economic losses will occur outside the agricultural sector. This means that both rural and urban households will be adversely affected. Per capita income consumption is projected to fall for both farm and nonfarm households (Pender 2008).

Nepal

Nepal's economy is largely dependent on climate-sensitive sectors, such as rain-fed agriculture. Its ecosystem is fragile, and its topography further makes it heavily prone to flooding. The country's political and economic institutions are weak and have been further eroded by years of civil turmoil. These insecurities make it difficult to initiate and implement economic and social policies in general and climate change in particular. Adaptation to climate change is costly, and financial resources are desperately needed to rebuild the country, reintegrate ex-insurgents, and maintain an effective civil service (Climate Change Fact Sheets 2008). This is particularly noticeable in the mountains, where the glaciers are melting, affecting the precipitation system and contributing to the formation of glacial lakes that could "burst" at any moment releasing torrents of water and debris.

According to the International Centre for Integrated Mountain Development (ICIMOD), average warming in Nepal is 0.6 degree centigrade higher than the global average (Shrestha, et al. 1999). The warming in the Himalayas was far greater at higher elevations. While it was significantly lower and often nonexistent in the Terai and Siwalik regions. "The mid-hills area used to be the main habitat of people and the impact of climate change was not very visible" but "those closer to the higher mountains seem to have greater awareness in terms of less snow coverage. The Sherpas are worried that if there is no snow in the winter, so will [they] get water?" (Schild 2008). An understanding of the course of evolution of the climate in the Himalaya's Hindu Kush (HKH) region has been difficult due to the fact that climate modeling was introduced in Nepal only recently. The existing global and regional models do not perform effectively due to the extreme topography of the region. Availability of reliable observed temperature and precipitation data in Nepal is also limited. Shreshtha et al. (1999; 2000) in their study reported that temperatures are increasing at a rather high rate and that the warming has been consistent and continuous since the mid-1970s. The average increase in annual temperature between 1977 and 2000 was 0.06°C per year. The warming temperatures have led to melting glaciers and change in the precipitation system.

In Nepal, over 80 percent of the population is dependent on agriculture. The main crops are rice, wheat, and maize. With an annual growth rate of 2.3 percent, the population of Nepal is rapidly increasing. It has doubled in the last twenty years from 13.5 million in 1975 to 27.1 million in 2005 (UNDP, 2007; UNDP. Human Development Report 2007/2008) putting tremendous pressure on the food sector. Further, the country is facing floods, soil erosion, and landslides that have greatly affected the availability of fertile soil, and climate change will likely worsen this situation. To meet these challenges the Government of Nepal proposes to increase the production of drought-resistant crop varieties by improving cropping practices to conserve water and by promoting crop diversification (MOPE 2004). Longer periods of drought will also drain natural resources in terms of quantity and quality. Thus, the collection of fuel wood, fodder and water, which are in general, women's responsibility, will probably take more time, thereby considerably increasing women's drudgery and affecting the entire family. Prevalent shortages of fuel, water, electricity, and gas all over the country are the result of a severe energy crisis, and they particularly affect women's work, even in the cities. The frequent absence of men for trading and herding puts greater household responsibilities on women; however, on the positive side women get more involved in many community activities.

MITIGATION AND ADAPTATION STRATEGIES TO DEAL WITH CLIMATE CHANGE

Adaptation to climate is the process through which people reduce the adverse effects of climate on their health and well-being, and take advantage of the opportunities that their climatic environment provides (Burton 1992; Burton 1997).

In economically constrained environments, planning of mitigation, and adaptation actions is a necessity, particularly given the large uncertainties about the scale, magnitude, and timing of the additional risks from climate change. The basic problem of policy making in the context of climate change is the uncertainty with regard to climate outcomes. Usually governments tend to allocate resources for more economically productive sectors, and give less attention to policies which are crucial to social issues but are less "productive" in the short range. Adaptation to cope with environmental damages suffers from these policy trends. Following is a brief discussion of these strategies adapted by Bangladesh and Nepal.

Bangladesh

The Bangladesh Climate Change Strategy and Action Plan was adopted by the government in 2009. It seeks to guide activities and programs related to climate change in Bangladesh. Until recently climatic risks have been poorly reflected in national policies and programs of Bangladesh. It launched the National Adaptation Program of Action (NAPA) in 2005, identifying fifteen priority activities which subsequently increased to forty-five programs in 2009. The first sectorial policy to unequivocally include climate change impacts and actions—the Coastal Zone Policy—was adopted in 2005. Also, climate change has been a key concern in the redrafting of the National Water Management Plan (NWMP). It identifies three climate hazards—tropical cyclones/storm surges, inland flooding, and droughts. The strategy contains forty-four programs formulated around six themes—food security/social protection/health, comprehensive disaster management, infrastructure, research/knowledge management, mitigation/low carbon development, and capacity building/institutional strengthening (Table 12.2). Of these, thirty-four programs listed under five themes are wholly or partially focused on adaptation. The government allocated $100 million of its own resources for fiscal year 2009–2010 and budgeted $100 million for the fiscal year 2010–2011 to implement these programs (World Bank 2010—EACC).

Jennifer et al. (2006), suggest livelihood diversification to influence adaptive capacities in rural Bangladesh. Effective climate change adaptation needs to be integrated (or mainstreamed) into other development initiatives. Bhaski Bhaskaran (2010: 13) observed that "while it is important to reduce emissions to limit global warming, it is also equally important to design credible climate policies to inform adaptation options as we have already committed to a certain level of warming over the next few decades." Ahmed and Alam (1998) have presented climate change scenarios, which were used for a number of subsequent national assessments. It was reported that the average increase in temperature would be 1.3°C and 2.6°C for the two projection years, 2030 and 2075. It was found that there would be a seasonal variation in changed temperature: 1.4°C change in the winter and 0.7°C in the monsoon months in 2030. For 2070 the variation would be 2.1°C and 1.7 °C for winter and monsoon, respectively. It was also reported that the winter rainfall would decrease at a negligible rate in 2030, while in 2075 there would not be any appreciable rainfall in winter. On the other

Table 12.2. Scope of Study in the Context of Programs and Activities Included in the Bangladesh Climate Change Strategy and Action Plan

Theme	Immediate	Short Term	Medium to Long Term
Food, security, social protection and health		Water and sanitation program in climate vulnerable areas Livelihood protection in ecologically fragile areas Livelihood protection of vulnerable socioeconomic groups (including women	Institutional capacity and research towards climate-resilient cultivars and their dissemination Development of climate-resilient cropping systems Adaptation against drought, salinity submergence and heat Adaptation in fisheries sector Adaptation in livestock sector Adaptation in health sector
Comprehensive disaster management	Improvement of floods forecasting and early warning system Improvement of cyclone and storm surge warning		Risk management against loss of income and property
Infrastructure	Repair and maintenance of existing flood embankments Repair and maintenance of existing cyclone shelter	Planning, design, and implementation of resuscitation of networks of rivers and khals through dredging and de-siltation work	Repair and maintenance of existing coastal polders Improvement of urban drainage Adaptation against floods Adaptation against future cyclones and storm-surges Planning, design, and construction of river training works
Research, and knowledge management	Establishment of a center for research, knowledge management, and training climate change Climate change modeling at national and sub-national levels	Preparatory studies for adaptation against sea level rise (SLR) and its impacts	Monitoring of ecosystem and biodiversity changes and their impacts Macroeconomic and sectoral economic impacts of CC Monitoring of internal and external migration of adversely impacted population and providing support to them through capacity building for their rehabilitation in new environment Monitoring of impact on various issues related to management of tourism in Bangladesh and implementation of priority action plan
Mitigation and low carbon development	Renewable energy development Management of urban waste Afforestation and reforestation programme	Rapid expansion of energy saving devices for example (CFL) Improving in energy consumption pattern in transport sector and options for mitigation	Improved energy efficiency in production and consumption of energy Gas exploration and reservoir Management Development of coal mines and coal fired power station(s) Lower emissions from agricultural land Energy and water efficiency in built environment
Capacity building and institutional strengthening	Revision of sectoral policies for climate resilience Mainstreaming climate change in national, sectoral and spatial development programs Strengthening institutional capacity for climate change management Mainstreaming climate change in the media	Strengthening human resource capacity Strengthening gender consideration in climate change management	
*Quantitative and Quantitative Analysis, Qualitative Analysis only, Not covered			

Source: EACC–World Bank 2010.

Table 12.3. Outputs of GCM Exercise using GFD 01 Transient

Year	Average Temperature			Temperature increase			Average Precipitation			Precipitation Increase		
	W	M	Ave	W	M	Ave	W	M	Ave	W	M	Ave
	(°C)			(°C)			Mm/month			Mm/Month		
1990	19.9	28.7	25.7	0.0	0.0	0.0	12	418	179	0	0	0
2030	21.4	29.4	27.0	1.3	0.7	1.3	18	465	189	+6	47	10
2075	22.0	30.4	28.3	2.1	1.7	2.6	00	530	207	-12	112	28

W stands for winter (i.e., December, January, and February: DJF) and M stands for Monsoon (i.e., June, July, and August: JJA).

Source: Ahmed, A.U., and Alam, M., 1998.

hand, monsoon precipitation would increase at a rate of 12 percent and 27 percent for the two projected years, respectively (Table 12.3).

There are a number of basic strategies that can be taken in response to climate change that categorize a range of ways to adapt to the altered situation. However, process that takes care of the local community and its adaptive capabilities is very critical; as climate change adaptation is generally context specific. Pender (2008) recommended mainstreaming climate change adaptation into development thinking and practices as a priority.

Nepal

In Nepal, it is essential to put a human face on climate change mitigation and adaptation strategies. Despite Nepal's better infrastructure, intervention plans and alert systems than Bangladesh, it is the people's response and commitment which will ultimately make these policies work. More efforts are needed at the grassroots level to help and support people to adapt their livelihoods to environmental insecurities and shield themselves against potential natural disasters. It is still more pertinent to involve women in this process by empowering them to participate in this process. Climate-induced disasters such as floods, landslides, and droughts have killed more than 4,000 people in Nepal over the last ten years. The economic losses caused by disasters add up to about USD 5.34 billion. Nepal has prepared its National Adaptation Programme of Action (NAPA) for adapting to extreme climate events and variability through an extensive country-driven consultative process. (MoE Nepal 2011). Nepal's commitment to reduce its emission of greenhouse gases represents an opportunity to provide the technologies for both reducing women's workload and improving health conditions, including the introduction of biogas, improved cooking stoves, and solar cookers. Those technologies must be made widely available to women. Nepal provides important eco-system services for the region, and those services will eventually benefit the country financially (Brigitte Leduc, 2008).

CONCLUSION AND POLICY PRESCRIPTIONS

The major strategies to combat the challenge of climate change are; global adaptation under the United Nations Framework Convention on Climate Change. It is also necessary to strengthen mechanisms and support for more practical and proactive adaptation under the convention by facilitating comprehensive national strategies and committing reliable funding for high priority implementation projects. National energy polices and industrial policies have to be implemented in order to contribute to climate mitigation by reducing or avoiding GHG emissions. To control the impacts of climate change, and limiting global GHG emissions will require deployment of currently available, and future low carbon technologies across a range of sectors on a global scale, along with other changes in economic activities. It should be noted that while energy is the main source of GHG emissions globally, land use change, forestry and agriculture currently account for almost 50 percent of GHG emissions in developing countries requiring greater attention to nonenergy and produced emissions. Reducing reliance on fossil fuels for energy is the essential factor for controlling climate change. Shifting from fossil fuels to renewable energy resources such as solar energy and hydropower, which do not emit CO_2 will help in mitigating climate change. Climate change polices have to be integrated into Sustainable Development Policies. Preventing deforestation would reduce carbon dioxide emissions and also help preserve biodiversity. Extensive tree plantation should be undertaken to reduce carbon dioxide concentration in the atmosphere. National and international laws and regulations that are favorable to country specific GHG mitigation have to be followed in "letter and spirit." The developing countries of Asia like Bangladesh, Nepal, and Pakistan, where impacts of climate change are likely to be felt more severely because of the topography, resource and infrastructure constraints, have to develop and implement incremental adaptation strategies and policies. Climate change mitigation and adaptation strategies have to be integrated while planning, designing and implementing development activities. A macro strategy consisting of sustainable and equitable development, which will increase education and technical skills; income levels; disaster preparedness and management and health care systems and reduce vulnerability, has to be envisaged. Contemplating a micro strategy involving the management of sectors that are most sensitive to the climate change is also equally important. In this regard, due importance must be given to developing new institutions or modifying existing ones to promote adaptation to climate change. Continued monitoring and analysis of variations and trends in key climatic elements has to be given priority. In all regions especially those which are typically prone to climate change effects workable weather forecasting systems must be in place. The role of nongovernmental organization (NGO), community and the public is crucial. They must be kept informed about of developments on risks of climate change. They should also be involved in planning, adaptation, and mitigation strategies. Climate polices have to be integrated into Sustainable Development Policies. Scientific research and technological development in key areas of climate change should be given priority.

Climate change is thus far and wide, and should be considered as one of the greatest challenges to modern human civilization. Therefore, it is essential to develop a portfolio of strategies that includes adaptation, mitigation, technological development, and research.

REFERENCES

Ahmed, A.U., Alam, M. and Rahman, A.A., 1998, "Adaptation to Climate Change in Bangladesh: Future Outlook," in *Vulnerability and Adaptation to Climate Change for Bangladesh*, S. Huq, Z. Karim, M. Asaduzzaman, and F. Mahtab (Eds.), Kluwer Academic Publishers, Dordrecht, pp. 125–143.

Allison, I. et al. 2009, *The Copenhagen Diagnosis, 2009: Updating the world on the Latest Climate Science.* Sydney: The University of New South Wales Climate Change Research Centre (CCRC).

Anjum Bari Farooqi, Azmat Hayat Khan*, Hazrat Mir (2005), *Pakistan Journal of Meteorology* Vol. 2: Issue 3.

Bhaski Bhaskaran, 2010, Climate modeling in Bangladesh: A model for capacity building in developing countries.

Brigitte Leduc 2008 Women's Environment and Development Organization. This study was edited by Tina Johnson and coordinated by Cate Owren, with additional support from Amanda Dentler.

Burton, I., 1992, *Adapt and Thrive*. Canadian Climate Centre unpublished manuscript, Downsview, Ontario.

Burton, I., 1997, "Vulnerability and Adaptive Response in the Context of Climate and Climate Change," *Climatic Change* 36, 185–196.

Christian Aid, 2007, Adapting to inevitable changes in the climate. Climate Change: Policy and Approach. (A Global Advocacy Briefing). London: Christian Aid.

Climate Change and Its Impact With Special Focus In Pakistan *Symposium on "Changing Environmental Pattern And Its Impact With Special Focus On Pakistan"* Zia Mustafa Symposium Vol 33 Paper No 290.

Climate Change Fact Sheets series 2008/1 www.ias.unu.edu).

Interview of Andreas Schild, Director General of ICIMOD, page 30, Mountain Forum 2008 Natural Reso·ırces: Women, Conflicts and Management *Mountain Forum Bulletin.*

IOM, 2010, Assessing the Evidence Environment, Climate Change and Migration in Bangladesh.

IPCC (Intergovernmental Panel on Climate Change). 2007a. *Climate Change 2007: Impacts, Adaptation and Vulnerability: Summary for Policymakers.* Working Group II Contribution to the Intergovernmental Panel on Climate Change Fourth Assessment Report. Geneva: IPCC).

IPCC, 2007, Climate Modeling in Bangladesh A model for capacity building in developing countries IPCC Fourth Assessment Report: Climate Change.

Jennifer Pouliotte, Nazrul Islam, Barry Smit and Shafiqul Islam Tiempo Issue 59 April 2006 pp. 18–22.

LEAD, 2008 Pakistan's Options For Climate change Mitigation and Adaptation, Study carried out for the British High Commission, Islamabad.

MoE Nepal, 2011 Adaptation to Climate Change NAPA to LAPA Government of Nepal Ministry of Environment.

MOPE, 2004. [The Ministry of Population and Environment (MOPE) became the Ministry of Environment, Science and Technology (MoEST) in 2005 MOPE. Initial National Communication of Nepal under the United Nations Framework Convention on Climate Change. Ministry of Population and Environment (MOPE), Kathmandu. 2004.

NIDOS, 2009 (Network of International Development Organisations in Scotland) www.nidos.org.uk, July.

Pender, J.S., 2008, What Is Climate Change? And How It Will Effect Bangladesh. Briefing Paper. (Final Draft). Dhaka, Bangladesh: Church of Bangladesh Social Development Programme.

Schild, Andreas, Natural Resources: Women, Conflicts and Management, *Mountain Forum Bulletin* 8, 30. www.ias.unu.edu/resourcecenter.

Shrestha A. B., Wake C. P., Mayewski P. A. and Dibb J. E., 1999, Maximum Temperature Trends in the Himalaya and Its Vicinity: An Analysis Based on Temperature Records from Nepal for the Period 1971–94 *Journal of Climate* 12, 2775–87.

Shrestha, A. B., Wake, C. P., Dibb, J. E., And Mayeski P. A., 2000, Precipitation Fluctuations in the Nepal Himalaya and its Vicinity and Relationship with Some Large Scale Climatological Parameters. *International Journal of Climatology,* 20, pp. 317–327.

The Kyoto Protocol: 2012, "Hot air." *Nature.* www.nature.com/news/the-kyoto-protocol-hot-air-1.11882., 29 November 2012.

The World Bank Group 2010 EACC Report.

UNDP, 2007 UNDP. Human Development Report 2007/2008: Fighting Climate Change: Human Solidarity in a Divided World. United Nations Development Programme (UNDP), New York.

UNFCCC Secretariat Press Release unfccc.int/files/press/news_room/press_releases_and_advisories/application/pdf/121208_final_pr_cop18_cf.pdf, p.2).

World Bank, 2010, *EACC Publications and Reports* Country Case Studies: Bangladesh.

Conclusion

Dhirendra K. Vajpeyi

Despite their geographical location, levels of social, economic and political development all the case studies in this collection of essays point out several common trends related to climate change and its impact on respective regions/countries. These common themes include changes in rainfall patterns (chapters 5, 6, 7, 10), frequency of droughts (5, 8, 10), rising CO_2 (2, 5, 6, 7, 10, 11), acidification and rise in sea levels (7, 8, 10), coastal inundation, agricultural production (5, 6, 7, 8, 10, 11), change in average temperatures (5, 6, 7, 9, 10, 11), health (3, 7, 11) and displacement of humans from their ancestral habitats (4, 5, 6, 7, 11). Similar concerns were voiced by the World Bank Report *Turn Down the Heat* (November 2012: 1–8). Another common point which percolates through these es says is the varying degrees of concerns, public awareness, and uneven impact of climate change on different regions and subregions of their countries. Some countries such as Russia (11) might benefit from mild climate change especially its cold regions such as Siberia, while other countries might suffer (5, 6, 7, 8, 10). The role of NGOs and mass media is also an important factor in raising public awareness, and reaction by political elites to climate change issues (5, 6, 7). These studies also point out that despite unsettled controversies, and uncertainties due to lack of "sufficient" visible scientific data both circumstantial evidence, from early migration of hummingbirds, to past climate variations provide a connection between climate change and increase in carbon dioxide in the atmosphere (Perremine 2013). However the debates and controversies related to climate change will continue. The doubters are right that uncertainties are rife in climate science. They are wrong when they present that as a reason for inaction (*The Economist*, 2010, March 20: 86) but at the same time a "lack of action on climate change . . . risks putting prosperity out of reach of millions of people in the developing world" (Kim 2012: VI).

Keeping in view the research presented in this anthology following are several policy observations:

- There is a broad consensus that burning of fossil fuels, and coal is dysfunctional to the environment, hence efforts should be made to clean up and regulate coal based power plants which inject a massive proportion of pollution in the atmosphere. It has been suggested that a coal tax be imposed. Like all taxes a coal tax will be resisted specially by China and India, but a long range cost-benefit analysis should mitigate economic risks for these countries. It should be noted that "when people are imposing serious risks on themselves (and society) it is not freedom of choice and ignore the consequences" (Sunstein 2013: 8). If a "paternalistic intervention (nudges)" benefits greater good of the society it is worth taking.
- Need to establish a strong review agency to monitor the progress of international agreements. An agency like International Atomic Energy Agency (IAEA).
- Research also shows that human displacement due to climate change is taking place but there is no international agency to assist climate refugees. The United Nations High Commission on Refuges (UNHCR) could help.
- Reduce deforestation, the clearing and burning of tropical forests which account for 15 percent of the world's carbon emissions.
- It has been observed that since all politics are local there is a greater need to strengthen and implement climate change laws at the local/domestic levels. "Global action is not going to stop climate change" (*The Economist*, 2010 November 27–December 3). If past actions/inactions (Copenhagen; Kyoto etc.) are any indications the above argument seems plausible, however, climate change and environmental issues are international in their nature and scope and do not require a passport to cross national borders. These issues require collective action. ". . . the issue is no longer on of rich versus poor. Some developing countries are now major emitters, and the developing world accounts for more than half of all current greenhouse gases. To be effective responses to climate change must address developing countries' needs, including their right to development, but develop-

ing countries are not so big that they must also actively contribute to remedies" (Birdsall 2013: IX–X).

- More research is needed to develop appropriate technology to (green technology) for adaptation and mitigation efforts especially in the developing world.
- More than one author in this anthology has lamented about the disconnect between national, and regional government policies and local needs. Top down policy prescription (including from international agencies) do not work. To succeed there must be a better coordination between the public, various sectors of civil society and the governmental policies. A strong meaningful partnership between various stakeholders must exist.
- Do not rely on self-regulation. Self regulation is self-deception.
- Strong political and economic commitment from society and policy makers is needed. Both must rise above domestic, international, and ideological politics, and stop giving excuses related to economic growth. Economic growth at the cost of overall human security and quality of life "threatens to roll back decades of sustainable development (Kim 2012: VI).

President Barak Obama's promise that "we will respond to the threat of climate change, knowing that failure to do so would betray our children and future generations" should not fall victim to partisan politics and narrow corporate interests. Diamond (2005) points out several contributing factors in the "Collapse" of societies, and wondered if the same things could happen to us. In *The World Until Yesterday* (2012: 283) Diamond cites "Wayne Gretsky Principle" (the legendary Ice hockey player) "if one takes no shots then one will miss no shots but one is also guaranteed to score no goals." It is time to take well aimed shots to meet the perils and dangers of climate change. If we fail to stare in the face of this potential catastrophe then it would confront and haunt our planet and the civilization for future generations. It is time to stop shadow dancing around these issues, and not treat them as short range inconveniences. It is this message which the authors of this collection of scholarly essays in *Climate Change, Sustainable Development and Human Security* seem to give us.

REFERENCE

Birdsall, Nancy (2013) "Preface Mattoo, Aaditya and Subramanian, Arvind, 2013" *Greenpont: A New Approach to Cooperation on Climate Change*, Washington D.C. Center for Global Development.

Diamond, Jared (2012). *The World Until Yesterday—What can we learn from Traditional Societies*? New York: Viking.

Diamond, Jared (2005). *Collapse—How Societies Choose to Fall or Succeed*, New York Viking.

Kim, Jim Yong (2012). Foreword, *Turn Down the Heat: Executive Summary*, Washington, D.C.: The World Bank.

Obama, Barak (2013). *Second Inaugural Address* January 21.

Perremine, Frederic (2013). "It's for your Good; Review Essay of Conly, Sarah 2013 *Against Autonomy: Justifying Coercive Paternalism*, in *The New York Review of Books* March 7: 8: 11.

Sunstein, Cass R. (2013). "Review Essay" *The New York Review of Books* March 7 of Sarah Conly's *Against Autonomy: Justifying Coercive Paternalism*, U.K: Cambridge University Press.

Selected Bibliography

"Atmospheric Carbon Dioxide and Global Surface Temperature Trends," *Center for Climate and Energy Solutions*, (www.c2es.org/facts-figures/trends/co2-temp).

Affholder, F.; G. C. Rodrigues; E. D. Assad, 1997. "Modelo Agroclimático para avaliação do comportamento do milho na região dos cerrados." *Pesquisa Agropecuária Brasileira*, v. 32, n.10, 993–1002.

Aggarwal, P. K. and A. K. Singh, (2010): Implications of global climatic change on water and food security. In: *Global Change: Impacts on Water and Food Security* [Ringler, C., A. K. Biswas, and S. Kline (eds.)]. Springer-Verlag, Berlin and Heidelberg, Germany, pp. 49–63.

Aggarwal, P. K., and Sinha, S. K., (1993) "Effect of Probable Increase In Carbon Dioxide and Temperature on Productivity of Wheat In India." *Journal of Agricultural Metrology.*

Ahmad Jamil and Haseen et al. (2011) Impact of Climate Change on Agriculture and Food Security in India *International Journal of Agricultural, Environment, and Biotech. Vol. 4, No. 2*: 129–137.

Ashrit, R. G., K. Rupa Kumar and K. Krishna Kumar (2001): "ENSO-Monsoon Relationships in a Greenhouse Warming Scenario," *Geophysical Research Letters,* 28(9): 1727–30.

Assad, E. D.; Hilton Silveira Pinto; J. Zullo Jr., F.R. Marin, G.Q. Pellegrino. 2008. "Mudanças Climáticas e a Produção de Grãos no Brasil: Avaliação dos Possíveis Impactos." *Revista Plenarium,* V. 5, p. 96–117.

Assad, E. D.; Luchiari Jr., 1989 "A future scenario and agricultural strategies against climatic changes: the case of tropical savannas." In: *Mudanças Climáticas e Estratégias Futuras*. USP. 30–31 October 1989. Brazil: São Paulo. SP.

Astill, James 2012 "The Vanishing North," *The Economist*, June 16–22, 13.

Ayers, J. and Huq, S. 2009 "Supporting Adaptation to Climate Change: What Role for Development Assistance?" *Development Policy Review* (27): 6, 657–92.

Bae, Chang-Hee C. 2005. "Tijuana-San Diego: Globalization and the transborder metropolis," in *Globalization and Urban Development*. Edited by H. W. Richardson and C.-H. C. Bae. Berlin: Springer, 181–197.

Bakewell, O. 2008 "'Keeping them in their place': The ambivalent relationship between development and migration in Africa," Third World Quarterly (29): 7, 1341–58.

Banyan 2012 "Snow Dragons," *The Economist*, September 1, 46.

Barnett, J. and Adger, W. N. 2007 "Climate change, human security and violent conflict," *Political Geography* 26: 639–655.

Bates, D. C. 2002 "Environmental Refugees? Classifying Human Migrations Caused by Environmental Change" Population and Environment 23: 5, 465–77.

Bauder, H. 2008 "Media Discourse and the New German Immigration Law." *Journal of Ethnic and Migration Studies* 34(1): 95–112.

Bayentin, L. Ouarda El Adlouni, P. Gosselin, et al. 2010 "Spatial Varability of Climate Effects on schemic heart disease hospitalization rates for the period 1989–2006 in Quebec, Canada." *International Journal of Heath Geographics* 9:10 1186/1476-072X-9–5.

Beck, U. 1992 Risk Society: Towards a New Modernity London: Sage.

Beck, Ulrich. (1992) *Risk Society: Towards a New Modernity*. London: Sage Publications.

Bell, D. 2004 "Environmental Refugees: What Rights? Which Duties?" *Res Publica* (10): 135–52.

Beniston, Martin, and David B. Stephenson. 2004. "Extreme Climatic Events and Their Evolution Under Changing Climatic Conditions." *Global and Planetary Change*. 44, no. 1: 1. g01.cnitc.cn/uploadfile/file/□□/□□□□□/1995-2005/2012010620521617.pdf.

Berkes, F. and Folke, C. 1998. "Linking social and ecological systems for resilience and sustainability," In: Berkes, F., Folke, C. (Eds.), *Linking Social and Ecological Systems* Cambridge: Cambridge University Press.

Biermann, F. and I. Boas 2008a. "Protecting climate refugees: The case for a global protocol," *Environment* 50 (6), 8–16.

Biermann, F. and I. Boas 2008b. "Response to Hulme," *Environment* 50 (6).

Bogardi, J. and K. Warner 2009 "Here comes the flood," *Nature Reports Climate Change* (3), 9–11.

Bohle, H. G. et al. 1994. "Climate Change and Social Vulnerability." *Global Environmental Change* 4 (1).

Boyd, E., N. Grist, S. Juhola and V. Nelson 2009. "Exploring Development Futures in a Changing Climate: Frontiers for Development Policy and Practice," *Development Policy Review* (27): 6, 659–74.

Brauch H. G., Spring U. O. et al. (editors) 2009. "Facing global environmental change: environmental, human, energy, food, health and water security concepts." Berlin: Springer.

Brooks, N., N. Grist and K. Brown 2009. "Development Futures in the Context of Climate Change: Challenging the Present and Learning from the Past," *Development Policy Review* (27): 6, 741–65.

Brown, Lori 2006 *State of the World: A year in Review*, New York: W. W. Norton.

Brown, Oli, Anne Hammill and Robert McLeman 2007 "Climate change as the 'new' security threat: implications for Africa," *International Affairs* 83: 6 (2007) 1141–1154.

Buainain, Antonio Marcio e Jr. Ruiz Garcia. 2013. "Pobreza rural e desenvolvimento do Semiárido Nordestino: resistência, reprodução e transformação." In *A Nova Cara da Pobreza Rural: desenvolvimento e questão regional*. Brasilia, IICA, 217–285. Série Desenvolvimento Rural Sustentável, volume 17.

Buainain, António Márcio, Claudio Dedecca, and Henrique Neder. 2012. *A Nova Cara da Pobreza Rural no Brasil: desafios para as políticas públicas*. Brasilia, IICA, Série Desenvolvimento Rural Sustentável, volume 16.

Buchner B., Dall'Olio S. 2004 Russia: The Long Road to Ratification. Internal Institution and Pressure Groups in the Kyoto Protocol's Adoption Process. "Fondazione Eni Enrico Mattei." Working Paper 151.

Burton, I. 2004 [2009 "Climate change and the adaptation deficit," E. L. F. Schipper and I. Burton (Eds.) *The Earthscan Reader on Adaptation to Climate Change* London: Earthscan, 89–98.

Byravan, S. and S. C. Rajan 2006 "Providing new homes for climate change exiles," *Climate Policy* (6), 247–252.

Cai, Jianming and Victor F. Sit. 2005. "Measuring world city formation—The case of Shangai," in *Globalization and Urban Development*. Edited by H. W. Richardson and C.-H. C. Bae. Berlin: Springer, 239–250.

Calfalnoieri, U. et al. 2007 "Human Health" *Climate Change*.

Caney, S. 2005 "Cosmopolitan justice, responsibility, and global climate change," *Leiden Journal of International Law* (18): 747–775.

Caney, S. 2008 "Human rights, climate change, and discounting," *Environmental Politics* (17): 4, 536–555.

Cao, J. et al. 2009 "Diurnal temperature range is a risk factor for Coronary heart disease death," *Journal of Epidemiology* 19: 328–332.

Castro, Josué de. 1946. *A Geografia da Fome: o Dilema brasileiro: pão ou aço*. Brazil: Rio de Janeiro, Antares ed. 1980.

Chaudhry, Anita and Aggarwal, P. K., (2007), "Climate Changes and Food Security in India," Indian Agriculture Research Institute, New Delhi.

Chen, Guangyao 2001: "China's Nongovernmental Organizations: Status, Government Policies, and Prospects for Further Development," *The International Journal of Not-for-Profit Law*, 3, March 2001.

Chen, Lincoln C., Tim G. Evans, and Richard A. Cash. 1999. Health as a Global Public Good. In Inge Kaul, Isabelle Grunberg, and Marc Stern, eds., *Global Public Goods: International Cooperation in the 21st Century*. New York, NY: United Nations Development Programme and Oxford University Press.

China National Committee for the Implementation of the UN Convention to Combat Desertification 1996: "China National Action Program to Combat Desertification (Abstract)," Beijing, at <www.unccd.int/actionprogrammes/asia/national/2000/china-eng.pdf (01.10.2009).

China's National Climate Change Program, 2004, Initial National Communication on Climate Change. Beijing, China.

Clark.; Moser, Schwarz et al. (1998). Assessing The Vulnerability of Coastal Communities to Extreme Storms: The Case of Revere, MA., USA. *Mitigation and Adaptation Strategies for Global Change*, 3, 59–82.

Claussen Eileen 1995 "Environmental and Security: The Challenge of Integration." *Environmental and Security Debates: An Introduction*, Washington DC The Woodrow Wilson Center, 40–43.

Climate Change 2007: The Physical Science Basis-Summary for Policymakers. Contribution of Working Group I to the Fourth Assessment Report of the Intergovernmental Panel on Climate Change. 2007. (www.ipcc.ch/pdf/assessment-report/ar4/wg1/ar4-wg1-spm.pdf).

Cohen, Jillian Clare and Patricia Illingworth. 2003. The Dilemma of Intellectual Property Rights for Pharmaceuticals: The Tension between Ensuring Access of the Poor to Medicines and Committing to International Agreements. *Developing World Bioethics* 3: 251–264.

Cohen, M. J. et al. (2008) *Impact of climate change and bio energy on nutrition*. International Food Policy Research Institute.

Collin, J., K. Lee and K. Bissell. 2002. The Framework Convention on Tobacco Control: The Politics of Global Health Governance. *Third World Quarterly* 23 (2): 265–282.

Conde de Saro, Rafael, Spanish Ambassador. 2011. Email Correspondence. 5 October.

Couldrey, M. and Herson, M. 2008 (Eds.) "Climate change and displacement," *Special Issue of Forced Migration Review* Issue 31, October 2008.

Craig, M. 1999. A Climate-Based Distribution Model of Malaria Transmission in Sub-Saharan Africa. *Parasitology Today* 15: 105–111.

Dale, Jiajun Wen, 2009, "Climate Change and China: Technology, Market and Beyond: A Report for Focus on the Global South." Focus On The Global South, Chulalongkorn University, Bangkok, Thailand.

Danilov-Danilyan V. I. 2003 "Kiotskij protokol: kritika kritiki." ESKO, №1.

Danoploulos, C. P., Dhirendra Vajpeyi, and Amir Bar 2004 (eds). *Civil Military Relations, Nation-Building and National Identities, Comparative Perspectives,* London: Praeger.

De Sherbinin, A. K. Warner and C. Ehrhart 2011a. "Casualties of Climate Change," *Scientific American* January, 2011: 32–39.

DeBardeleben J., Heuckroth K. 2001. Public attitudes and ecological modernization in Russia. In: "The Struggle for Russian Environmental Policy." 49–76.

DeLaet, Debra L. and David E. DeLaet. 2012. *Global Health in the 21st Century: The Globalization of Disease and Wellness.* Boulder, CO: Paradigm Publishers.

Deschenes, O. et al. 2009 "Extreme weather events, mortality, and migration," *The Review of Economics and Statistics* 91: 659–681.

Dessler, Andrew, and Edward A. Parson 2006 *The Science and Politics of Global Climate Change—A Guide to the Debate,* Cambridge, UK: Cambridge University Press.

Detraz, Nicole 2009 "Climate Change and Environmental Security: For Whom the Discourse Shifts," *International Studies Perspectives,* 10 pp. 303–320.

Deudney, Daniel 1991. "Environment and Security: Muddled Thinking." *Bulletin of the Atomic Scientist*, vol. 47 no. 3 April.

Deudney, Daniel and Richard Matthews (eds) 1995. *Contested Ground: Security and Conflict in the New Environmental Politics*, Albany: SUNY Press.

Diamond, Jared 2005. *Collapse—How Societies Choose to Fail or Succeed*, New York: Viking.

Dwivedi, O. P. 2008. "The Well-Being of Nations: Reflections on Integrating the Human and Ecosystem well-being (eds) Dhirendra Vajpeyi and Renu Khator *Globalization, Governance and Technology: Challenges and Alternatives*, New Delhi: Deep and Deep pp. 60–75.

Easterling D. R., G.A . Meehl, C. Parmesan, S. A. Changnon, T. R. Karl, and L. O. Mearns. 2000 (a). "Climate Extremes: Observations, Modeling, and Impacts." *Science* (New York, NY). 289, no. 5487: 2068–74. clivac.eri.ucsb.edu/classes/Tropical_meteorology/classes_ucsb/climate_change/articles_climate_change_tropics/Easterling_et_al_2000%20extreme%20events.pdf.

Easterling, D. R., J. L. Evans, P. Ya Groisman, T. R. Karl, K. E. Kunkel, and P. Ambenje. 2000 (b). "Observed Variability and Trends in Extreme Climate Events: A Brief Review." *Bulletin of the American Meteorological Society*. 81, no. 3. (g01.cnitc.cn/uploadfile/file/□□/□□□□□/2000/2012010714282449.pdf).

Economy, Elizabeth 2004: *The River Runs Black: The Environmental Challenge to China's Future* (New York: Cornell University Press).

Elbe, Stefan. 2006. Should HIV/AIDS be Securitized? The Ethical Dimensions of Linking HIV/AIDS and Security. *International Studies Quarterly* 50: 119–144.

Epstein, Paul R., Dan Farber, and Jeffrey Sachs. 2011. *Changing Planet, Changing Health: How the Climate Change Crisis Threatens our Health and What We Can Do about It*. Berkley, CA: University of California Press.

Fernandez et al. (2012) World Bank report Impacts of Climate Change on Brazilian Agriculture [P118037].

Fidler, David P. 2004. Germs, Governance, and Global Public Health in the Wake of SARS. *The Journal of Clinical Investigation* 113: 799–804.

Field, Christopher, Barros, Vicente, Stocker, Thomas F., and Dahe, Qin. Managing the Risks of Extreme Events and Disasters to Advance Climate Change Adaptation Special Report of the Intergovernmental Panel on Climate Change. Cambridge University Press, 2012.

Flannery, Timothy Fridtjof. The Weather Makers: The History and Future Impact of Climate Change. London: Allen Lane, 2006.

Flavin, Christopher 2005 "Redefining Global Security," in *State of the World* 2005, *Redefining Global Security* (ed.) Michael Renner et al. New York: W. W. Norton.

Ford, J. *The Role of Typansomiases in African Ecology: A Study of Tsete Fly Problem.* Claredon Publishers.

Gaby, Keith 2011 "Climate 411 New Report on Climate Change Says Wilder Weather is Headed Our Way" *Climate Central*, November 18.

Gasper, Des. 2005. Securing Humanity: Situating 'Human Security as Concept and Discourse" *Journal of Human Development* 7(2): 221–245.

Gasques, J. G.; E. T. Bastos; C. Valdes; M. R. P. Bacch. 2012. *Produtividade da Agricultura Brasileira e os Efeitos de Algumas Políticas. Revista de Política Agrícola*, v. n.3, p. 83, 2012.

Gilbert, Alan. 2005. "Globalization and Latin America: Understanding the links of Colombia's capital" in *Globalization and Urban Development*. Edited by H.W. Richardson and C.-H.C . Bae. Berlin: Springer, 165–180.

Gobal Warming Deniers are full of . . . Hot Air. 2011. *The Union of Concerned Scientists Fact Check*. Cambridge, MA: www.ucsusa.org.

Gosain, A. K. et al. (2006): "Climate Change Impact Assessment on Hydrology of Indian River Basins," *Current Science*, 90(3): 346–53.

Goswami, B. N. (2005), *Interdecadal variability, in The Asian Monsoon*, Edited by B. Wang, Springer, New York. pp. 295–328.

Graizbord, Boris, Allison Rowland, and Adrian Guillermo Aguillar. 2005. "Mexico City as a peripheral global player: The two sides of a coin," in *Globalization and Urban Development*. Edited by H. W. Richardson and C.-H. C. Bae. Berlin: Springer, 147–164.

Grubb, Michael 1995 "Seeking Fair Weather: Ethics and International Debate on Climate Change," *International Affairs*, 71 (3) 463–496.

H.-M. Füssel and R. J. T. Klein. (2006). Climate Change Vulnerability Assessments: An Evolution of Conceptual Thinking. *Climatic Change*, 75:301–329, 2006.

Hayter, T. 2003. "No borders: the case against immigration controls," *Feminist Review* 73, 6–18.

Henk, Dan 2005. "Human Security: Relevance and Implications," *Parameters*, Summer pp. 91–106.

Henry L., Douhovnikoff V. 2008. Environmental Issues in Russia. "The Annual Review of Environment and Resources." Vol. 33.

Hertsgaard, Mark 2011. *Hot-Living Through the Next Fifty years on Earth*, New York: Mifflin Harcourt.

Ho, Peter 2001: "Greening Without Conflict? Environmentalism, NGOs and Civil Society in China," *Development and Change*, 32, 893–921.

Ho, Peter and Eduard B. Vermeer (eds.) 2006: *China's Limits to Growth: Greening State and Society* (Malden et al.: Blackwell Publishing).

Holbig, Heike 2009: "Ideological reform and political legitimacy in China: Challenges in the post-Jiang era," in: Gunter Schubert (ed.): *Regime Legitimacy in Contemporary China: Institutional Change and Stability* (Oxon, New York: Routledge), 13–34.

Homer-Dixon, Thomas F. 1994 "Environmental Scarcities and Violent Conflict: Evidence from Cases," *International Security* 19(1) pp. 5–40.

Houghton, J. T. Climate Change 2001: "The Scientific Basis: Contribution of Working Group I to the Third Assessment Report of the Intergovernmental Panel on Climate Change." Cambridge: Cambridge University Press, 2001. www.acrim.com/%5C/Reference%20Files/CLIMATECHANGE%202001%20-%20The%20Scientific%20Basis.pdf.

Houghton, J. T., G. J. Jenkins, J. J. Ephaums, eds. 1990. *Climate Change: The IPCC Scientific Assessment*. UK: Cambridge University Press.

Hulme, M. 2008. "Commentary—Climate refugees: Cause for a new agreement?" *Environment* 50 (6).

IPCC 2012. *Managing the Risks of Extreme Events and Disasters to Advance Climate Change Adaptation*. Cambridge: Cambridge University Press.

IPCC, 2007: "Climate Change 2007: The Physical Science Basis. Contribution of Working Group I to the Fourth Assessment Report of the Intergovernmental Panel on Climate Change-Frequently Asked Questions." [Solomon, S., D. Qin, M. Manning, Z. Chen, M. Marquis].

Janik, Laura. 2011. "Understanding Hepatitis B and the Global Failure to Act." *Global Health Governance* IV (2).

Jayaraman, Rajshri and Ravi Kanbur. 1999. International Public Goods and the Case for Foreign Aid. In Inge Kaul, Isabelle Grunberg, and Marc Stern, eds., *Global Public Goods: International Cooperation in the 21st Century*. New York, NY: United Nations Development Programme and Oxford University Press.

Jiang, Hong 2009: "China's Great Green Wall Proves Hollow: Tree planting damages environment in northern China," *The Epoch Times*, 29.07.2009, at www.theepochtimes.com/n2/china-news/desertification-in-china-20291.html (17.12.2012).

Jodha, Narpat S. et al. (eds.) 2002: *Poverty Alleviation in Mountain Areas of China*. Proceedings of the International Conference held from 11–15 November, 2002, in Chengdu, China (Feldafing: InWEnt gGmbH).

Johnson, C. 2012. "Governing climate displacement: The ethics and politics of human resettlement," *Environmental Politics* (21): 2, 308–28.

Karl, Thomas. Weather and Climate Extremes in a Changing Climate Regions of Focus: North America, Hawaii, Caribbean, and U.S. Pacific Islands. Washington, DC: U.S. *Climate Change Science Program*, 2008. purl.access.gpo.gov/GPO/LPS110676.

Kassiola, Joel and Sujian Guo (eds.) 2010: *China's Environmental Crisis: Domestic and Global Political Impacts and Responses* (New York: Palgrave Macmillan).

Kaul, Inge, and Ronald U. Mendoza. 2003. Advancing the Concept of Public Goods. In Inge Kaul, Pedro Conceicao, Katell de Goulven, and Ronald Mendoza (eds.) *Providing Global Public Goods: Managing Globalization*. New York, NY: United Nations Development Programme and Oxford University Press.

Keohane, O. Robert, Kal Raustiala 2008. "Toward a Post-Kyoto Climate Change Architecture: A Political Analysis," *Law-Econ Research Paper*, No. 08–14, pp.1–26.

Keohane, Robert and Joseph Nye. 1977. *Power and Interdependence*. Longman Publishing.

Khon V. C., Mokhov I. I., Latif M., Semenov V. A., Park W. 2010. "Perspectives of Northern Sea Route and Northwest Passage in the twenty-first century." *Climatic Change*, Vol. 100, No. 3. 757–768.

King, Gary and Christopher Murray 2001. "Rethinking Human Security," *Political Science Quarterly*, pp. 589–590.

Kirton, John J., editor. 2009. *Global Health*. Burlington, VT: Ashgate Publishing.

Knowlton, Kim, Miriam Rotkin-Ellman, Linda Geballe, Wendy Max, and Gina M. Solomon. 2011. Six Climate Change–Related Events In The United States Accounted For About $14 Billion In Lost Lives And Health Costs. *Health Affairs* 30 (11): 2167–2176.

Koremenos, Barbara, Charles Lipson, Duncan Snidal. 2001. The Rational Design of Institutions. *International Organization* 55 (4): 761–799.

Korppoo A., Vatansever A. A. 2012 "Climate Vision for Russia: From Rhetoric to Action." Carnegie Endowment Policy Outlook, August 2012.

Kothawale D. R., Munot A. A., Krishna Kumar K. (2010). "Surface Air temperature variability over India during 1901–2007 and its association with ENSO." 42: 89–104.

Koty Reddy T., 2010, "Impact of Environment on Poverty in India." *The Indian Economy Review*, Vol. Vii, March Issue, p. 28.

Kumar, K. and J. Parikh (1998) "Climate Change Impacts on Indian Agriculture: The Ricardian Approach" in A Dinar, R Mendelsohn et al. (ed.), *Measuring the Impact of Climate Change on Indian Agriculture*, World Bank Technical Paper No 402 (Washington DC: The World Bank).

Kumar, K. K., B. Rajagopalan et al. (2006): "Unravelling the Mystery of Indian Monsoon Failures during El Nino," *Science*, 314: 115–19.

Kyle, J. L. and E. Harris 2008 "Global Spread and Persistence of Dengue," *Annual Review of Microbiology* 62: 71–92.

Lacerda, F. F. and Paulo Nobre. 2010. Aquecimento Global: conceituação e repercussões sobre o Brasil. *Revista Brasileira de Geografia Física*, v. 3, p. 14–17.

Lacerda, F.F.; H. Dias; E. D. Assad; J. M. O. Assis; M. S. B. Moura. 2010. "Extremos e Variabilidade Climática no Nordeste Brasileiro e em Pernambuco." In: II Workshop de Mudanças Climáticas e Recursos Hídricos de Pernambuco, Recife. Mudanças Climáticas e Recursos Hídricos de Pernambuco.

Lacis, A. A. et al. 2010 "Atmospheric CO_2: Principal Control Knob Governing Earth's Temperature," *Science* October 356–359.

Li, Ding 2008: *An Empirical Research on the System of Comprehensive Intervention of Ecological Poverty Alleviation in Arid Regions* (Lanzhou: Lanzhou University Press) [in Chinese].

Li, Hanlin and Fenzi Wei 2005: "Research Report on Minqin County Ecological Migration," *Northwest China and Development Forum Collection* [in Chinese].

Li, Mei et al. 2010: "Evolution of Water Supply and Demand in the Shiyang River Basin over nearly 40 Years," *Arid Land Geography*, 33 (4), 651–658 [in Chinese].

Liao, Jie 2007: *Research on the LUCC in Minqin based on Remote Sensing and GIS* (Lanzhou: Lanzhou University Press) [in Chinese].

Lima, R. C. C., Cavalcante, A. M. B., e Marin, A. M. P. (editors). 2011. *Desertificação e Mudanças Climáticas no Semiárido Brasileiro*. Campina Grande, INSA, 2011.

Lipton, M. 1983 "Why poor people stay poor," J. Harriss (Ed) *Rural development: theories of peasant economy and agrarian change*. London: Hutchinson, 66–81.

Liverman, D. M. 1994 "Vulnerability to Global Environmental Change," in Cutter S. L. (ed.) *Environmental Risks and Hazards*, New Jersey, Prentice-Hall.

Lu, Ding and William A. W. Neilson (eds.) 2004: *China's West Region Development: Domestic Strategies and Global Implications* (Singapore et al.: World Scientific).

Lu, Yiyi 2009: *Non-governmental Organizations in China* (London, New York: Routledge).

Ma, Jinzhu and Michael W. Edmunds 2006: "Groundwater and lake evolution in the Badain Jaran Desert ecosystem, Inner Mongolia," *Hydrogeology Journal*, 14, 1231–1243.

Ma, Jinzhu et al. 2005a: *The Shiyang River Drainage Area: The Evolution of the Water Environment and Sustainable Use of the Water Resources* (Lanzhou: Lanzhou University Press) [in Chinese].

Ma, Quanlin et al. 2004: "Environmental Problems of the Shiyang River Basin, its Root Causes and Rehabilitation Measures," *Journal of Safety and Environment*, 4 (5), 64–68 [in Chinese].

Marcotullio, Peter J., Sarah Rothemberg and Miri Nakahara. 2005. "Globalization and urban environmental transitions: Comparison of New York's and Tokyo's experiences" in *Globalization and Urban Development*. Edited by H.W . Richardson and C.-H. C. Bae. Berlin: Springer, 289–310.

Marengo, J. A. 2006, *Mudanças Climáticas Globais e seus Efeitos Sobre a Biodiversidade: Caracterização do Clima Atual e Definição das Alterações Climáticas para o Território Brasileiro ao Longo do Século XXI*; Brasília: MMA, 2006; 201 p.

Marin, F. R.; J. W. Jones, A. Singels, F. Royce, E. D. Assad, G. Q. Pellegrino; F. Justino, 2012. Climate change impacts on sugarcane attainable yield in southern Brazil. *Climatic Change*, v. 1, p. 1, 2012.

Marin, Fábio Ricardo; Maria Leonor Lopes-Assad, E. D. Assad; C. E. Vian, M. C. Santos, 2008. Sugar cane crop efficiency in two growing seasons in São Paulo State, Brazil. *Pesquisa Agropecuária Brasileira*, v. 43, p. 1449–1455.

Marland G., Boden T. A., Andres R. J. 2011 "Global, Regional, and National Fossil-Fuel CO_2 Emissions." Carbon Dioxide Information Analysis Center. cdiac.ornl .gov/trends/emis/overview.html.

Matthews, Jessica T. 1989 "Redefining Security," *Foreign Affairs*, Spring pp. 162–177.

McAdam, J. and B. Saul 2010. "Displacement with Dignity: International Law and Policy Responses to Climate Change Migration and Security in Bangladesh," *University of New South Wales Faculty of Law Research Series 2010*. Working Paper 63 (December 2010).

McGranahan, G., Balk, D. and Anderson, B. 2007 "The rising tide: assessing the risks of climate change and human settlements in low elevation coastal zones," *Environment and Urbanization* (19): 1, 17–37.

McLeman, R. and B. Smit 2006. "Migration as adaptation to climate change," *Climatic Change* (76): 31–53.

McMichael A. J. et al. (editors) 2003 "Climate Change and Human Health: Risks and Responses." WHO. Geneva.

Mendelsohn, Robert, Ariel Denar and Larry Williams 2006. "The Distributional Impact of Climate Change on Rich and Poor Countries." *Environmental and Development Economics* 11.

Mertha, Andrew 2008: *China's Water Warriors: Citizen Action and Policy Change* (New York: Cornell University Press).

Nef, Jorge 2008 "Political Economy of Globalization," in Vajpeyi and Khator (eds) pp. 141–169.

Nelson, D. R., W. A. Adger and K. Brown (2007) "Adaptation to environmental change: Contributions of a resilience framework," Annu. Rev. Environ. Resour. (32) 395–419.

Nishimura, Motoki 2006 "The Impact of Global Warming to Agriculture and Food security," *Global Environmental Research*, Vol. 11, No. 1, pp. 35–42.

Nobre, P. 2011. "Mudanças Climáticas e desertificação: os desafios para o Estado Brasileiro." In Lima, Cavalcante e Marin. *Desertificação e Mudanças Climáticas no Semiárido Brasileiro*. Campina Grande, INSA/MCTI, 211p.

Nordas, R. and N. P. Gleditsch 2007 "Climate change and conflict," *Political Geography* (26), 627–638.

Nordhaus, William 2012. "In the Climate Casion: An Exchange," *The New York Review of Books*, August 16, 85.

Nordhaus, William 2012 "Why the Global Warming Skeptics are Wrong?" *The New York Review of Books*, March 22, 32–34.

O'Brien, Karen L. and Robin M. Leichenko 2000 "Double Exposure: Assessing the impacts of climate change within the context of Economic Globalization" *Global Environmental Change*, 10, 221: 232.

Oliphant R. 2011 Counting the Cost of Russia's Melting Permafrost. "The Moscow Times," Issue 4739, 6 October 2011.

Olivier J., Janssens-Maenhout G., Peters J. 2012 "Trends in global CO_2 emissions." 2012 Report. The Hague: PBL/JRC.

Olson, Mancur. 1965. *The Logic of Collective Action: Public Goods and the Theory of Groups*. Harvard University Press.

Oyama, M. D., C. A. Nobre, 2003. "A new climate-vegetation equilibrium state for Tropical South America." *Geophysical Research Letter*, 30, 23, 2199.

Paris, Roland 2001 "Human Security-A Paradigm Shift or Hot Air," *International Security*, Autumn 87–108.

Parry M. and T. Carter 1998. *Climate Impact and Adaptation Assessment*. London: Earthscan Publications.

Parry, M. L., Canziani, O. F., Palutikof, J. P., van der Linden, P. J., Hanson, C. E., 2007, *Climate Change 2007: Impacts, Adaptation and Vulnerability*, Cambridge University Press.

Perch-Nielson, S. L., Battig, M. B. and Imboden, D. 2008 "Exploring the link between climate change and migration," *Climatic Change* (91): 375–93.

Pinto, H. S., E. D. Assad, 2008. Global warming and the new geography of agricultural production in Brazil. EMBRAPA e UNICAMP. Brasilia, British Embassy, 2008. 82 pp.: www.cpa.unicamp.br/aquecimento_agricola_en.html consulted in 2013/01 www.agritempo.gov.br/climaeagricultura/ consulted in 2013/01.

Platt, Anne E. 1996. *Infecting Ourselves: How Environmental and Social Disruptions Trigger Disease*. Washington D.C. World Watch Institute.

Price-Smith, Andrew. 2002. *The Health of Nations: Infectious Disease, Environmental Change, and Their Effects on national Security and Development*. Boston, MA: The MIT Press.

Raghavan, S. and S. Rajesh, (2003): Trends in Tropical Cyclone Impact: A Study in Andhra Pradesh, India. *Bulletin of the American Meteorological Society*, 84, 635–644.

Rajiv K. and N. H. Ravindranath (2011). "Impact of climate change on Indian forests: a dynamic vegetation modeling approach," *Mitigation and Adaptation Strategies for Global Change*, Volume 16, Number 2, pp. 119–142.

Raleigh, Clionadh, and Henrik Urdal 2007 "Climate Change, Envrionmental Degradation and Armed Conflict," *Political Geography* 26: 674–694.

Randall, David A., Richard A. Wood et al. 2007. "Climate Models and Their Evaluation" in *Climate Change 2007: The Physical Science bases. Contribution of Working Group I to the Fourth Assessment Report of the Intergovernmental Panel on Climate Change*, Cambridge, U.K.: Cambridge University Press.

Ranger, N., S. Hallegatte et al., (2011) An assessment of the potential impact of climate change on flood risk in Mumbai. *Climatic Change*, 104, 139–167.

Rawls, John. (1971) *A Theory of Justice*. Oxford: Oxford University Press.

Reifen C. and R. Toumi 2009 "Climate Projections: Past Performance no guarantee of future skill?" *Geophysical Research Letters* 36: 10,1029GL038082.

Reuveny, R. 2007 "Climate change-induced migration and violent conflict," *Political Geography* (26), 656–673.

Reuveny, Rafael 2007. "Climate Change and Induced Migration." *Political Geography*, 26, pp. 656–673.

Richardson, Harry W. and Peter Gordon. 2005. "Globalization and Los Angeles," in *Globalization and Urban Development*. Edited by H. W. Richardson and C.-H. C. Bae. Berlin: Springer, 197–210.

Rodriguez-Puebla, C., S. M. Ayuso, M. D. Frias, L. A. Garcia-Casado, L. A. 2007. "Effects of climate variation on winter cereal production in Spain." *Climate Research* 34, 223–232.

Rosenberg, N. J. 2010 "Climate Change, Agriculture, Water Resources: What do we tell those that need to know?" *Climate Change*, 100.

Roy J., A. Ghosh, A. Majumdar, P. Roy, A. P. Mitra and C. Sharma, (2005), Socio-economic and physical perspectives of water related vulnerability to climate change: results of field study in India, *Science and Culture* 71, (8), pp. 239–259.

Ru, Xin et al. 2008: *Bluebook of China's Society: Society of China Analysis and Forecast*. (Beijing: Social Science Academic Press).

Sabonis-Helf T. 2006 "Climate Change Policy in Russia, Ukraine and Kazakhstan." Washington DC: The National Council on Eurasian and East-European Research.

Saich, Tony 2000: "Negotiating the State," *The China Quarterly*, 161, 124–141.

Sandler, Todd. 1992. *Global Challenges: An Approach to Environmental, Political, and Economic Problems*. Cambridge, UK: Cambridge University Press.

Schwartz, Jonathan and Shawn Shieh (eds.) 2009, *State and Society Responses to Social Welfare Needs in China: Serving the People* (London, New York: Routledge).

Shapiro, Judith 2012: *China's Environmental Challenges* (Cambridge, Malden: Polity Press).

Shimko, Keith. 2009. *International Relations: Perspectives and Controversies*, 3rd edition. Boston, MA: Wadsworth Publishing.

Shope, R. 1992. Global Climate Change and Infectious Diseases. *Environmental Health Perspectives* 96: 171–174.

Shukla P.R., Rana A., Garg A., Kapshe M. and Nair R. (2004) *Climate Change Assessment for India: Applications of Asia Pacific Integrated Model* (AIM), Universities Press, Hyderabad.

Singer, S. Fred 2012 "The Climate Contrarians," *The New York Review of Books*, August 16, 85.

Sinha, S. K. and Swaminathan, M. S. (1991), "Deforestation Climate Change and Sustainable Nutrients Security," *Climate Change* 16, pp. 33–45.

Sitarz, Daniel (ed.) 1993 *Agenda 21: The Earth Summit Strategy to Save our Planet*," Boulder, CO: Earth Press.

Sorensen, André. 2005. "Building world city Tokyo: Globalization and conflict over urban space" in *Globalization and Urban Development*. Edited by H. W. Richardson and C.-H. C. Bae. Berlin: Springer, 225–238.

Specter, Michael 2012 "The Climate Fixers—Is there a Technological Solution to Global Warming," *The New Yorker*, May 14, pp. 96–103.

Sugiura, Toshihiko 2006 "The Current Situation of Fruit Growing Influenced by Global Warming and Countermeasures," *Journal of Agricultural Science*, Vol. 61, No. 3, pp. 107–122.

Sugiura, Toshihiko, Yokozawa Masayuki 2004. "Impact of Global Warming on the Environments for Apple and Satsuma Mandarin Production Estimated from Changes of the Annual Mean Temperature," *Journal of the Japanese Society for Horitulcultural Science*, No. 73, pp. 72–78.

Sun, Yanfei and Dingxin Zhao 2008, Environmental Campaigns, in Kevin J. O'Brien (ed.): *Popular Protest in China* (Cambridge/London: Harvard University Press), 144–162.

Tangri, A. K. *(2003)*, "Impact of climate change on Himalayan glaciers" Proceedings of V and A Workshop on Water Resources, Coastal Zones and Human Health, NATCOM, *27–28 June*, Indian Institute of Technology (ITT), New Delhi.

Tolosa, Hamilton. 2005. "The Rio/São Paulo Extended Metropolitan Area" in *Globalization and Urban Development*. Edited by H. W. Richardson and C.-H. C. Bae. Berlin: Springer, 125–146.

Trenberth, K. (2005). Uncertainty in Hurricanes and Global Warming. *Science*. 308. Pp. 1753–1754. (yly-mac.gps.caltech.edu/reprintsyly/A_RecentPapers/li%20jiang_h2o/zjunk/early%20versions/ref/Trenberth_Hurricanes_Science 2010%20copy.pdf).

Tynkkynen N. 2010. Russia, a great ecologic Power in global climate policy? Framing climate change as a policy problem in Russian public discussion. "Environmental Politics," Vol. 19, No. 2. 179–195.

Underdal, A.; J. Hovi, S. Kabbelken, T. Skodvin. 2012. "Can conditional commitments break the climate change negotiation deadlock?" *International Political Science Review*, vol. 33, n. 4, September, 475–493.

United Nations Development Programme (UNDP). 1994. *New Dimensions of Human Security*. New York, NY: Oxford University Press.

Vajpeyi, Dhirendra, 2001, *Deforestation, Environment and Sustainable Development*, London: Praeger.

Vajpeyi, Dhirendra 2012 *Water Resource Conflicts and International Security—A Global Perspective*, New York: Lexington Books.

Van Aalst, Maarten K. 2006. "The Impacts of Climate Change on the Risk of Natural Disasters." *Disasters*. 30, no. 1: 5–18. onlinelibrary.wiley.com.proxy.lib.uni.edu/store/10.1111/j.1467-9523.2006.00303.x/asset/j.1467-9523.2006.00303.x.pdf?v=1&t=h29j55oe&s=cc973d62daaa4263381486c5adb3a5fa538973f2.

Venkateshwarlu, B. (2009). *Climate Change and Sustainable Agriculture: Securing the Small and Marginal Farmers in India*, The Human Impact of Climate Change: Policy Notes for Parliamentarians, CLRA; New Delhi.

Verweij, Marco. 2006. "Is the Kyoto Protocol Merely Irrelevant, or Positively Harmful, for the Efforts to Curb Climate Change?" In Marco Verweij and Michael Thompson, eds. *Clumsy Solutions for a Complex World: Governance, Politics and Plural Perceptions*. New York, NY: Palgrave Macmillan.

Victor, David G. et al. 2009 "Geoengineering Option—A last resort against global warming?" *Foreign Affairs*, March–April 64–76.

Victor, G. David 2004. *The Collapse of the Kyoto Protocol and the Struggle to Slow Global Warming*, Princeton University Press: New Jersey.

Whiteside, Alan. 2002. Poverty and HIV/AIDS in Africa. *Third World Quarterly* 23 (2): 313-332.

Woolings, T. 2010 "Dynamical Influences on European Climate: An Uncertain Future," *Philosophical Transactions of the Royal Society* A 368.

World Bank (June 2008) *Climate Change Impacts in Drought and Flood Affected Areas: Case Studies*, Report No. 43946-IN.

World Bank (2006). *Overcoming Drought: Adaptation Strategies for Andhra Pradesh, India*. Directions in Development: Environment and Sustainable Development. Washington, DC: World Bank.

World Bank. (2009) *South Asia: Climate Change Strategy*. Draft Report, World Bank: South Asia Region.

Wu, W. B., R. Shibasaki, P. Yang, G. X. Tan, K. I. Matsumura, K. Sugimoto. 2007. "Global-scale modelling of future changes in sown areas of major crops." *Ecological Modelling* 208, 378–390.

WWF-Russia. "WWF Russia Strategic Plan 2012-2017 2012" Moscow. www.WWF-Russia.ru.

Xie, Lei 2009: *Environmental Activism in China* (New York: Routledge).

Yang, Guobin 2005, Environmental NGOs and Institutional Dynamics in China, in: *The China Quarterly*, 181, 46–66.

Yang, Yifei 2009: *A Study of Ecological Poverty in Minqin County* (Lanzhou: Gansu Agricultural University Press) [in Chinese].

Yang, Yongchun 2003: "Analysis of the Environmental Changes in Oases at Lower River Reaches in Arid Regions and their Reasons: A Case Study of Minqin County at the Lower Reaches of the Shiyang River on the Hexi Corridor, Gansu Province," *Human Geography*, 18 (4), 42–47 [in Chinese].

Zeng, N. 2003. Drought in the Sahel. *Science* 302, pp. 999–1000. www.sciencemag.org/content/302/5647/999.short.

Zetter, R. 2007 "More labels, fewer refugees: Remaking the refugee label in the era of globalization" *Journal of Refugee Studies* (20): 2, 172–92.

Zhong, Huaping et al. 2002: "On Water Resources and Environmental Protection of the Downstream Shiyang River in Minqin," *Northwest Water Resources and Water Engineering*, 13 (1), 10–13 [in Chinese].

Zlobin V. I., Danchinova G. A., Suncova O. V., Badueva L. B. 2004 "Izmenenie klimata i zdorov'e Rossii v XXI veke." M.: Adamant. 121–124.

Zullo J. R., J., H. S. Pinto, E. D. Assad, A. M. H. de Avila, A. M. H. 2011. "Potential for growing Arabica coffee in the extreme south of Brazil in a warmer world." *Climatic Change*, v. 1, p. 1573–1480, 2011.

Index

adaptation, 18, 88–93, 329–332

Bangladesh, 325–335
Brazil, 169–199

carbon dioxide (CO2), 27, 37–46
China, 111–168; carbon dioxide, 37–46; climate refugees, 83–109; Minqin Oasis, 139–168; mitigation, 122–132
climate extremes, 29–37; monsoon, 31; precipitation, 30; tropical cyclones, 35–37
climate refugees, 83–109; National Adaptation Program of Actions, 69–94

desertification, 170–171, 182–185; agriculture, 172–176; deforestation, 171–172,182–185; regional climate models (AGCMs), 176–182; urbanization, 186–190

geo-engineering, 19
glaciers, 5–6

India: agriculture, 210–211; coastal areas, 213–215; food security, 211–212; human health, 216–217; natural ecosystems, 213; water resources, 215–216
integrated assessment (IA), 18–19
Intergovernmental Panel on Climate Change (IPCC), 4, 15, 27–36, 65, 66, 86, 87, 202, 213, 231, 281, 325
International Renewable Energy Agency (IRENA), 71–75

Japan, 227–247; adaptation, 239–244; floods, 231–234; heat strokes, 237–239; human security, 236–237; sea level, 233–236

Kyoto Protocol, 16, 52, 64–70, 267–268

mitigation, 18
Montreal Protocol, 39, 52, 61–64, 68–70

National Aeronautics Space Administration (NASA), 4, 28, 30, 31, 34, 42–43
Nepal, 325–335

oceans, 6–7, 37–46

Russian Federation (Russia), 248–279, 307–321; attitudes, 260–265; Dimitri Medvadev, 270, 309–310, 318, 321; domestic policy, 267–272; human security, 254–260; international efforts, 272–276; Russian Public Opinion Research Center (VCIOM), 261–264, Caucasus, 254, 257, 309, 316; Siberia, 251, 261, 309; Vladimir Putin, 318

Turkey, 281–305; adaptation, 298–300; agriculture, 288–289; bio-diversity, 293–294; human security, 290–293; industrialization, 285–286; Mediterranean, 281–305; Ministry of Environment and Urbanization (MEU), 284, 287, 288, 292, 293, 295, 300; National Climate Change Action Plan (NCCAP), 283–284, 294–295; socio-economic impact, 294–295; temperatures, 289–290; transportation, 286–287; treaties, 300–302

United Nations Development Program (UNDP), 207, 211, 294
United Nations Framework on Climate Change (UNFCC), 1, 16, 85, 88, 93, 208, 282

WORLD Health Organization (WHO), 52, 53, 55–56

About the Contributors

Eduardo Delgado Assad holds a master's degree and a PhD from University of Montpellier, France. From 1993 to 2006 he was the technical and scientific coordinator of National Agricultural Climate Risks at the Brazilian Ministry of Agriculture, head of the National Climate Change and Agricultural Network of Ministry of Science and Technology, head of the Climate Change Research Platform at EMBRAPA, and coordinator of the vulnerability impacts groups on the Brazilian Panel of Climate Change. In 2011 he was National Secretary of Climate Change for the Brazilian Ministry of Environment. He was also Embrapa coordinator to Brazilian Low Carbon Agriculture. Nowadays, he coordinates several national and international research projects on the impacts of climate change on agriculture and Low Carbon Agriculture Monitoring.

Igor N. Barygin is a professor of political science at St. Petersburg State University. His academic and research interests are ecological politics, comparative politics, area studies, and right-wing political parties in Europe. His main publications include, *Socialnaia baza evropeiskih kraine pravih* (Social basis of European right-wing radicals), *Politicheskie partii, dvijenia I organisazii sivremennoi Rossii* (Political parties, movements and organizations in contemporary Russia), *Politologia: slovar* (Political Science: Glossary), *The Russian right-wing political parties/Modern Europe after*

Fascism 1943–1980's, edited by S. Y. Larsen, Vol. I–II, and *Pravoradikalnie politicheskie partii i dvigjenia v sovremennoi Evrope* (Right-wing political parties and movements of modern Europe).

Maria Bondes received her MA from Hamburg University, Germany. She is a research fellow at the German Institute of Global and Area Studies (GIGA), and the Institute of Asian Studies, Hamburg, Germany. Her research focuses on China's environmental problems and activism. During her studies at the University of Hamburg she spent one year at Lanzhou University, in China's Western Gansu Province and has conducted extensive field research on the region's climate change and ecological problems as well as environmental activism in the area. Right now she is back in China conducting more research. She has presented several research papers at international seminars and conferences.

Antonio Marcio Buainain received his PhD in economics from the University of Campinas (1999). He is presently a professor of economics at the Institute of Economics of the University of Campinas—Unicamp, in Campinas, Brazil. Between 1989–1994 he worked as an economist of the Latin American and the Caribbean Unit of the Food and Agriculture Organization (FAO), in Rome, as an advisory and consultant to several Brazilian ministries, including Ministry of Agriculture, Ministry of Rural and Agrarian Development and Ministry of Science and Technology, IDB, World Bank, UNISCO, WIPO. He has published forty-eight papers in national and international journals, organized, authored, and co-authored thirty-four books (author / co-author of fifteen and main editor of nineteen titles) and forty chapters. He has published 236 articles in leading Brazilian newspapers, and had 117 full papers published in congresses, conference, and seminars proceedings.

Vanessa Duarte is a PhD student at the University of Évora, Portugal, and a member of CEGAFE (Centre for Advanced Studies in Management and Economics of the University of Évora). She specializes in innovation and decision sciences. Her research has been supported by a BJC scholarship from the University of Évora. Presently she is affiliated with CEFAGE-UE.

Natalia Eremina is an assistant professor at the School of International Relations Saint-Petersburg State University, Russia. She is a member of the research group on "Ecological issues as factors of political destabilization." Her academic and research interests are, ecological policy, comparative politics, political radicalism, nationalism, ethno-political conflicts, and right-wing parties and movements. She has a PhD in political science. Some of her main publications include, "Ethnic Policy in the European

Union's states (the example of the United Kingdom)" (2007), "The reforms of the European Union: Political, Legal and Economic Aspects," "Federalism: Financial and Economic Aspects (Russian-British parallels)," "Ethnic Fragmentation of the European Society in the Globalization and European Integration Context," "Trends, Prospects, Challenges of Globalization," "Regionalism and regionalization as the Grounds of the European Integration," "The issue of Adaptation of the EC Regional Policy in the Context of Russia," "Regional Political Culture as the Factor of State Development: Integration versus Disintegration," and "Modernization of Russia: The Territorial Dimension ."

Cheng Fang-Ting is a PhD student (ABD) at the Graduate School of Law and Politics, University of Tokyo, Japan. Her academic research areas include comparative environmental policies and international security. Her MA thesis (2009 University of Tokyo) was on "The Kyoto Protocol and Japanese Environmental Diplomacy (1997–2002): Japan's Response towards the U.S. Withdrawal." She has several published articles and ten scholarly presentations at international conferences. Her publications and research presentations include two major papers, and eight presentations on Japanese ecological issues.

Leonid M. Gregoryev received his PhD from Moscow State in economics. He is currently the Chair of World Economy, Dept. of World Economy and Politics, National Research University—Higher School of Economics, Moscow; Deputy Director General—Chief of Research of Russian Energy Agency; chairman of the Board of WWF-Russia since 2006; Deputy Minister of Finance and Economy; chairman of Foreign Investment Committee RF in 1991–1992; World Bank in 1992–1997; General Director of Bureau of Economic Analysis in 1997. He has published more than 300 publications.

Laura L. Janik is an assistant professor of political science at the University of Northern Iowa. Her teaching and research interests include global health governance, political economy, environmental politics, nonstate actors, and international organization. Her most recent publications can be found in *Global Health Governance*, *The Qualitative Report*, and *Simulation and Games*. Laura received her PhD from the University of Connecticut in 2009. She is the acting co-director for the Center for International Peace and Security Studies at the University of Northern Iowa and currently serves as the secretary for the International Studies Association Midwest.

Craig Johnson is an associate professor of political science at the University of Guelph, Canada. He holds a PhD in International Development from the London School of Economics (2000), and has taught at the London School of Economics, the School of Oriental and African Studies, University College London, and the University of Oxford. Between 2000 and 2002, he was a post-doctoral fellow with the Overseas Development Institute in London, where he helped lead a three-year study of liberalization and livelihood diversification in India, Nepal, and Bangladesh. In 2009, he was an ESRC-SSRC visiting fellow with the Tyndall Centre for Climate Change Research and the Environmental Change Institute, both at the University of Oxford. Dr. Johnson is currently taking forward new research on forced displacement and environmental change, exploring the politics of rights, vulnerability, and development in the South Asian context. He is the author of *Arresting Development: The Power of Knowledge for Social Change* (Routledge, 2009) and coeditor (with John Farrington, Priya Deshingkar and Daniel Start) of *Policy Windows and Livelihood Futures: Prospects for Poverty Reduction in Rural India* (Oxford University Press, 2006).

Palakodety Sita Janaki Kreshina received her PhD from Andhra University, India (1989) in cytogenetic studies of some telo aneuplorids in Pearl Millet. Presently she is an associate professor at the Institute of Public Enterprise Hyderabad, India. She has presented forty-eight scientific papers at national and international seminars and conferences in the Netherlands, Kenya, Russia, and India, and has two published books on Biotechnology.

Norman Laws is a research associate at Leuphana University, in Lunbeburg, Germany. His publications include *Climate Change: Implications for Europe's Security, Defense Policy: How Climate Change affects the Security of Europe*, and *AV Akademikerverlag: Saarbrucken* (2011).

Ding Li is an associate professor of human geography at Lanzhou University's College of Earth and Environmental Sciences, Gansu Province, China. His research focuses on the sustainable development of western China's arid and semi-arid regions. He has led various projects focusing on sustainable development and poverty alleviation in western China, including Minqin County, and has published extensively on the region in Chinese.

Jian Li is an associate professor of anthropology at the University of Northern Iowa. He has conducted ethnographic field research in China, Thailand, Laos, and Myanmar. He has published research articles in *Human Organization, Culture and Agriculture, China's Ethnic Groups, The Chinese Journal of Schistosomiasis Control, Anthropology News, Journal of SESA*, and the *Journal of Chinas Central University for Nationalities*. His current research focuses on development and health, medical anthropology, energy, reemerging infectious diseases, and China's environmental issues.

Igor A. Makarov received his PhD from the National Research University Higher School of Economics (NRU-HSE), Moscow, Russia. He is a research fellow at the NRU-HSE, in the Center for Comprehensive European and International Studies (CCEIS). He is also a lecturer at the NRU-HSE, Faculty of World Economy and International Affairs, and in the Department of World Economy. His research interests include, environmental economics, the economics of climate change, climate policy, economics of water scarcity, economics of foods and food insecurity, and international political economy. His main publications include "Food Crisis 2007–2008 in the Context of the Problem of Climate Change," "To Bread and Water," "Global Problems of Climate, Water, and Food in Public Goods Theory," "International Economics," "Towards the New Understanding of Sustainable Development," "Global Climate Change and World Economy," "World Economy and International Relations," "The Role of the USA in Coping With Global Climate Change," and "USA-Canada: Economics, Politics, Culture." He also participated in the Symposium on Methodology, in Reading, UK.

Rebecca McGuire is a freelance researcher on environmental policy-making.

Ramkumar Mishra obtained his PhD in finance from the University of Rajasthan, India (1973). He was a Commonwealth Fellow, British Council Fellow and has a SDA from Milan, Italy. He has forty-five books, and hundreds of research papers published. At present he is director and senior professor at the Institute of Public Enterprise, Osmania University, Hyderabad, India.

Roopinder Oberoi received her PhD from the University of Delhi, India. She specializes in political science and public administration. She was awarded a post doctorate fellowship from the University Grant Commission in 2009 and has contributed several chapters in the field of corporate governance, management, and sustainable development in various books, journals, and has presented seventeen research papers at national and international conferences. Presently she is an assistant professor at Delhi University.

Hilton Silveira Pinto is a professor at the Institute of Biology and former director of the Center for Research in Meteorology Applied to Agriculture at the State University of Campinas—Unicamp. He has a

PhD in agronomy and has spent time as a post doctorate and invited professor at the University of Guelph, Canada. He was scientific advisor to the Ministry of Science and Technology, to the Ministry of Industry and Commerce, and he was S. Paulo State Secretary of Science and Technology. He has coordinated twenty-eight research projects with national and international financing in the areas of risk zoning and climate change in agriculture. He has published seventy-one papers in national and international scientific journals, forty-nine book chapters, and has received three awards for scientific merit.

Alla Salmina has a PhD in Sociology. At present she is a researcher at the National Research University—Higher School of Economics (HSE), lecturer in the Department of Sociological Research Methods at the National Research University Higher School of Economics (HSE). She has several publications on Russian economy.

Mahendra Prasad Singh has a PhD in political science from the University of Alberta, Edmonton, Canada (1975). He has held high administrative positions in various academic institutions in India. His research interests and major publications (twelve Books and twenty-five chapters and articles) are in Indian politics. At present he is an honorary senior fellow and director of research at the Centre for Multilevel Federalism Institute of Social Science, Delhi, India.

Miguel Rocha de Sousa has a PhD in Economics from Nova School of Business and Economics, Lisbon, Portugal. Presently he is an assistant professor at the University of Evora (UE). His areas of research are international organizations, political economy, land reform, economic growth and development, macroeconomics of the euro, co-integration and economic literacy. He is an associate of SEDES, and RC35 of Technology and Development. He has extensively published in his area of research interests.

Clifton White is a graduate student in Public Policy at the University of Northern Iowa with research interest in environmental issues.

Paul Williams received his PhD in political science from the University of California, Los Angelis (1992). Presently he is an assistant professor at Bilkent University, Ankara, Turkey. He and Ali Tekin are the coauthors of *Geo-Politics of the Euro-Asia Energy Nexus: The European Union, Russia, and Turkey* (N.Y.: Palgrave-Macmillan 2011). Dr. Williams has published seventeen chapters and has presented twenty-four research papers in regional and international conferences. His areas of academic and research interests include international environmental policy making.

ABOUT THE EDITOR

Dhirendra Vajpeyi is a professor of political science at the University of Northern Iowa. He has authored, coauthored, and edited sixteen books and numerous chapters and articles. They include *Environmental Policy making in the Third World; Technology and Development; Local Government and Politics in the Third World; Indira Gandhi's India; Deforestation: A Comparative Analysis, Civil-Military Relations; Nation-Building and National Identity; Local Democracy and Politics in South Asia; Modernizing China; Water Resource Management: A Comparative Perspective; Law, Politics and Society in India; Politics, Technology and Bureaucracy in South Asia; Globalization, Governance and Technology: Challenges and Alternatives*; and *Water Resource Conflicts and International Security: A Global Perspective.* Dr. Vajpeyi has lectured at Universities in Moscow, St. Petersburg (Russia), Peoples Republic of China, Chile, and Klagenfurt (Austria). He was a senior research fellow at the Hover Institute, Stanford University (California). Currently he is the chair of the IPSA Research Committee 35 on "Technology and Development." He holds a PhD in political science from Michigan State University.